Jerzy Maćkowiak

Fluid Dynamics of Packed Columns

Jerzy Maćkowiak

Fluid Dynamics of Packed Columns

Principles of the Fluid Dynamic Design of Columns for Gas/Liquid and Liquid/Liquid Systems

Translated by Claudia Hall
with the cooperation of Anna Ługowska-Czok

Springer

Dr.-Ing. Jerzy Maćkowiak
ENVIMAC Engineering GmbH
Im Erlengrund 27
46149 Oberhausen
Germany
j.mackowiak@envimac.de

ISBN 978-3-540-88780-5 e-ISBN 978-3-540-88781-2
DOI 10.1007/b98397
Springer Heidelberg Dordrecht London New York

Library of Congress Control Number: 2009926972

Cover design: WMXDesign GmbH, Heidelberg

Printed on acid-free paper

Springer is part of Springer Science+Business Media (www.springer.com)

Foreword by Prof. Górak, Technical University of Dortmund

Columns for distillation, absorption and extraction have been applied for years in chemical, petrochemical, food, energy and electronic industry. They are often equipped with different kinds of packings. Understanding of principles for packing design, internal traffic of phases within the column and implementing of theoretical models into industrial practice is of vital interest of both – students and practitioners. The book of Dr. J. Maćkowiak gives a deep insight into the fluid dynamics of packed columns. It is based on personal experience of the author and on a tremendous experimental data base on pressure drop (10,000 points), flooding points (1,2000 points) and liquid hold-up (1,100 points) measured for about 200 different types of random and structured packings. This is the biggest experimental data base published ever.

The big value of this book is also the theoretical model, called "Suspended Bed of Droplets" which is bases on the dependency between the flow resistance coefficient and packing shape. This model is predictive, i.e. it allows calculation of pressure drop and flooding gas velocity for unknown packing geometry. The model is valid for pressures from vacuum up to 100 bar and for specific liquid loads up to 250 $m^3\ m^{-2}h^{-1}$. In this book the reader will find the sound theoretical analysis of two phase flow in packed column, extensive correlation of existing data for traditional and modern packings as well as practical equations which can be directly used in academic teaching courses and industrial applications.

Dr. J. Maćkowiak has committed to paper his 20 years experience as practitioner while operating company ENVIMAC GmbH and as researcher, publishing his results in journals and conference papers. Calculation methods of packed columns, presented in the book will increase the design accuracy of distillation, absorption and extraction which cause up to 60% of total processing costs in chemical industry.

Dortmund, February 2009

Prof. Dr.-Ing. A. Górak
Technical University of Dortmund
Department of Biochemical and Chemical Engineering
Laboratory of Fluid Separations

Foreword by Prof. Dr. Ing. A. Mersmann, Technical University of Munich

For many years now, packed columns have been used in various domains, such as diffusional separation technology, environmental technology and biotechnology, providing a mass transfer contact area for gases and liquids. In recent decades, scientific as well as economic factors have continually led to new and more efficient packings being designed. The industry is striving to accurately dimension modern packings made with new materials and innovative geometric designs and to operate them in a reliable manner. This requires first of all the compilation and general presentation of experimental data, a difficult and time-consuming task, which the author has been undertaking for many years.

This work is based on more than 10,000 experimental pressure drop data, in excess of 1,100 liquid hold-up data and more than 1,200 flooding point data for approx. 160 different random and structured packings made of various materials commonly used in practice. The majority of this data was compiled by the author in an accurate and reproducible manner whilst working for the Institute of Thermal Separation Processes at Bochum University. This book exceeds all publications worldwide in terms of its volume of data, and the industry will be grateful to the author as well as Springer Publishing for its publication.

It is my hope and wish that, for the benefit of mankind, the comprehensive and accurate results and conclusions of this work will lead to an improvement in the design and operation of packed columns.

Alfons Mersmann

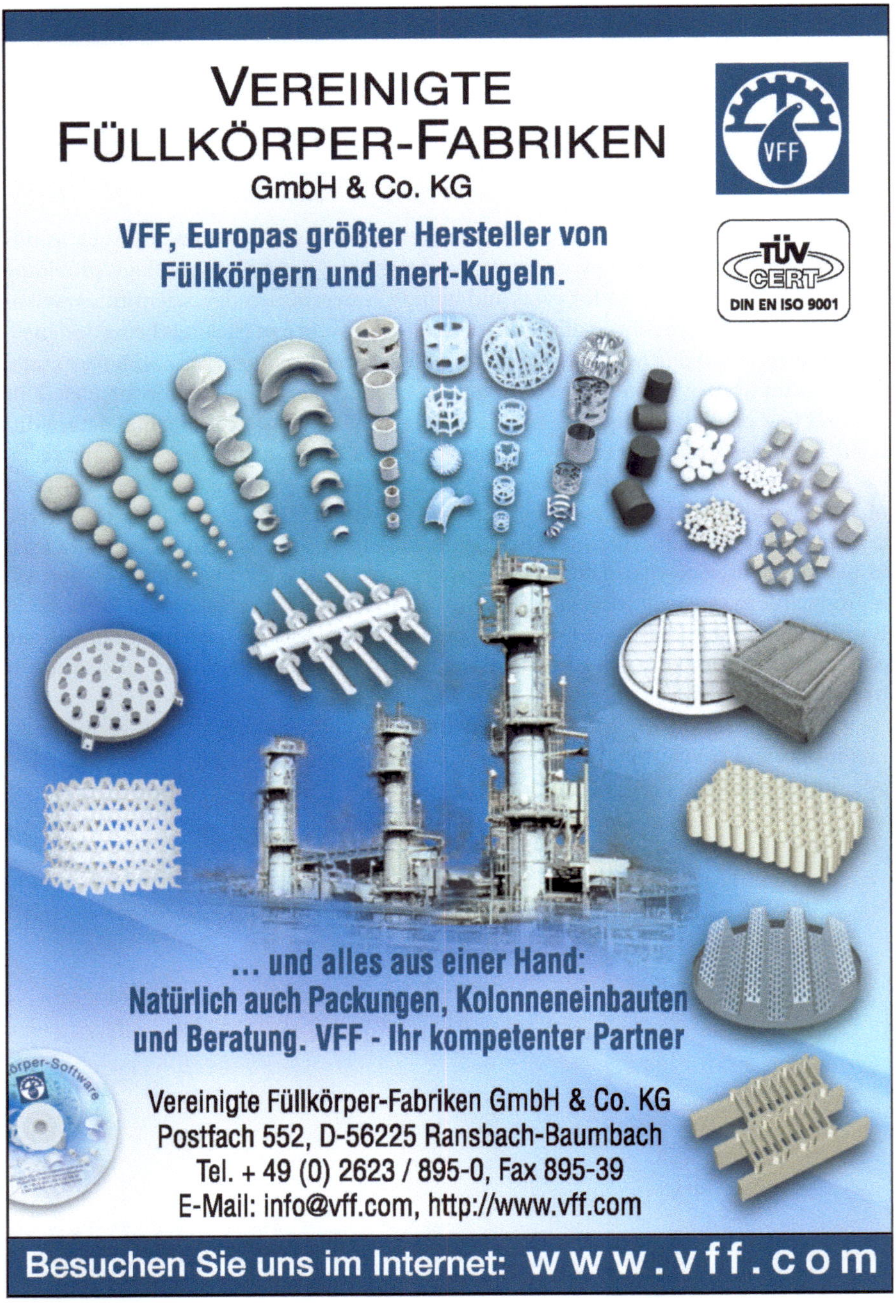
VEREINIGTE
FÜLLKÖRPER-FABRIKEN
GmbH & Co. KG
VFF
TÜV CERT
DIN EN ISO 9001
VFF, Europas größter Hersteller von
Füllkörpern und Inert-Kugeln.
... und alles aus einer Hand:
Natürlich auch Packungen, Kolonneneinbauten
und Beratung. VFF - Ihr kompetenter Partner
Vereinigte Füllkörper-Fabriken GmbH & Co. KG
Postfach 552, D-56225 Ransbach-Baumbach
Tel. + 49 (0) 2623 / 895-0, Fax 895-39
E-Mail: info@vff.com, http://www.vff.com
Besuchen Sie uns im Internet: www.vff.com

Preface

The first German edition of the book "Fluid dynamics of packed columns with modern random and structured packings for gas/liquid systems" was published in 1991. It sold out within a few years. Added to this were numerous enquiries, in particular within the industry, prompting me to publish a second, extended edition.

A packed column remains the core element of any diffusional separation process. This underlines the need for basic design principles for packed columns, which enhance the design process by making it more accurate and reliable.

The SBD (suspended bed of droplets) model introduced in the first German edition of the book was well received by the experts and is now used by a large number of companies in the industry, as it offers improved reliability in the fluid dynamic design of packed columns. For the purpose of facilitating the design process, the SBD model was integrated into the simulation programme ChemCAD. The software programme FDPAK, which is available for Windows, has certainly contributed to the widespread use of the SBD model. The programme is very user-friendly and the calculation results are presented in tabular as well as graphic form, showing flood load, pressure drop and hold-up diagrams in the entire operating range.

The first German edition concentrates on the description of the fluid dynamics of columns with random and structured packings in the vacuum and normal pressure range of up to approx. 2 bar and for specific liquid loads of up to 100 $m^3m^{-2}h^{-1}$ for gas-liquid systems. This range covers a majority of the applications and tasks relevant in the absorption and desorption of highly and/or moderately soluble gases as well as in rectification under vacuum and normal pressure.

The importance of absorption and rectification under high pressure has steadily increased in the past 10 years, calling for an upgrading of the existing model. Fortunately, new publications emerged during the last 15 years, presenting experimental data on pressure drop, flooding point and hold-up parameters for high-pressure systems. Based on this, it was possible to and expand the validity of the correlation derived in the first German edition for determining the hold-up at flooding point to include the range of high liquid loads and therefore higher hold-up parameters (Chap. 2).

Using the SBD model, it is now possible to describe operations under higher pressure, which is very significant in practice, as it is linked to high liquid loads and low gas velocities. The SBD model has been verified using experimental data taken at high pressure of up to 100 bar. There are some practical numerical examples at the end of each chapter, which provide an insight into the application of the model.

The current edition will introduce a generally valid procedure of the single pressure drop calculating based on the knowledge of the form factor (P and an additional model for calculating the pressure drop of irrigated structured and random packings, based on the knowledge of the law of resistance $\psi_{LV} = f(Re_L)$ for two-phase counter-current flows and of the liquid hold-up $h_L{}^0$ in the entire operating range up to flooding point. This model is suitable for applications, in which the only available data for determining the law of resistance is experimental pressure drop data for a two-phase system (no given pressure drop data for dry random packings), or in which the pressure drop above the loading line for low viscous mixtures needs to be determined more accurately.

A large amount of experimental data has shown that this model generates satisfactory results up to flooding point for laminar $Re_L < 2$ as well as for turbulent liquid flow $Re_L > 2$.

The correlation for determining the gas velocity at flooding point introduced in the first edition has been modified further and can now also be applied to any type of structured packing, tube columns with regularly stacked Pall rings, Hiflow rings and Białecki rings and regularly stacked layers of Pall rings, Raschig rings, Hiflow rings and Białecki rings. Following the latest findings, it has been possible to mathematically compute various loading capacities of structured column internals of types X and Y with flow channels at different gradients. This has also been taken into account in the expanded general correlation for calculating the gas velocity at flooding point, which makes this correlation applicable to any type of column internal.

Chapter 7 introduces for the first time the basic fluid dynamics principles of packed columns for liquid/liquid extraction. The previously mentioned SBD model for gas/liquid systems is transferable to liquid/liquid systems. The method used to calculate the gas velocity at flooding point of the disperse and continuous phases will be explained by means of some numerical examples.

The guiding idea behind this book was to develop a closed, consistent concept for designing packed columns for gas/liquid and liquid/liquid systems, in order to make the calculation of individual parameters more transparent and create a basis for objective comparisons between different column internals.

In contrast to other studies, this book uses a different approach for logging processes within packed columns, which is based on the specific flow behaviour of droplet systems.

The occurrence of droplet systems in packed columns was confirmed by Bornhütter and Mersmann in 1991. Hence, despite the highly complicated processes of two-phase flows in packed columns, it was possible to form straight-forward, user-friendly correlations, which are ideal for practical applications when it comes to developing solutions for a wide range of tasks. The simple correlations are particularly advantageous when comparing a large number of different columns internals. In addition, this book should help scientists as well as students to gain a better understanding of flow processes in gas/liquid and liquid/liquid systems.

As opposed to other studies, this book draws on the publications of other authors to support and expand the application ranges of the SBD model. However, it does not use them for the illustration and comparison of different calculation models. This work is based on more than 10,000 experimental pressure drop data, in excess of 1,200 flooding point data and approx. 1,100 liquid hold-up data for nearly 200 different types of packings. Compared to the 1st edition, the figures have more or less doubled.

What is particularly significant from a scientific point of view, is the knowledge that the experimental work and test systems under vacuum and high pressure previously required for the research and development of new types of column design can be reduced to just a few steps, thus allowing the user to determine the model parameters of the SBD model fast and with minimal experimental effort using the air/water test system. It is worth noting that tests using single-phase air flows under ambient conditions are sufficient to determine the system-independent law of resistance $\psi = f(Re_V)$. In the case of simple types of packing, it is possible to determine the resistance factor of single-phase flow simply based on the geometric configuration without requiring experimental evidence, as described in Chap. 3. It is therefore possible to transfer the entire fluid dynamics of one type of column internal to any application in diffusional separation technology.

J. Maćkowiak

Summary

For many years now, there has been a constant demand for low pressure drop column internals in rectification, absorption, stripping and liquid/liquid extraction. The prevailing trend in the chemical industry is to replace tray columns with those containing modern structured and random packings. When planning the design of packed columns, it therefore particularly important to use reliable methods for predicting the mass transfer and hydrodynamic behaviour of the two-phase flow.

For this reason, the book is aimed at illustrating the basic principles of the fluid dynamic design of columns with modern random and structured packings, using a new design concept which can be applied to any type of packing.

One of the author's priorities was the practical aspect of the work, since the standard approach to rectification, absorption and extraction as the core areas of thermal process engineering is purely empirical.

The design data for gas/liquid and liquid/liquid systems presented in the current edition are strongly interlinked and based on the SBD model for both systems.

The design process for gas/liquid systems is based on correlations which have been experimentally derived and verified. The process allows the calculation of the flooding point, pressure drop and hold-up for gas/liquid systems virtually in the entire operating range up to flooding point. The following features of the process should be highlighted:

The model parameters are calculated for a given packing density using the water/air system under normal conditions. The experiments conclude that these model parameters for the separation of mixtures in rectification technology in the vacuum to high pressure range are applicable for the entire operating range. Hence, the experimental work can be minimised to just a few steps using only the air/water system under normal conditions.

The model also allows the calculation of flooding point and hold-up parameters for liquid/liquid systems virtually in the entire operating range up to flooding point.

Structure

The content of this book is divided into seven chapters. The first six chapters deal with gas/liquid systems (Part 1); Chapter 7 is dedicated to liquid/liquid extraction. The structure of the first five chapters largely matches the calculation steps applied in the fluid dynamic design of packed columns for gas/liquid systems. Chapter 1 briefly describes the structure of packed columns and their significance for the separation of mixtures under vacuum conditions as well as for absorption and desorption. It also outlines the design processes commonly used today. This is followed by an analysis of the hydraulic behaviour of packed columns and their relevant parameters, with Chap. 2 examining the flooding point of the lower operating range and Chap. 3 determining the pressure drop of dry random packings. Chapters 4 and 5 deal with the pressure drop of irrigated random packings and their liquid hold-up parameters.

The correlations derived in Chaps. 2, 3 and 4 can be used to calculate the apparatus diameter and determine the pressure drop and liquid hold-up in a given operating range and at flooding point.

For the purpose of illustrating the topics, each chapter begins with a brief overview of the most relevant work on the respective topic as judged by the author, followed by the presentation of the chosen approach.

The results of this work are summarised in Chap. 6. The Tables 6-1a–c contain the geometric column data, such as packing density N_0, specific geometric surface of packing elements a_0, void fraction ε_0, specific weight G of 1 m^3 according to the manufacturer. In addition, the chapter contains all numerical values of the parameters required for the fluid dynamic design of approx. 200 randomly and regularly stacked packing elements well as tube columns and structured packings.

By using the equations derived for the calculation of each parameter, it was possible to condense the extensive research material, which is discussed in the summary of the results of this work in Chap. 6. The end of each chapter contains example calculations to illustrate the individual correlations for determining the vapour load factor at the flooding point of the liquid hold-up as well as the pressure drop of irrigated and dry random packing elements. The numerical examples are practice-oriented and explain the correlations mentioned before, based on the examples of different packings.

The references relating to the individual chapters are listed at the end of each chapter.

The book boasts comprehensive tables and charts with information on experimental data and test conditions, highlighting the enormous volume of tests as well as the scope of application offered by this process. Following Chap. 6 is a description of FDPAK, the well-known and extensively used software programme for the fluid dynamic design of packed columns.

The correlations most commonly used in this work are compiled and explained in a separate list of symbols at the beginning of the book.

The first German edition of the book was written between 1988 and 1990 and the second one evolved between 1997 and 2002. The first English addition is based on experimental data stemming from more than 200 – mostly modern – random and structured packings between 1965 and 2008.

The tests were mainly carried out by the author using distillation plants with diameters $d_S$of 0.15/0.22/0.5 m and absorption plants with diameters of 0.15/0.22/0.3 m and 0.45/0.6 m as well as industrial plants with diameters of 0.8/1.2/1.6/1.8/3 m owned by ENVIMAC Engineering GmbH. A considerable amount of test data on rectification ($d_S = 0.5$ m) and absorption ($d_S = 0.3$ m) originates from the author's work as Scientific Assistant at the Institute for Chemical Engineering, Technical University of Wrocław/Polen (1970–1976). In co-operation with Dr. Ing. S. Filip, Dr. Ing. Z. Ługowski, Dr. Ing. S. Suder and Dr. Ing. habil. A. Kozioł from the Technical University of Wrocław, the author carried out a number of studies on modern packings at industrial plants, which are also referred to in this monograph. Some of the test results, in particular the rectification data for metal Pall rings (15–80 mm), stem from the findings of Prof. Billet. Further test data on extraction and rectification processes were taken from a number of scientific studies, which were personally conceived and overseen by the author at the Institute for Thermal Separation Processes at Bochum University, under Prof. Dr. Ing. R. Billet.

In 1990, a database was created for the purpose of evaluating all experimental data, including literature data. It currently holds more than 1,200 experimental flooding point data, in excess of 1,100 liquid hold-up data and more than 10,000 pressure drop data for irrigated random and structured packings. The number of test mixtures is 32. The result is a comprehensive data pool, which is constantly being updated with the addition of new experimental data.

The progress in terms of the accuracy in designing packed columns for gas/liquid and liquid/liquid system is considerable. The results of this work have led to accurate predictions of the fluid dynamic behaviour of random packing elements, simply based on their geometric form without requiring any chemical engineering tests.

Acknowledgments

A special thank you goes to *Prof. Dr. Ing. A. Mersmann*, Technical University of Munich, for many fruitful discussions and valuable suggestions for the draft of the first German edition, *Prof. Dr.-Ing. A. Górak*, Technical University of Dortmund, for reviewing the first English edition, as well as to my friend *Prof. Dr. Ing. habil. A. Kozioł*, Technical University of Wrocław and *Dr. Ing. J. Szust*, ENVIMAC Engineering GmbH, for his contribution towards the development of the FDPAK programme and evaluating the flooding point data.

My son, *Dipl. Ing. Jan Maćkowiak*, proof-read the second German edition thoroughly and converted it to the correct format for printing.

Mrs Dipl. Ing. Anna Ługowska-Czok prepared the first English edition for printing and produced the complex graphs showing flooding point and pressure drop parameters in order to illustrate the spread of the experimental data.

I would also like to thank *Mrs Claudia Hall* for the translation of this book as well as her co-operation and valuable contributions to the English edition.

I would particularly like to thank *my wife Nathalie and my children Jan and Anna* for their patience, and all those who have supported me in writing this book.

Jerzy Maćkowiak, 2009

Contents

Part 1
Principles of the Fluid Dynamic Design of Packed Columns for Gas/Liquid Systems

Part 2
Principles of the Fluid Dynamic Design of Packed Columns for Liquid/Liquid Systems

Part 1

Principles of the Fluid Dynamic Design of Packed Columns for Gas/Liquid Systems

List of Symbols for Part 1

Formula Variables, Latin Letters

a	$m^2\ m^{-3}$	geometric surface area of packing per unit volume
a'	$m^2\ m^{-3}$	effective surface area of dry packing per unit volume
a_o	$m^2\ m^{-3}$	standard geometric surface area of packing per unit volume
a_e	$m^2\ m^{-3}$	effective interfacial surface area of packing for gas-liquid contact per unit volume
A, B, C	–	Antoine constants for calculating the pressure of saturated vapours of pure substances, with index "1" for more volatile components and index "2" for less volatile components
A	m^2	total non-perforated wall surface area of individual packing element
A_0	m^2	perforated wall surface area of individual packing element
A_1	m^2	surface area of individual packing element
A_e	m^2	total effective mass transfer area of packing for gas-liquid contact
A_F	m^2	surface area of packing
A_i	–	constant
A_w	m^2	surface area of column wall
A_s	m^2	cross-sectional area of column
B_L	–	dimensionless liquid load
C, C_i	–	constants
C_B	$s^{2/3} m^{-1/3}$	model parameter for determining the pressure drop of an irrigated packed bed for $Re_L \geq 2$
$C_{B,0}$	–	dimensionless parameter in Eq. (4-41), $C_{B,0} = C_B \cdot g^{1/3}$, below the loading line $C_{B,0} = 0.8562$
$C_{C,0}$	–	dimensionless parameter in Eq. (4-44), below the loading line $C_{C,0} = 1$
C_C	$s^{2/3} m^{-1/3}$	parameter for determining the pressure drop of an irrigated packed bed for $Re_L < 2$
C_{Fl}	–	constant for calculating the vapour velocity at flooding point, Eq. (2-50)
$C_{Fl,0}$	–	universal flooding-point constant for packing in flood correlation, Eq. (2-52)

C_H	€/t	costs of heating steam
C'_0	–	shear factor
C_0, $C_{0,B}$	€/h	operating costs of vacuum and normal pressure distillation column
C_P	–	constant for determining the total liquid hold-up of packings (for turbulent liquid flow)
C_T	–	constant for droplet diameter d_T
d	m	nominal packing diameter
d_i	m	inside packing diameter
d_h	m	hydraulic diameter of packed bed
d_p	m	particle diameter
d_R	m	tube diameter, inside
d_S	m	column or tower diameter
d_T	m	droplet diameter acc. to Sauter
d_T^*	–	dimensionless particle diameter, Eq. (2-25)
D	kgs^{-1}	mass flow rate of distillate
$\dot{D}$	$kmols^{-1}$	molar flow rate of distillate
D_V, D_L	m^2s^{-1}	diffusion coefficient for more volatile component in vapour mixture resp. in liquid
E_K		kinetic energy
f, f_i with i = 1,2,. . ..		function
F	kgs^{-1}	mass flow rate of feed
$\dot{F}$	$kmols^{-1}$	molar flow rate of feed
F_p	m^{-1}	packing factor of dry packed bed $F_p = a/\varepsilon^3$
$F_{P,exp}$	m^{-1}	experimentally determined packing factor of dry packed bed at flooding point for two-phase flow
$F_{P,0}$	m^{-1}	packing factor of dry packed bed for standard packing density, $F_{P,0} = a_0/\varepsilon_0{}^3$
F_V	$(m/s)\sqrt{kg/m^3}$ $Pa^{1/2}$ $m^{-1/2}kg^{1/2}s^{-1}$	gas or vapour capacity factor
$F_{V,Fl}$	$(m/s)\sqrt{kg/m^3}$ $Pa^{1/2}$ $m^{-1/2}kg^{1/2}s^{-1}$	gas or vapour capacity factor at flooding point
$F^*_{V,Fl}$	ms^{-1}	flood load factor
$F_{V,0}$	$(m/s)\sqrt{kg/m^3}$ $Pa^{1/2}$ $m^{-1/2}kg^{1/2}s^{-1}$	gas or vapour capacity factor at upper loading line

$F_{V,U}$	$(m/s)\sqrt{kg/m^3}$ $Pa^{1/2}$ $m^{-1/2}kg^{1/2}s^{-1}$	gas or vapour capacity factor at lower loading line
$F_V/F_{V,Fl}$	–	relative gas or vapour capacity
g	ms^{-2}	local acceleration due to gravity
G	kgm^{-3}	specific packing weight per unit volume
h	m	height of individual packing element
$h_L = V_L/V_S$	m^3m^{-3}	total liquid hold-up based on empty column
h_L^0	m^3m^{-3}	liquid hold-up based on free column volume, $h_L^0 = h_L/\varepsilon$
$h_{L,S}$	m^3m^{-3}	liquid hold-up above the loading line and below flooding point
$h_{L,Fl}^0$	m^3m^{-3}	liquid hold-up at flooding point, based on free column volume
h_{st}, h_d, h_H	m^3m^{-3}	static, dynamic, adhesion liquid hold-up
Δh_V	$kJ\,kmol^{-1}$	enthalpy of evaporation
H	m	height of packed bed
HETP	m	height equivalent to a theoretical stage
HTU_{0V}	m	height of an overall transfer unit related to vapour phase
HTU	m	height of a transfer unit
$(\Delta H)_i$	m	height of individual packed bed section
i		variable
k	–	proportionality factor
K	–	wall factor
K_A	N	lifting force
K_g	N	gravitational force
K_ψ	N	resistance force
K_0	N	surface force
K_R	N	shear force
K_η	N	viscosity force
K_1, K_2	–	constant K_1 and exponent K_2 for determining the resistance coefficient for single-phase flow of gas in packing for $Re_V < 2100$ and for Eq. (3-14)
K_3, K_4	–	numerical value of K_1 and K_2, for $Re_V \geq 2100$
K_P	–	parameter in Eq. by Teutsch [16], see Chap. 4
$K_{\rho V}$	–	correction factor for gas density
L	kgs^{-1}	mass flow rate of liquid
$\dot{L}$	$kmols^{-1}$ $kmolh^{-1}$	molar flow rate of liquid
$\dot{L}/\dot{V}$	–	molar flow ratio
m	–	exponent or parameter in Eq. (2-46)
M	$kgkmol^{-1}$	molar weight
n_i	–	number of test points in Eq. (2-31)

n_t	–	number of theoretical stages
n_t/H	m^{-1}	theoretical separation efficiency, number of theoretical stages per 1 meter of packing height
N, N_0	m^{-3}	packing density of any type of packed bed, standard packing density according to manufacturer's specifications
NTU_{OV}	–	number of overall transfer units
p	mbar	pressure
p_T	mm Hg	vapour pressure of pure component, Eq. (1-1)
Δp	Pa	pressure drop of irrigated packed bed
Δp_0	Pa	pressure drop of dry packed bed
$\Delta p/H$	Pam^{-1}	pressure drop per 1 m of irrigated packed bed
$\Delta p_0/H$	Pam^{-1}	pressure drop per 1 m of dry packed bed
$\Delta p/n_t$, $\Delta p/NTU_{0V}$	Pa	specific pressure drop
$\Delta p/\Delta p_0$	–	pressure drop quotient
$r = \dot{L}/\dot{D}$	–	reflux ratio
r_{min}	–	minimum reflux ratio
s	m	wall thickness of packing element
t	°C	temperature
$\bar{u}_T$	ms^{-1}	effective falling velocity of individual droplet in packed bed
u_F	ms^{-1}	mean film velocity
u_0	ms^{-1}	effective gas velocity, at which a droplet is kept in suspension
u_T	ms^{-1}	reduced gas velocity of individual droplet in packed bed
u_0	ms^{-1}	gas velocity in packed bed based on an empty column at flooding point for $h^0_{L,Fl} \rightarrow 0$
u_K	ms^{-1}	characteristic droplet velocity
u_L	$m^3m^{-2}h^{-1}$ $m^3m^2s^{-1}$	specific liquid load based on cross-sectional area of an empty column
$u_{L,U}$	$m^3m^2s^{-1}$	specific liquid load at lower loading line
u_R	ms^{-1}	relative velocity of phases
u_V	ms^{-1}	gas or vapour velocity, based on the cross-sectional area of an empty column
$\bar{u}_V$	ms^{-1}	mean effective gas or vapour velocity
$u_{V,Fl}$	ms^{-1}	gas or vapour velocity at flooding point, based on the cross-sectional area of an empty column
V	kgs^{-1}	mass flow rate of vapour
$\dot{V}$	$kmols^{-1}$	molar flow rate of vapour
V_L	m^3s^{-1}	volumetric flow rate of liquid
V_V	m^3s^{-1}	volumetric flow rate of vapour or gas
V_F	m^3	packing volume
V_L	m^3	liquid volume

V_S	m^3	empty column volume
V_1	m^3	volume of individual packing element
$\dot{W}$	$kmol s^{-1}$	molar flow rate of bottom product
W	$kg s^{-1}$	mass flow rate of bottom product
x	kmol $kmol^{-1}$	mole fraction of the more volatile component in liquid
x	–	contraction coefficient, Eq. (4-9)
X_{Fl}	–	flow parameter at flooding point
y	kmol $kmol^{-1}$	mole fraction of the more volatile component in vapour
y_m	kmol $kmol^{-1}$	mean mole fraction of the more volatile component
$y*$	kmol $kmol^{-1}$	mole fraction of the more volatile component in vapour at equilibrium state
z, z_{Fl}	–	quotient, $z = h_{L,S}/h_L$, $z_{Fl} = h_{L,Fl}/h_L$

Formula Variables, Greek Letters

α	deg	flow channel angle in packed bed, see Fig. 1-2
α	–	relative volatility
$\delta(i)$	%	relative error, based on experimental value of i
$\bar{\delta}_i$	%	mean relative error
δ_L	m	mean film thickness
ε, ε_0	$m^3 m^{-3}$	relative void fraction of any type of packing, standard void fraction
η	mPas $kg m^{-1} s^{-1}$	dynamic viscosity
λ	–	resistance coefficient in Eq. (3-2)
λ_L	–	resistance coefficient for two-phase flow, Eq. (4-1)
λ_0	–	phase flow ratio at flooding point, $\lambda_0=(V_L/V_V)_{Fl}$
μ	–	form factor of irrigated packing
ν	$m^2 s^{-1}$	kinematic viscosity
ρ	$kg m^{-3}$	density
σ_L	$mN m^{-1}$ $N m^{-1}$	surface tension of liquid for gas/liquid systems
τ	s	contact time
ψ	–	resistance coefficient for single-phase flow (vapour or gas flow) through packed bed, see Eq. (3-8)
ψ_0	–	resistance coefficient for single-phase flow for non-perforated packing elements, see Eq. (3-28)
ψ_{Fl}	–	resistance coefficient for single-phase flow of gas phase at flooding point

ψ_R, ψ_0	–	resistance coefficient for droplet swarm or for individual droplet falling in air packed bed
ψ_{VL}	–	resistance coefficient for two-phase flow
φ	–	ratio of diameters, d_S/d
φ_P	–	form factor of dry packing in Eq. (3-21)
$\Delta\rho$	kgm^{-3}	density difference, $\Delta\rho = \rho_L - \rho_V$
Φ	–	efficiency factor of dry packed bed $\Phi = a/a$
Σ	–	total (sum)

Dimensionless Numbers

$$B_L = \left[\frac{\eta_L}{\rho_L \cdot g^2}\right]^{1/3} \cdot \frac{u_L}{\varepsilon} \cdot \frac{1-\varepsilon}{\varepsilon \cdot d_P}$$ dimensionless liquid load

$$C_L = \frac{\rho_L \cdot \sigma_L^3}{\eta_L^4 \cdot g}$$ liquid number

$$Fr_{Fl}^* = \frac{u_{V,Fl}^2}{d_T \cdot g} \cdot \frac{\rho_V}{\Delta\rho}$$ extended Froude number

$$Fr_L^0 = \frac{u_L^2}{\varepsilon^2 \cdot g \cdot d}$$ Froude number of liquid

$$Fr_L' = \frac{u_L^2}{g \cdot d}$$ Froude number of liquid

$$Fr_L = \frac{u_L^2 \cdot a}{g}$$ Froude number of liquid

$$\mathrm{Re}_L = \frac{u_L}{a \cdot \nu_L}$$ Reynolds number of liquid

$$Re_L' = \frac{u_L \cdot d}{\nu_L}$$ Reynolds number of liquid

$$\mathrm{Re}_T = \frac{u_T \cdot d_T}{\nu_V}$$ Reynolds number of droplet

$$\mathrm{Re}_V = \frac{u_V \cdot d_P}{(1-\varepsilon) \cdot \nu_V} \cdot K$$ modified Reynolds number of vapour or gas

$$Re_V' = \frac{u_V \cdot d}{\nu_V}$$ Reynolds number of vapour or gas

$$We_{crit} = \frac{d_T \cdot u_V^2 \cdot \rho_V}{\sigma_L}$$ critical Weber number of droplet

$$\frac{We_L}{Fr_L} = \frac{\rho_L \cdot g}{\sigma_L} \cdot \left[\frac{\varepsilon \cdot d_p}{1-\varepsilon}\right]^2$$ ratio of Weber number to Froude number

$$T_L \cong 0.9 \cdot \left[\frac{F_V}{F_{V,Fl}}\right]^{2.8}$$ shear stress number

Indices

1	more volatile component
2	less volatile component
calc.	calculated value
crit	critical value
A	stripping section
D	distillate
e	extrapolated value
exp.	experimentally determined value
eff.	effective
F	feed
Fl	at flooding point
i	value of variable, e.g. $\Delta p/H$, $HTU_{0\,V}$, $\Delta p/n_t, \rho_L, \eta_V \ldots$ in height section, ΔH_i, $i=1 \ldots n$
j	$j = 1 \ldots n$
L	liquid
m	mean value
max	maximum value
min	minimum value
O	upper (at upper loading line)
0	based on standard packing density N_0
P	valid for packing
R	relative
S	valid for load range (above the loading line and below the flooding point)
T	top of the column
u, U	lower; lower limit
V	vapour or gas
W	bottom for mole fractions x, y or for water with physical characteristics ρ, η, σ

Mathematical Operator Symbols

$\in$	within the range $(\ldots)$
∂	partial differential

Abbreviations

[A], [ ]	Author or other sources of experimental data
Fig.	figure
BR	Białecki ring
BR-S	stacked Białecki rings
Fi	internal data – S. Filip, TU Wrocław I-13
Eq.	equation

Chap.	chapter
IS	Intalox saddle
Lu	internal data – Z. Ługowski, TU Wrocław I-13
max. Re_L	valid up to maximum Reynolds number of liquid, cp. Table 4-4 and Table 5-1a–5-1c
Mc	internal data – J. Maćkowiak, TU Wrocław I-13 (1975)
mK	with collars
MP	test point no....
NSW	packing produced by Norddeutsche Seekabelwerke (Nor-Pac)
oK	without collars
PR	Pall ring
RA	seal ring
Re/Chr/Mc	internal data – Rebhan, R. Chromik, J. Maćkowiak, RUB Bochum
TC	tube column filled with stacked packing
RR	Raschig ring
S	stacked packing elements
SBD	model of suspended bed of droplets

Material Designation

C	ceramic (stoneware or porcelain)
M	metal
PE	polyethylene
PP	polypropylene
PVDF	polyvinylidene fluoride

CHAPTER 1

Introduction

1

1.1 General Information on Packed Columns

In thermal process engineering, packed columns as well as tray columns are often used for heat and mass transfer processes in rectification, absorption and extraction as well as for the cooling of gases and liquids and wastewater and groundwater treatment. They are mainly used for counter-current gas/liquid flow. Figure 1-1 shows the schematic structure of a packed column with random packing elements.

Figures 1-2a and 1-2b show conventional as well as modern random and structured packings which are used in separation technology today.

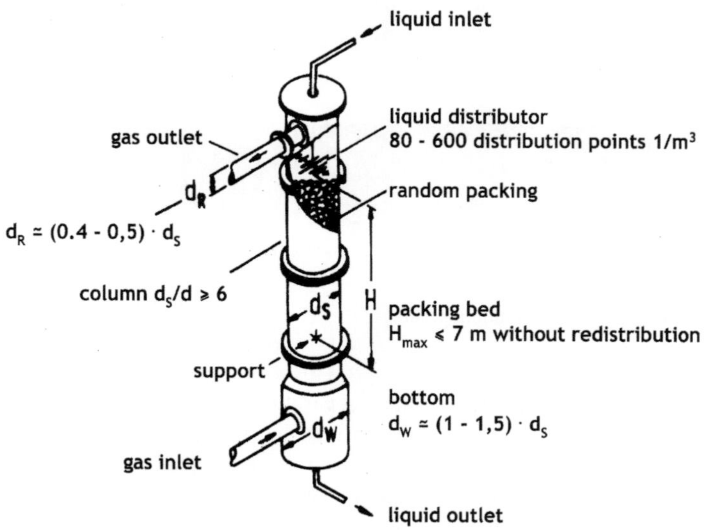

Figure 1-1. Schematic representation of packing

J. Maćkowiak, *Fluid Dynamics of Packed Columns*,
Chemische Technik/Verfahrenstechnik, DOI 10.1007/b98397_1,

Figure 1-2a. Commonly used random packings

Packed columns belong to the group of separation plants, in which the liquid, driven by gravity, flows down through the random or structured packing in the form of a trickle film or droplets. They are characterised by low pressure drop and a high operating range.

The simple design of packed columns makes them suitable for the most important basic operations in thermal process engineering – rectification, absorption, desorption and extraction – using various column internals, including the more conventional metal and ceramic ones as well as packing elements made of plastic.

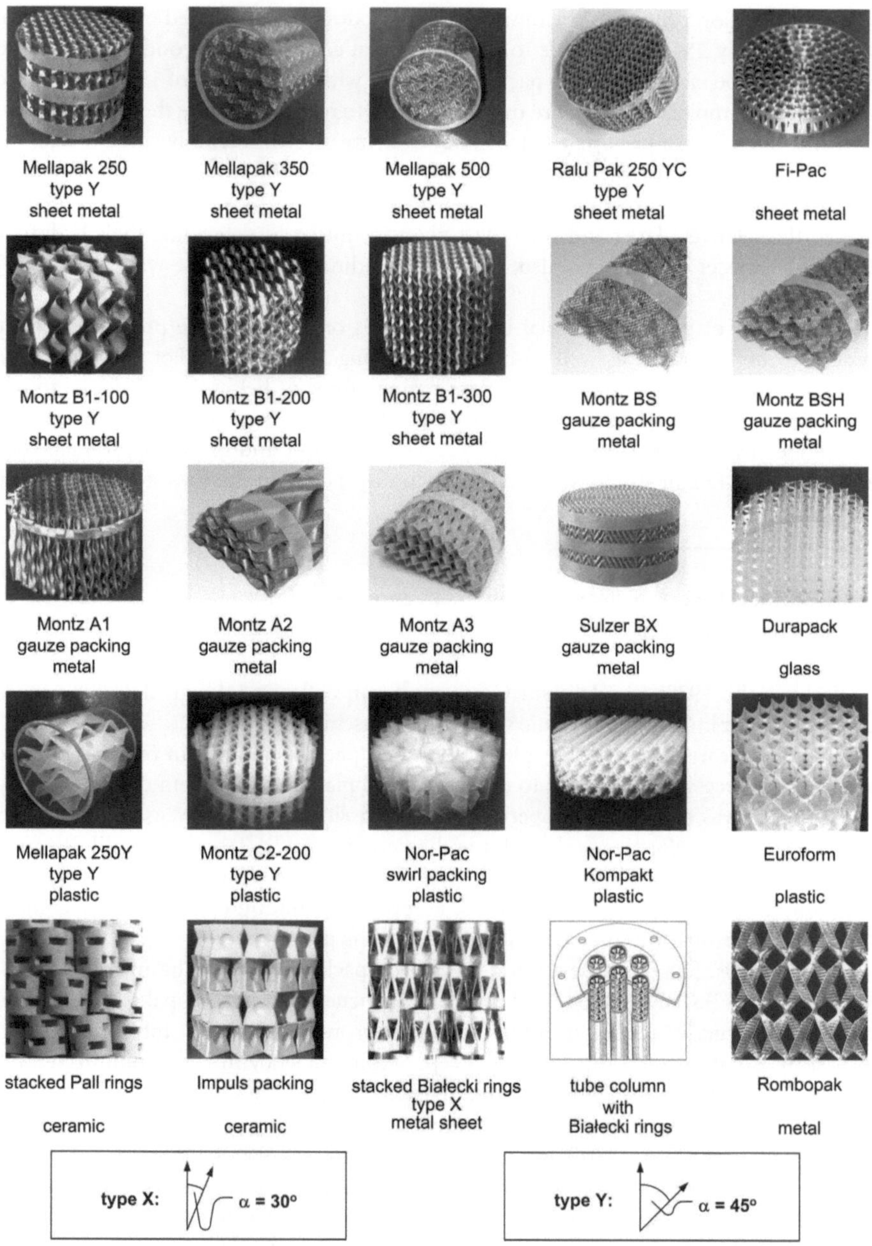

Figure 1-2b. Various structured packings, stacked packings and tube columns

In rectification processes, the dimensions of random packings used on an industrial scale are mostly 25–80 mm, due to the low capital cost and their good hydraulic and mass transfer behaviour. Smaller packing elements with dimensions of 15–25 mm used in smaller columns $d_S \leq 0.8$ m are more expensive to operate, despite their high separation efficiency. Larger packing elements, however, are not considerably cheaper than the 80 mm rings, yet their efficiency is significantly lower. In addition, there is little research material available on larger packing elements with dimensions between 50 and 90 mm, due to the relatively large and costly test plants required to determine their hydraulic and mass transfer parameters. Also, plants of this dimension are not available in every research centre.

The choice of packing material largely depends on the corrosive properties of the system to be separated as well as on the operating temperature. For the separation of aqueous alcohol solutions at moderate temperatures of below 100°C, for example, polypropylene or PVDF would be a comparatively cost-effective material. In the case of very aggressive substances, materials such as nickel, ceramic and/or plastic products such as PP, PVDF, ETFE or PTFE, would be more appropriate than stainless steel.

1.2 Development of Packed Columns and Their Significance in Rectification and Absorption Technology

As early as the 1930s, the US and Germany began collecting design data for packed columns, albeit initially for ceramic spheres and Raschig rings only.

For many years up to the late 1960s, the use of packed columns in rectification and absorption processes was limited to relatively small plants with column diameters of up to 1 m. This was due to the properties of Raschig rings, commonly used at the time, which became less effective as the column diameter increased. It was not until Pall rings were introduced by BASF in the 1960s, that these limitations were partly reduced. In addition, little consideration was given to an even distribution of liquid at the column top and the pre-distribution of gas on inlet below the packed bed.

In the 1960s, Sulzer set new standards in the packing design. The introduction of gauze packings BX and CY, followed by the sheet metal packing Mellapak 250 Y, opened up new constructive possibilities for the design of modern column internals, characterised by a high loading capacity, good separation efficiency in large columns as well as a low pressure drop throughout the operating range. These characteristics set the trend for the development of new, modern types of random and structured packings in Germany in the late 1970s and early 1980s, such as Nor-Pac, Hiflow rings, Top-Pak, Envipac, Dtnpac, VSP ring as well as Intalox metal packing, structured Montz packing, structured Ralupak and Rombopak etc. These developments were the result of intensive co-operation between industry and universities, in particular the Institute of Thermal Separation Processes of Bochum University in Germany.

The 1990s were marked by further developments in the area of random metal packings, such as Raschig Super rings [33] and Mc-Pac packing [34] as well as ceramic packings R-Pac and SR-Pac and structured glass packings such as Durapak.

In the same decade, the manufacturers Sulzer and Montz increased the range of structured packings, with specific packing ranges from 100 to 750 $m^2\ m^{-3}$ and channel angles of 30°-type X and 45°-type Y. They launched the newly developed, high-efficiency structured Optiflow packing as well as Mellapak Plus [30, 31].

As primary energy sources had been subject to high costs since the 1970s, this trend satisfied market demands for column internals that were characterised by low pressure drop and high loading capacity, a trend that is still prevalent today.

The advantages of using modern lattice-type and structured packings are particularly significant in the vacuum rectification of thermally unstable mixtures or separation processes with a high number of theoretical stages [1, 2] as well as in absorption technology. In the 1990s, packings were also successfully used for applications in pressure absorption and pressure rectification in the oil industry, where tray columns were being upgraded to structured packings.

A comparison between random packings and trays at identical top pressure p_T shows that, due to the low total pressure drop $\Delta p = p_T - p_W$ of packed columns, the temperature at the column bottom t_W at a given number of theoretical stages n_t is also lower.

The separation of thermally unstable mixtures requires low temperatures at the column bottom t_W, leading to an increase in the temperature difference $(t_H - t_W)$ between the heating medium t_H and the liquid mixture t_W. In addition, a low temperature at the column bottom allows for the use of low pressure steam to heat the reboiler, which is more cost-effective.

The rectification of mixtures under vacuum ensures that the product purities achieved at the top and at the bottom of the column are equally high at a lower reflux ratio, compared to normal pressure rectification, since the relative volatility α of the mixtures tends to increase as the top pressure drops.

The use of lower reflux ratios for the separation of mixtures under vacuum conditions results in a lower consumption of heating steam, compared to rectification under normal pressure.

Columns operating under vacuum could lead to energy savings of up to 25% [1], as shown in the two examples in Fig. 1-3 a and b.

The optimum operating pressure p_T at the column top, under which maximum energy efficiency can be achieved, is the minimum of the function: energy cost ratio of top pressure $C_0/C_B = f(p_T)$ see Fig. 1-3.

Where C_0 represents the operating costs of normal pressure rectification. Depending on the difference between the temperatures at the bottom t_W and at the top t_D of the column, and on the price of the heating steam C_H, the optimum operating temperature $t_{D,opt}$ of a condenser operated with cooling water is in the range of 40–55°C [1]. Once the temperature at the column top $t_{D,opt}$ has been defined, it is possible to determine the top pressure p_T, assuming that a pure top product is generated, based on the known Antoine equation:

$$p_T = \exp\left[A_1 - \frac{B_1}{(t_D + 273.15) + C_1}\right] \tag{1-1}$$

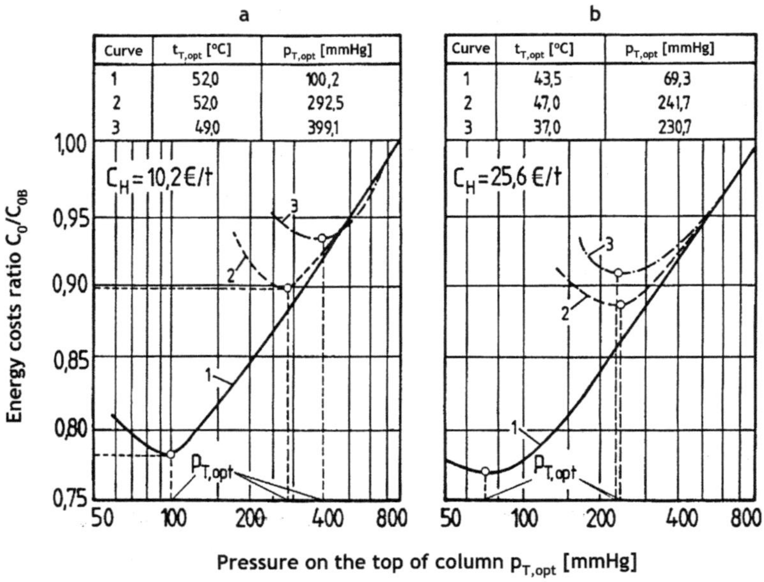

Figure 1-3. Energy cost ratio C_0/C_{0B} relating to column operation under normal pressure p_{TB}, as function of pressure at the column top p_T for various mixtures acc. to Billet, Ługowski, Maćkowiak [1]. Curve 1: System: toluene/ethylbenzene; Curve 2: System: benzene/toluene; Curve 3: System: methanol/ethanol. Assumption: Concentration at the top $x_D = 1$, at the bottom $x_W = 0$, in feed $x_F = 0.5$. Reflux ratio $r/r_{min} = 1.2$, heat steam cost $C_H = 10$ and 25 €/t, cooling water costs $C_K = 0.05$ €/t

where A_1, B_1 and C_1 are the Antoine constants for higher-volatility components. A list of numerical values for the constants A_1, B_1, C_1 for tested systems as well as additional references can be found, amongst others, in the well-known monographs by Gmehling and Kolbe [35] and Gmehling, Onken [36].

The monograph published by Billet [2] includes detailed information on additional energy saving measures, such as optimum feed concentration, use of heat pumps etc.

Absorption and desorption are the main range of industrial applications for random plastic packing elements, as those separation processes mostly occur within a moderate temperature range. Particular significance must be accorded to modern, randomly filled packing elements with lattice structures, which are characterised by extremely low pressure drop and high loading capacity, thus resulting in a smaller construction and lower operating costs. The discrepancy between random and structured packings in terms of their hydraulic and mass transfer behaviour has been decreasing in the last 10 years. Flue gas scrubbers are an exemplary application of modern Hiflow packing elements, e.g. in the power plant Ludwigshafen and Buschhaus, with column diameters of up to 9.4 m. Random Envipac packings, sizes 2 and 3, produced by ENVIMAC are used in Lurgi plants for direct gas cooling towers with column diameters of up to 8 m.

Applications of this kind, using random and structured packings in columns with diameters between 5.5 and 7.0 m, have now become common practice. Hence, there is continued demand for accurate design data for these columns.

The accurate design of random and structured packings with low specific pressure drop and high loading capacity for a number of separation tasks used in absorption, in rectification under vacuum and normal pressure as well as in the high pressure range can therefore be regarded as a significant contribution towards energy saving.

1.3 Brief Overview of Existing Monographs and/or Complex Reviews on Packed Column Design

The design of packed columns using random and structured packing is based on the determination of the column diameter d_S and the packing height H as well as the total pressure drop Δp at the operating point and at the flooding point. For this purpose, the following parameters for each packing element must be known:

(a) operating range, i.e. flooding point and lower loading line
(b) pressure drop of irrigated packing $\Delta p/H$ per 1 m of packed bed
(c) liquid hold-up h_L
(d) theoretical separation efficiency as n_t/H and/or height of transfer unit HTU_{OV} for the vapour phase or HTU_{OL} for the liquid phase

In addition, the number of theoretical stages n_t required for the separation of a mixture is necessary in order to calculate the packing height.

The following section outlines the most important studies, from the author's point of view, which have made a significant contribution towards the modelling of the fluid dynamics of packed columns.

The first important monograph was written by Kirschbaum [3], with its fourth and last edition published in 1969. It was the first work to introduce a simple, closed process for the dimensioning of 8–50 mm ceramic Raschig rings. Based on experimental data as well as simple models, it presented correlations for calculating the upper loading line, the pressure drop of two-phase flow as well as the number n_t/H, which is today known as theoretical separation efficiency, both at the upper loading line and at approx. 65% of the upper loading line.

Very early on, Kirschbaum [3] recognised the relationship between the separation efficiency n_t/H and the pressure drop $\Delta p/H$ per 1 m of packing height and developed the following correlation, based on approx. 300 experimental data items:

$$\frac{n_t/H}{\psi_L} \cong 0.13 \quad \left[m^{-1}\right] \tag{1-2}$$

which gave the total resistance coefficient ψ_L of two-phase flow according to Eq. (1-3) for a given packing diameter d, known pressure drop $\Delta p/H$ of the irrigated packing and vapour velocity u_V.

$$\frac{\Delta p}{H} = \psi_L \cdot \frac{u_V^2}{2} \cdot \frac{\rho_V}{d} \tag{1-3}$$

This concept was based on the assumption that the pressure drop Δp/H above 65% of the upper loading line for a packing size is system-independent.

Linking the pressure drop Δp/H and the separation efficiency n_t/H to avoid having to ascertain many unknown parameters was an idea that was further pursued by Billet [15] as well as other publications [16–18]. Kirschbaum's monograph [31] is a standard work, which is virtually as up to date now as it was when it was first published.

Mersmann [4, 5] evaluated the results published in other studies on rectification and absorption investigations using the air/water test system, and compiled design data for the 15–35 mm packing elements used in the 1960s, namely Intalox saddles, Pall rings and ceramic Berl saddles. Based on these studies, it is possible to determine the column diameter as well as the packing efficiency at the loading point for small packings with d < 35 mm. The derived correlations for the hydraulic behaviour of packed beds are based on a film model applicable to laminar and turbulent gas and liquid flows. Graphic correlations for flooding point, loading point, pressure drop of irrigated packings, liquid hold-up and height of transfer unit were also introduced – not in the operating range below the loading line, but in the maximum of the function $HTU_{OV} = f(u_V)$.

From today's point of view, the work published by Mersmann [4, 5] is an exemplary demonstration of how to derive general considerations on the hydraulic behaviour of random packings using the model of film flow in packed columns. It has maintained its scientific relevance until today [10, 11, 14, 19].

Beck [6] used a large amount of research material, primarily based on experimental data taken by Billet [7], which was published, amongst others, in BASF reports in the 1960s, and developed empirical correlations for calculating flooding point, pressure drop above the loading line as well as separation efficiency at the minimum of the curve $n_t/H = f(u_V)$. The advantages of the method developed by Beck lie mainly in the simple use of the correlations, which have a practical significance in the design of packed columns. The following points should be noted in connection with this method:

1. The separation efficiency can only be determined at the minimum of the function $n_t/H = f(u_V)$ – not in the entire operating range. The minimum of the function $n_t/H = f(u_V)$ is typical of random packings with small nominal dimensions and is partly determined by the system's hydraulic properties and its physical properties.
2. The correlation for determining the separation efficiency does not take into account the diffusion properties of the systems to be separated.

Weiß et al. [8] found that their measured pressure drop values Δp/H correlated satisfactorily, i.e. with a ±20% accuracy, to Beck's formula [6] for 35 mm metal Pall rings. In the case of 25 mm ceramic Pall rings, however, the measured values were 100% higher than the result of Beck's calculation [6].

Schmidt [9] introduced a concept for the design of columns packed with randomly filled 8–50 mm ceramic Raschig rings. He performed elaborate rectification tests in a column with a diameter of 0.3 m and a packing height of 2 m, using 7 binary mixtures at top pressures ranging between 66 and 1000 mbar, and used the results of his measurements to develop two empirical equations for calculating the mass transfer coefficient in the gas and liquid phases. In addition, he introduced correlations for determining the upper and lower loading line as well as the pressure drop and the liquid hold-up throughout the operating range.

The capacity diagrams developed by Schmidt [9] are still used as practical working diagrams today. A valuable contribution for practical application was also the introduction of the term "lower loading line" as well as the correlations from which the "lower loading line" can be analytically derived.

Brauer [10] as well as Brauer and Mewes [11] gave a detailed and critical overview of the individual publications which had been released up to 1967, and recommended various correlations which allowed the calculation of flooding point, pressure drop and mass transfer coefficient in the gas and liquid phases for Raschig rings, Pall rings, Intalox saddles and Berl saddles commonly used in the past. It should be noted that the correlations for the fluid dynamic design presented in these monographs were based on and/or directly taken from studies by Mersmann [4, 5] and Teutsch [12]. The entire international literature on mass transfer in the liquid phase was studied and presented in the form of dimensionless numbers. The calculation equations with an accuracy of approx. ±30% were derived on the basis of absorption measurements using the systems CO_2−water/air and/or O_2−water/air.

Using dimensionless numbers, Brauer and Mewes [11] were able to summarise approximately the results of experimental studies (literature data) on mass transfer in the gas phase in irrigated columns packed with Raschig rings, spheres and Berl saddles. There is no further information available on the accuracy of the formula; however, based on the attached diagrams, it can be assumed that the measured values are given with a relative error of ±20−50%. This correlation can be found even today, particularly in the latest work of Zech and Mersmann [13, 14].

A monograph published by Billet [15] elaborates on the hydraulic behaviour and the efficiency of random metal packings (Raschig rings and Pall rings) and, for the first time, of structured gauze packings BX, CY in rectification mostly under vacuum and normal pressure. The measurements used by Billet [15] were taken at large-scale BASF pilot plants with different distilliation systems, which is why they are particularly valuable for the development and verification of new calculation methods. Billet [15] modified the graphic Sherwood correlation for determining the gas velocity at the flooding point of 15−50 mm metal Pall and Raschig rings for rectification systems under vacuum, normal and high pressure.

The following formula for calculating the packing height is suggested on the basis of Billet's [15] introduction of the term *specific pressure drop* $\Delta p/n_t$:

$$H = n_t \cdot \left(\frac{\Delta p}{n_t}\right) / \left(\frac{\Delta p}{H}\right) [m] \tag{1-4}$$

The measurements taken by Billet [15] at a BASF distillation plant are presented by means of the correlations $\Delta p/H = f(F_V)_{\dot{L}/\dot{V}=1}$, $n_t/H = f(F_V)_{\dot{L}/\dot{V}=1}$, and $\Delta p/n_t = f(F_V)_{\dot{L}/\dot{V}=1}$. This method of presenting experimental results is still used in literature today.

This book also takes into account Billet's [15] experimental data on the hydraulic behaviour of packed columns for the development of a new method for the standard presentation of the fluid dynamics of columns with any type of packing design. The basic principles of the method for determining the separation efficiencies of packed columns have already been publicised in joint presentations and individual publications by the author and Billet [16–18].

Following years of co-operation with the author, Billet [21] has published a new monograph, in which he pursues another model for the fluid dynamics of random and structured packings under vacuum and normal pressure, based on film flow.

Reichelt's professorial dissertation [19] is a comprehensive and valuable literature study on the subject of random packings, focusing on the fluid dynamics of the two-phase flow for a number of different random ceramic packings. Based on the experimental results using the air/water test system, he developed empirical correlations for determining the pressure drop of single-phase flow. He presented a modified version of Mersmann's flooding point diagram and summarised and discussed the results of the most important studies previously published as well as the main equations, including their ranges of validity.

Zech and Mersmann [13, 14] were the first who attempted to systematically analyse the possibility of transferring absorption data to rectification processes. Using the theory of instationary diffusion for short contact times, it was possible for the first time to separate the volumetric mass transfer coefficient known from experiments into the mass transfer coefficient and specific mass transfer area. The comparison of measurements found in literature with the calculation based on the method developed by Zech and Mersmann [13, 14] showed good concurrence, which suggests that it is possible to predict the separation efficiency of distillation columns using absorption data. The analysis was carried out in the operating range below the loading line. The loading line was monitored using another modfied correlation by Mersmann [4]. As a result, the studies by Zech and Mersmann [13, 14] presented a closed concept for the dimensioning of ceramic Raschig rings, Berl saddles, Pall rings as well as metal Raschig rings.

The monograph by Strigle [20], published in 1987, provided correlations, working diagrams and a large amount of empirical data from a practical point of view, which are very useful for the design of packed columns in a diverse range of applications, such as rectification, absorption, desorption, liquid/liquid extraction, gas cooling etc.

Bornhüter [22] analysed the flow behaviour of a liquid in a column with a diameter of $d_S = 1$ m filled with modern lattice-type packings with nominal dimensions of 25–90 mm. He found that the liquid flowed in each packing as a stream of droplets, in addition to films and runlets, depending on the specific liquid load. Bornhütter [23] modified Mersmann's flooding capacity diagram (1965) and developed correlations for calculating liquid hold-up, mass transfer coefficient in the liquid and gas phases as well as effective mass transfer area for the lattice-type packings which he analysed.

Krehenwinkel [23] investigated the fluid dynamics and the mass transfer of packed columns in a range of high pressures using classic random packings, whereas Spiegel [24] and Ghelfi et al. [25] looked at structured packings. Their studies contain valuable experimental data, which has been incorporated in the SBD model described in this work.

Krehenwinkel's work [23] includes empirical correlations for calculating the gas velocity at the flooding point, the pressure drop and the volumetric mass transfer coefficient in the gas and liquid phases.

In 1990/1991, Maćkowiak [26, 27] described an additional method for determining the pressure drop of packed columns with any type of column internal throughout the entire operating range up to flooding point, based on the knowledge of the resistance factor ψ_{LV} for two-phase counter-current flow. The equations and numerical examples can be found in Chap. 5.

Finally, mention must be made of numerous studies carried out by SULZER-Winterthur, which have contributed enormously to the development of structured packings, such as Mellapak, gauze packings BX, CY, Optiflow as well as Mellapak Plus [30, 31]. They have also led to ground-breaking applications in the field of thermal separation processes in rectification in the vacuum, normal and high pressure range as well as in high-pressure absorption. Further information on this topic can be found, amongst others, in the publications [24, 28–32].

1.4 Conclusion Chapter 1

The above publications show that it is easy to determine the main dimensions of columns with classic packing elements, such as Pall rings, Raschig rings, Intalox saddles based on today's good theoretical knowledge of hydraulic and mass transfer principles. The current methods for the design of packed columns, however, cannot always be applied to any separation task, unless they have been verified by experiments, in particular when it comes to new types of lattice and structured packings. There are numerous experimental results available for conventional types of packings, such as Raschig rings, Intalox saddles and Pall rings made of ceramic and/or metal. In addition, there is a considerable amount of experimental data available on modern types of packings with large dimensions $d = 50-80$ mm and on sheet-metal packings, which are more suitable for industrial applications today. The data taken from experiments using thin-walled Pall rings, Hiflow rings, Nor-Pac packing elements, Envipac, Dtnpac, Tellerette, Glitsch CMR rings, VSP rings, Białecki rings, Raschig Super rings [33], Mc-Pac [34], R-Pac, sheet-metal packings produced by Sulzer, Montz and Raschig, Durapak, Rombopak, Glitsch made of metal, plastic and ceramic etc., shown in Figs. 1-2a and 1-2b, is summarised in this book and presented in a uniform manner, using existing and new theoretical knowledge on fluid dynamic principles of two-phase counter-current flow.

References Chapter 1

1. Billet R, Ługowski Z, Maćkowiak J. Optimaler Druck bei der Rektifikation. Chemie Technik, vol. 12 (1983) p. 69/76
2. Billet R. Energieeinsparung bei thermischen Stofftrennverfahren . Dr. A. Hüthig-Verlag, Heidelberg (1983)

3. Kirschbaum E. Destillier- und Rektifiziertechnik. Springer-Verlag, Berlin/Göttingen/Heidelberg 4.Edition (1969)
4. Mersmann A. Zur Berechnung des Flutpunktes in Füllkörperkolonnen . Chem.-Ing.-Techn., vol. 37 (1965) no. 3, p. 218/226
5. Mersmann A. Die Trennwirkung von Hohlfüllkörpern. Chem.-Ing.-Techn., vol. 37 (1965) no. 7, p. 672/680
6. Beck R. Ein neues Verfahren zur Berechnung von Füllkörperkolonnen . Published by VFF, Ransbach/Baumbach, Westerwald (1969)
7. Billet R. Recent Investigations of Metal Pall Rings. Chem. Eng. Prog., vol. 63 (1967) no. 9, p. 53/65
8. Weiß S, Schmidt E, Hoppe K. Obere Belastungsgrenze und Druckverlust bei der Destillation in Füllkörperkolonnen. Chemische Technik, vol. 27 (1975) no. 7, p. 394/396
9. Schmidt R. Zweiphasengegenstrom und Stoffaustausch in Schüttschichten – Beitrag zur Vorausberechnung von Füllkörpersäulen. VDI-Forschungsheft 550, Düsseldorf (1972)
10. Brauer H. Grundlagen der Einphasen- und Mehrphasenströmungen . Verlag Sauerländer, Aarau u. Frankfurt/Main (1971)
11. Brauer H, Mewes D. Stoffaustausch einschließlich chemischer Reaktionen . Verlag Sauerländer, Aarau u. Frankfurt/Main (1971)
12. Teutsch T. Druckverlust in Füllkörperkolonnen bei hohen Berieselungsdichten. Chem.-Ing.-Techn., vol. 36 (1964) no. 5, p. 496/503
13. Zech JB. Flüssigkeitsströmung und Stoffaustausch in berieselten Füllkörperschüttungen. Dissertation TU München (1978)
14. Mersmann A. Thermische Verfahrenstechnik. Springer Verlag, Berlin, New York (1980)
15. Billet R. Industrielle Destillation. Chemie Verlag, Weinheim (1973)
16. Billet R, Maćkowiak J. Allgemeines Verfahren zur Berechnung der Trennwirkung von Füllkörperkolonnen für die Rektifikation. Chem.-Ing.-Techn., vol. 55 (1983), no. 3, p. 211/213
17. Billet R, Maćkowiak J. Neues Verfahren zur Auslegung von Füllkörperkolonnen für die Rektifikation. Verfahrenstechnik vt, vol. 17 (1983) no. 4, p. 203/211
18. Billet R, Maćkowiak J. How to use the absorption data for design and scale-up of packed column. Fette, Seifen, Anstrichmittel, vol. 86 (1984) no. 9, p. 349/358
19. Reichelt W. Strömung und Stoffaustausch in Füllkörperapparaturen bei Gegenstrom einer flüssigen und einer gasförmigen Phase . Verlag Chemie (Reprotext), Weinheim, Bergstraße (1974)
20. Strigle RF. Random packings and packed Towers. Gulf Publishing Company, Houston, Texas, London (1987)
21. Billet R. Packed Towers –in Processing and Environmental Technology. VCH -Chemie Verlag, Weinheim (1995)
22. Bornhütter K. Stoffaustauschleistung von Füllkörperschüttungen unter Berücksichtigung der Flüssigkeitsformen. Dissertation: TU München, Dezember (1991)
23. Krehenwinkel H. Experimentelle Untersuchungen der Fluiddynamik und der Stoffübertragung in Füllkörperkolonnen bei Drücken bis zu 100 bar. Dissertation TU Berlin, Dezember (1986)
24. Spiegel L. Mellapak für die Hochdruckrektifikation und –absorption. Swiss Chem. vol. 8 (1986) no. 2a, p. 23–28
25. Ghelfi L, Kreis H, Alvarez JA, Hunkeler R. Structured Packing in Pressure Columns. Poster – Int. Conference and Exibition on Destillation and Absorption. Birmingham, England, 7–9 September (1992)
26. Maćkowiak J. Bestimmung des Druckverlustes berieselter Füllkörperschüttungen und Packungen. Staub-Reinhaltung der Luft vol. 50 (1990) no. 12, p. 455–463
27. Maćkowiak J. Pressure Drop in Irrigated Packed Columns. Chem. Eng. Processing vol. 29 (1991) no. 5, p. 93–105
28. Plüs RC, Ender Ch. A new high-efficiency packing „Mellaring – VSP". Achema 1991, presentation, conference volume (09.06.1991). Frankfurt a.M., Dechema-Verlag Sulzer Bros AG, Schweiz
29. Suess A, Spiegel L. Hold-up of Mellapak structured packings. Sulzer Bros AG (Schweiz) Ltd. – Separation Column, Winterthur
30. Süess Ph, Meier W. Optiflow-Improved efficiency in distillation and absorption columns. Technical information – Separation Columns- by Sulzer Chemtech, Winterthur
31. Mellapak Plus – A new generation of Structured packings. Technical information – Separation Columns – by Sulzer Chemtech, Winterthur

32. Spiegel L, Meier W. Structured packings – Capacity and pressure drop at very high liquids loads. cpp-chemical plants + processing, no. 1 (1995)
33. „Ein Hochleistungsfüllkörper stellt neue Maßstäbe" – Raschig-Super-Ring Nr. 1 Metall und Nr. 2 Metall. Technical information by Raschig, DFM 295 13607.3/30.05. (1996)
34. Maćkowiak J. Mc-Pac-ein neuer metallischer Füllkörper für Gas/Flüssigkeitssysteme. Chem.-Ing.-Techn. vol. 73 (2001) no.: 1+2, p. 74–79
35. Gmehling J, Kolbe J Thermodynamik 2. Edition, VCH-Verlag, Weinheim 1992
36. Gmehling J, Onken U Vapor-Liquid Equilibrium Data Collection, Part IVb, Part VId, Part VIe, Part VIIa, Part VIIb, Part VIIIa. DECHEMA Chemistry Data Series, Frankfurt from 1998 on

Two-Phase Flow and Operating Range 2

2.1 Hydraulic Processes in Packed Columns

The accurate design of packed columns requires a knowledge of the hydraulic characteristics of the respective packing element throughout the entire operating range. The gas velocity at the flooding point $u_{V,Fl}$ is a particularly important parameter in this context, as it is required for the calculation of the maximum loading capacity at a given specific liquid load u_L. The higher the loading capacity of the column in the case of countercurrent flow of the phases, i.e. the higher the gas velocity at the flooding point u_V, the smaller the required cross-section and diameter of the column.

When designing a packed column, it is also important to calculate the pressure drop of the irrigated packing $\Delta p/H$ per 1 m packed bed height and the respective liquid hold-up h_L. In recent years, these have become important technical-economic parameters. They allow the assessment of packed columns in terms of their suitability for the rectification of temperature-sensitive mixtures and for all gas/liquid systems, i.e. for absorption, desorption, cooling and preheating of gas as well as for absorption using slow chemical reactions, where the residence time of the liquid must be known.

If there is only gas flowing through the packing, i.e. $u_L = 0$, a pressure drop occurs in the column Δp_0, Fig. 2-1, and increasingly so, as the gas capacity F_V rises. The pressure drop also depends on the dimension and shape of the packing and is directly proportional to the packing height H. If the packing is irrigated by the downward countercurrent flow of the liquid, i.e. $u_L > 0$, there are mostly films and runlets forming on the individual packing elements, assuming these are small and non-perforated. In the case of highly perforated packing elements of any size, the so-called lattice work packings, which are used more and more frequently, there are not only films and runlets, but also a large number of droplets and sprays, which increases with the size of the packing. On the basis of today's knowledge, this applies to any type of packing of any shape and material, i.e. plastic, ceramic and metal, in particular in the case of very small and moderate liquid loads. In two-phase flow, once the gas has reached a certain velocity, the formation of additional droplets can occur through separation from films. The total liquid hold-up h_L of such droplets, runlets and films reduces the free cross-section available for the gas flow.

J. Maćkowiak, *Fluid Dynamics of Packed Columns*,
Chemische Technik/Verfahrenstechnik, DOI 10.1007/b98397_2,

Figure 2-1. Dependencies (**a**) $h_L = f(F_V)$, (**b**) $\Delta p/H = f(F_V)$, (**c**) $n_t/H = f(F_V)$ as examples of qualitative illustration of activities in packed columns

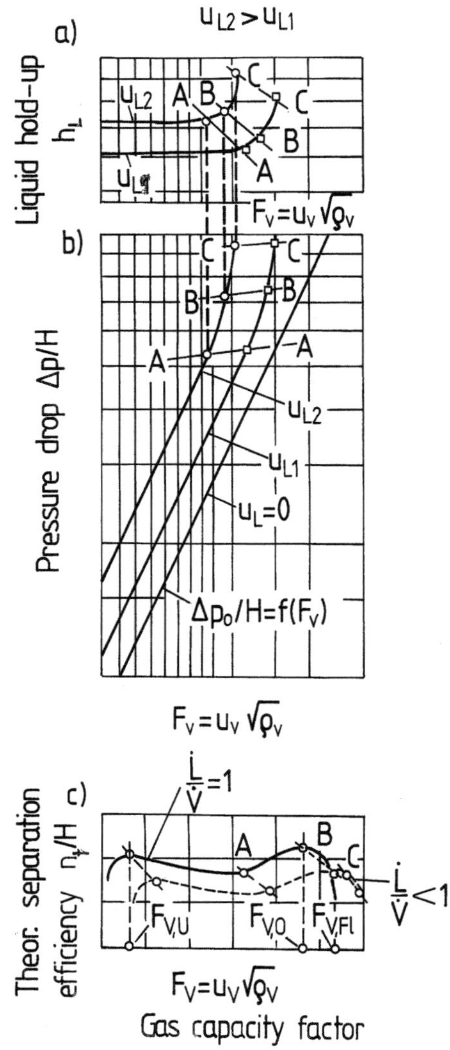

The effective void fraction of the packing (ε-h_L) decreases, which results in an increase in pressure drop $\Delta p/H$, based on the accepted law of resistance acc. to Darcy and Weisbach. As the specific liquid load u_L increases and the gas velocity u_V remains constant, the liquid hold-up h_L and the pressure drop of the irrigated packing $\Delta p/H$ also increase (see Fig. 2-1a,b).

Figure 2-1b shows a qualitative, double-logarithmic representation of the dependency of the pressure drop per 1 m packing height for irrigated random or structured packings $\Delta p/H$ on the gas capacity factor F_V, whereas Figs. 2-2a and 2-2b is a quantitative representation, applicable to randomly filled 25 mm metal Białecki rings, with the

Figure 2-2a. Hydraulic characteristics of random 25 mm metal Białecki rings, valid for the system air/water under normal conditions. Pressure drop $\Delta p/H$ as a function of the gas velocity u_V or gas capacity factor F_V, u_L with the specific liquid load u_L as a parameter

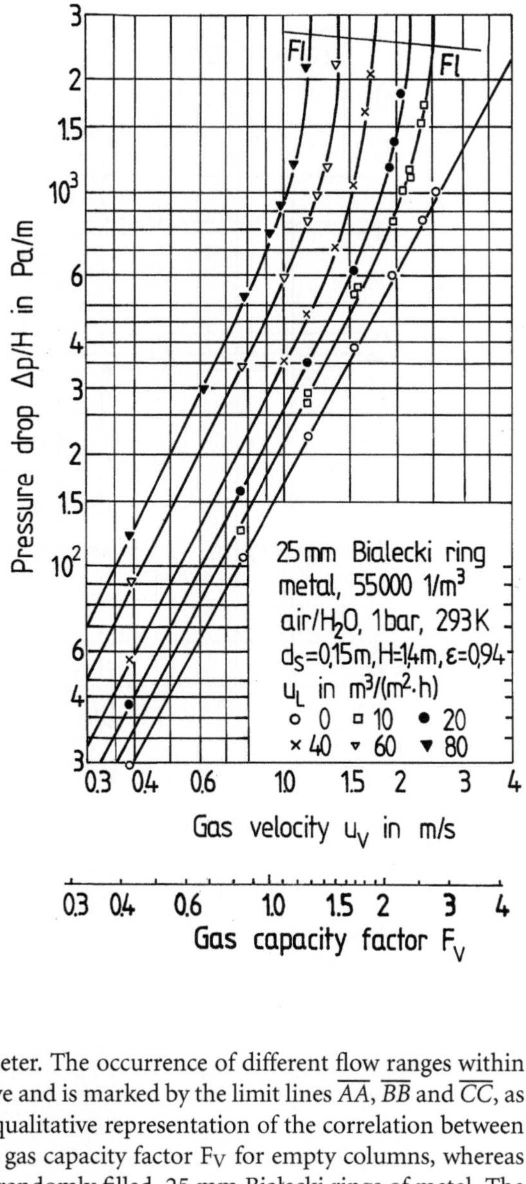

specific liquid load u_L as the parameter. The occurrence of different flow ranges within the operating range is very distinctive and is marked by the limit lines $\overline{AA}$, $\overline{BB}$ and $\overline{CC}$, as shown in Fig. 2-1. Figure 2-1a is a qualitative representation of the correlation between the total liquid hold-up h_L and the gas capacity factor F_V for empty columns, whereas Fig. 2-2b shows the correlation for randomly filled, 25 mm Białecki rings of metal. The diagrams are similar for other random or structured packings, analogous to Figs. 2-2a and 2-2b. The combined representation of the correlation between the pressure drop $\Delta p/H$ and liquid hold-up $h_L$and the vapour capacity factor F_V highlights the characteristic relationship between the parameters shown here.

Up to the first limit line $\overline{AA}$, the so-called loading line, the pressure drop curves $\Delta p/H$ of irrigated packings at small and/or moderate liquid loads $u_{L,1}$ and $u_{L,2}$ run parallel

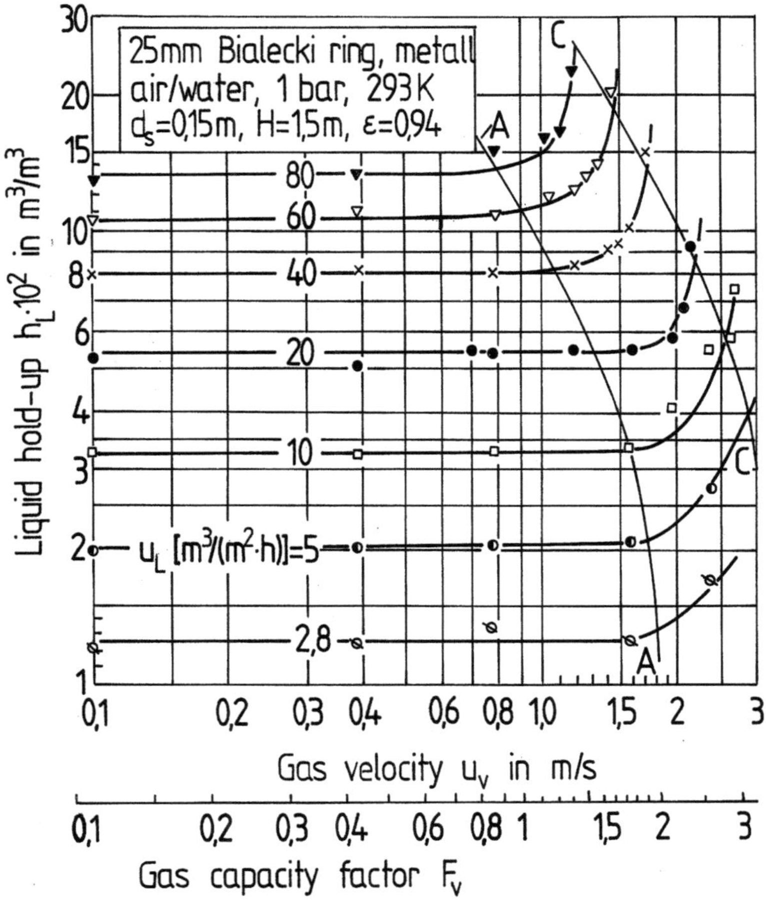

Figure 2-2b. Hydraulic characteristics of random 25 mm metal Białecki rings, valid for the system air/water under normal conditions. Liquid hold-up h_L as a function of the gas velocity u_V or gas capacity factor F_V, u_L with the specific liquid load u_L as a parameter

with the pressure drop curve $\Delta p_0/H$ of dry packings with $u_L = 0$, based on experimental research on modern types of packings. The loading line is at approx. 65% of the flood load factor $F_{V,FL}$ pertaining to the flooding line $\overline{CC}$. In this operating range, the liquid flow has practically no influence on the gas flow. Hence, the liquid hold-up h_L does not depend on the gas capacity factor F_V and/or the gas velocity u_V, but it increases with the liquid load u_L.

A further increase in the gas velocity u_V above the loading line $\overline{AA}$ results in an increase in the liquid hold-up h_L, Fig. 2-2b, and in the pressure drop, Fig. 2-2a. In this operating range, the shearing forces of the gas prevent the liquid from flowing off.

The pressure drop curves in the range between the loading line $\overline{AA}$ and the upper loading line $\overline{BB}$ rise steadily from 1.95 to 2.95 [2, 3, 12], Fig. 2-2a. At the same time, the liquid hold-up h_L increases, leading to an increase in the effective mass transfer area, see Fig. 2-2b. Packed columns are usually operated at gas velocities u_V between 30 and 80% of the gas velocity at the flooding point $u_{V,Fl}$.

The term "packed columns" refers to columns with randomly filled packings, with stacked or structured packings and tube columns filled aligned with packing elements.

2.2 Flooding Point

2.2.1 Flooding Mechanisms

Depending on the liquid load, the type and size of the packing and the physical properties, there are two different flooding mechanisms which can occur in packed columns:

(a) flooding at high phase flow ratios

$$\lambda_0 = \left(\frac{u_L}{u_V}\right)_{Fl} \tag{2-1}$$

where columns with large-surface packing elements are being filled with liquid with phase inversion

(b) flooding at small phase flow ratios λ_0, due to droplet entrainment in the gas phase

In packed columns, the liquid flows in counter-current to the gas, in the form of films and runlets. This applies in particular to small, non-perforated packing elements with low void fraction, e.g. $\varepsilon = 0.4-0.6$ m^3 m^{-3}. The void space of the packing elements contains dead space, which is filled with more and more liquid as the liquid load u_L increases. Flooding occurs when the entire column is filled with liquid. In the case of high specific liquid loads u_L and very low gas velocities u_V, i.e. at very high phase flow ratios at the flooding point λ_0, the so-called phase inversion occurs through the formation of bubbles, whereby the gas phase is dispersed and the liquid now forms the continuous phase.

The flooding mechanism is characterised by a steep increase in the pressure drop curve $\Delta p/H = f(F_V)$ as the gas capacity factor F_V increases, line $u_{L,3}$ in Fig. 2-1b.

The last 30 years have seen a new development in the packings design, as more and more metal and plastic packings with an open structure are being produced – the so-called lattice packings. Today, the use of perforated packing elements with dimensions of $d \geq 0.050$ m is common practice for industrial applications. The void fraction of these metal and plastic packing elements with $d \geq 0.025$ m is in the range of $\varepsilon = 0.92-0.98$ m^3 m^{-3}. They do not contain any dead space that can be filled with liquid. The operating range of modern "lattice packings" is therefore greater than that of full-surface packing elements, which means that the influence of the gas on the liquid flow is different, compared to the first type of packing.

In the case of highly perforated packing elements, or lattice packings, the formation of droplets occurs as a result of liquid dripping from the individual packing elements or from films and runlets, and can be entrained by the upward gas flow. This process is shown in Fig. 2-3, acc. to Bornhütter and Mersmann [66, 87]. The amount of droplets in the total hold-up ranges between 5 and 42%, depending on the type and size of the packing. As a rule, the number of droplets in the packing increases, the higher the liquid load u_L and the larger the size of the packing elements. Low phase flow ratios at the flooding point λ_0 are characteristic of columns operated in the vacuum and normal pressure range. This mechanism of droplet formation has been verified by specific studies carried out by Bornhütter and Mersmann [66, 87].

As a result, smaller as well as larger perforated packing elements are characterised by a different flooding mechanism, as droplets occur in liquid-fluidised beds, compared to

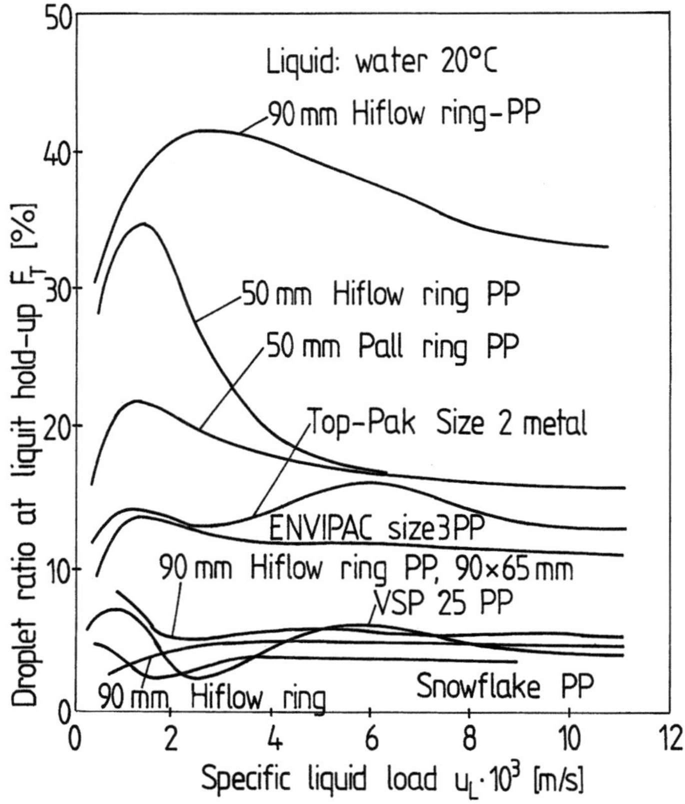

Figure 2-3. Droplet hold-up for various packings during water irrigation acc. to Bornhütter/Mersmann [66, 87]

classic, non-perforated packings, where flooding is caused by droplets being entrained upwards, in particular in the case of smaller phase flow ratios at the flooding point λ_0. The latter is characterised by a constant increase in pressure drop $\Delta p/H$, as the gas capacity factor F_V rises, Fig. 2-1b, lines $u_{L,1}$ and $u_{L,2}$.

Accurate gas velocities at the flooding point can only be obtained by means of pressure drop curves, as shown in Fig. 2-1b. The visual determination of the flooding point, on the other hand, can cause large errors.

2.2.2 Droplet Formation in Packed Columns

Droplet Formation

Droplet formation at lattice packing elements can occur without the influence of gas, as a result of the following: dripping from the sharp-edged packing elements, break-up of sprays and threads (runlets) at a certain distance from the individual packing element and through contact with the ledges of the packing elements. In the case of higher gas velocities, droplet formation can occur through transfer of the gas kinetic energy to runlets and films [66, 87].

The following considerations initially refer to the second case only and are based on the physical notion that a large number of droplets are sheared from the films and runlets of the liquid and entrained by the upward gas flow. As a result, the proportion of droplets to the total hold-up is even higher than shown in Fig. 2-2b. By shearing off droplets from the runlets, the gas transfers an impulse onto the droplets. The shear force of the gas K_R

$$K_R = C_0' \cdot \frac{\pi \cdot d_T^2}{4} \cdot \frac{\rho_V}{2} u_V^2 \tag{2-2}$$

is partly converted during droplet formation as a result of the surface tension σ_L. (C_0' is the so-called shear factor [58]). The surface force K_0 is given as:

$$K_0 = \pi \cdot d_T \cdot \sigma_L. \tag{2-3}$$

While in suspension, the displaced droplet has to overcome the gravitational force minus the lifting force.

$$K_g = \frac{\pi \cdot d_T^3}{6} \cdot (\rho_L - \rho_V) \cdot g = \frac{\pi \cdot d_T^3}{6} \cdot \rho_L \cdot g \quad \text{for } \rho_L >> \rho_V. \tag{2-4}$$

The balance of forces

$$K_R = K_0 + K_g \tag{2-5}$$

leads to the following correlation (2-6), acc. to Covelli, Mülli [58], based on the solution of Eq. (2-5) for a minimal effective gas velocity, at which it is just possible to separate droplets from runlets:

$$\bar{u}_{V,crit} = \left[\frac{16 \cdot 9\sigma_L \cdot \Delta\rho \cdot g}{3C_0'^2 \cdot \rho_V^2}\right]^{1/4} = 1.55 \cdot \frac{1}{\sqrt{C_0'}} \cdot \left(\frac{\sigma_L \cdot \Delta\rho \cdot g}{\rho_V^2}\right)^{1/4}. \tag{2-6}$$

For $C_0' = 1$, the formula (2-6) is identical to the one developed by Mersmann [38] and Levich [57] for droplets falling in gases and liquids.

This permissible velocity $\bar{u}_{V,crit}$ can be used to determine a critical droplet diameter $d_{T,crit}$, based on Eq. (2-5):

$$d_{T,crit} = \frac{6}{16} \cdot \frac{C_0' \cdot \rho_V \cdot \bar{u}_{V,crit}^2}{(\rho_L - \rho_V) \cdot g} = 2 \cdot \left[\frac{3\sigma_L}{2 \cdot g(\rho_L - \rho_V)}\right]^{1/2} \Rightarrow$$

$$d_{T,crit} = 2.44 \cdot \left[\frac{\sigma_L}{\Delta\rho \cdot g}\right]^{1/2} \tag{2-7}$$

Larger droplets with diameters greater than $d_{T,crit}$ break down into smaller droplets. Covelli, Mülli [58] presented the following theory for the shear factor C_0' in Eq. (2-2):

$$C_0' = 6.53 \cdot \sqrt{\frac{\sigma_L \cdot g \cdot (\rho_L - \rho_V)}{\rho_V^2 \cdot u_V^4}} \cdot \left(1 + 245.4 \cdot \mathrm{Re}_T^{-0.842}\right) \tag{2-8}$$

It was developed on the basis of data, taken at a heated tube for the test system water/steam at 1 bar, and applies to Reynolds numbers of droplets Re_T, acc. to Eq. (2-9), in the range of $Re_T = 300$ to $Re_T = 8000$:

$$Re_T = \frac{\bar{u}_V \cdot d_T}{\upsilon_V} \tag{2-9}$$

Based on Eq. (2-9), it is possible, for a given gas velocity $\bar{u}_V$, to calculate the shear factor C_0', acc. to Eq. (2-8), followed by the critical effective gas velocity $\bar{u}_{V,crit}$, acc. to Eq. (2-6), if the Reynolds number Re_T is known.

If the following is true:

$$\bar{u}_V \geq \bar{u}_{V,crit} \tag{2-10}$$

then droplet formation occurs in the packing by shearing from films and runlets. In the case of high gas velocities, the shear factor C_0' is expected to be minimal, acc. to Eq. (2-8). These conditions are most likely to trigger droplet formation.

The formation of droplets is usually described by means of the so-called critical Weber number We_{crit}:

$$\frac{\bar{u}_{V,crit} \cdot d_T \cdot \rho_V \cdot C_0'}{\sigma_L} = \frac{F_{V,crit}^2 \cdot d_T \cdot C_0'}{\varepsilon^2 \cdot \sigma_L} = We_{crit} \cdot C_0' / \varepsilon_2 = const. \Rightarrow$$

$$We_{crit} \cong \frac{12 \cdot \varepsilon^2}{C_0'} \tag{2-11}$$

which is based on the ratio of the shear force K_R, acc. to Eq. (2-2), to the surface force K_O, Eq. (2-3). Equation (2-11), developed by Wallis and quoted by Stichlmair [55], applies to low-viscosity mixtures ($\eta_L \approx 1$ mPas).

If the critical Weber number is not reached, there is no droplet formation from runlets and films. This is expected to be the case for high liquid loads, for random packings with low void fraction as well as for systems with extremely low surface tension σ_L, i.e. for pressure rectification.

Droplet Entrainment

The droplets forming in the packing can only be entrained if the gas velocity is sufficiently high. The gas velocity can be ascertained from the balance of forces acting on a suspended droplet. The thrust force of the gas K_ψ must be equated to the gravitational force K_g minus the lifting force K_A. The following correlation applies:

$$K_\psi = K_g - K_A \Rightarrow$$

$$\psi_{Fl} \cdot \frac{\bar{u}_V^2 \cdot \rho_V}{2} \cdot \frac{d_T^2}{4} = \frac{\pi \cdot d_T^3}{6} \cdot (\rho_L - \rho_V) \cdot g \tag{2-12}$$

Solving Eq. (2-12) for $\bar{u}_V$ gives Eq. (2-13)

$$\bar{u}_V = \sqrt{\frac{4}{3} \cdot \frac{d_T \cdot g}{\psi_{Fl}}} \cdot \sqrt{\frac{\rho_L - \rho_V}{\rho_V}} \tag{2-13}$$

Estimating the Lower Limit, Which Allows Droplet Formation from Films and Runlets in Packed Columns

It is possible to give an approximate estimation of the lowest effective gas velocity $u_{V,Fl,min}$ for the test system air/water at 20 °C and 1 bar, which allows droplet entrainment from the packing. This is based on the assumption that the packing elements are sufficiently large to allow the substitution of the same resistance coefficient $\psi_{Fl} \cong \psi_R$ into Eq. (2-13) that applies to a droplet swarm in an empty column. It can be calculated using Chao's formula (2-14), quoted by Soo [59]:

$$\psi_R = \frac{32}{Re_T} \left[1 + 2 \cdot \frac{\eta_L}{\eta_V} - 0.314 \cdot \frac{1 + 4 \cdot (\eta_L / \eta_V)}{\sqrt{Re_T}} \right] \tag{2-14}$$

For the test system air/water, the droplets in packed columns reach an average diameter of $d_T \cong 0.003$ m. u_V is now substituted for the value $u_V \cong 1.5\ ms^{-1}$. The objective is to find out whether, given the above gas velocity, it is still possible for droplets of this size to be entrained from the packing. Equation (2-9) gives the value $Re_T = 268.95$ and Eq. (2-14) leads to a resistance coefficient ψ_R of 12.68, which is then used to calculate the gas velocity u_V, acc. to Eq. (2-13), as $1.5\ ms^{-1}$. The example shows that, at a gas velocity of $1.5\ ms^{-1}$ or higher, it is still possible for droplets to be entrained in columns containing larger packing elements of, e.g., $d \geq 0.050$ m. Given a packing size of $d \cong 0.090$ m and $u_V \approx 1.5\ ms^{-1}$, the liquid load at the flooding point is in excess of $100\ m^3m^{-2}\ h^{-1}$ [27, 28]. The maximum operating range for larger, perforated packing elements with $d \geq 0.050$ m, at which droplet formation and entrainment can occur, is relatively wide. However, it decreases with the size of the packing elements. It is difficult to provide a more accurate estimation of the lower limit of the model's validity, which assumes that flooding occurs through droplet entrainment by the upward gas flow. This is due to the fact that no numerical values are currently available for the resistance coefficient in connection with droplet fall in the packing.

In addition, it is not yet possible to estimate the proportion of droplets formed through dripping from the packing edges and those formed through shearing. The lower value of the gas velocity in the air/water system is estimated to be $u_V \geq 1-1.5\ ms^{-1}$. In the case of F_V factors, it is in the region of 1.1–1.65 or considerably lower.

Based on Eq. (2-13), it is possible to calculate the gas velocity u_V that allows the suspension of droplets with a diameter of d_T. If this value is lower than the gas velocity $u_{V,Fl}$, it is no longer possible for droplets with d_T to be entrained by the upward gas flow. In such cases, it is recommended to describe the flooding mechanism in packed columns using the *film gas thrust shear force model* or *film model* developed by Mersmann [2, 3], see Sect. 2.2.3.

Bornhütter (1991) [66, 87] carried out experiments to analyse the formation of droplets in a column with a diameter of 1 m containing modern packing elements, such as Hiflow ring, Pall ring, VSP ring, Envipac, Snowflake, with diameters ranging between 25 and 90 mm and using liquids with viscosities between 1 and 27 mPas. He found that the proportion of droplets to the hold-up ranged between 5 and 45%, see Fig. 2-3. What all types of packings had in common was the fact that the proportion of droplets remained largely constant as the specific liquid load increased, whereas the proportion increased if the loads were reduced, falling below $u_L = 0.003\ ms^{-1}$.

Based on the theoretical considerations illustrated above and the experimental data found in literature [66], it can be assumed that, due to the constant occurrence of droplets in the packing, the flooding point mechanism in packed columns is defined by the entrainment of droplets.

2.2.3
Literature Overview – Status of Knowledge

The analysis of most representative correlations for evaluation of flooding gas velocity in packed column was provided by Kister [88], Piche [73], Kuźniewska-Lach [74], Grabbert [92] et al.

The methods described in the relevant literature for determining the gas velocity at the flooding point $u_{V,Fl}$ can be divided into two groups. The first group includes graphic correlations developed by Sherwood, Shipley and Holloway [4], Lobo, Leva [6], Billet [5], Eckert, Kafarov, Planowski [1, 50], Kirschbaum [19], Strigle [31] etc., who assume droplet formation at the flooding point. Table 2-1 shows the individual developmental stages of the most important correlations developed by this group (no. 1–7). The second group of methods assumes film flow in the packing, see correlations 8–14 in Table 2-1 [1, 4, 5, 6, 8, 9, 37–89, 94].

The first group of flooding point correlations is based on the assumption that, in accordance with visual observations, droplets are formed in the void spaces of the packing elements, which then fall down into the packing elements situated below, see Sect. 2.2.1.

At the flooding point, the balance of forces acting on a suspended droplet leads to correlation (2-12), which is then transposed to give the following equation for the flood load factor $F^*_{V,Fl}$:

$$F^*_{V,Fl} = \frac{F_{V,Fl}}{\sqrt{\rho_L - \rho_V}} = \sqrt{\frac{4}{3} \cdot \frac{d_T \cdot g}{\psi_0}} \quad \left[\mathrm{ms}^{-1}\right]. \tag{2-15}$$

The term $\sqrt{(4/3) \cdot d_T \cdot g \cdot \psi_0^{-1}}$ in Eq. (2-15) is summed up to the so-called flood load factor $F^*_{V,Fl}$, which can only be determined experimentally for packed columns. Here, the quotient, Eq. (2-15), is given by dividing the gas capacity factors $F_{V,Fl}$, known from experiments, by the root of the density difference $(\rho_L - \rho_V)$ and applied to the operating conditions at the flooding point using the term $(L/V)_{Fl} \cdot \sqrt{\rho_V/\rho_L}$ (2-16).

$$F^*_{V,Fl} = f(X_{Fl}) \quad \text{with } X_{Fl} = \left(\frac{L}{V}\right)_{Fl} \cdot \sqrt{\frac{\rho_V}{\rho_L}} \tag{2-16}$$

The term X_{Fl} in Eq. (2-16), known as the flow parameter, is the root of the ratio of the kinetic energy of the liquid $E_{K,L}$ to the kinetic energy of the gas $E_{K,V}$. This type of capacity diagram, based on Eq. (2-16), was adopted by Böden [1, 4, 8, 55] and is still used today [40–48], mostly for the air/water system.

Figure 2-4a shows a typical capacity diagram for 15 mm metal Pall rings, based on Eq. (2-16). Figures 2-4b and 2-4c show further examples of capacity diagrams for various random and structured packings, which were developed on the basis of recent experiments. The figures clearly show to what extent the flooding point is influenced by the size, type and material of the packing. These variables were initially described by introducing the quotient a/ε [1] (Eq. (2-17)) and later by using the packing factor $F_P = a/\varepsilon^3$ of the dry packing [4], see Eq. (2-18).

$$\frac{u^2_{V,Fl}}{2 \cdot g} \cdot \frac{a}{\varepsilon} \cdot \frac{\rho_V}{\rho_L} = f\left[\left(\frac{L}{V}\right)_{Fl} \cdot \sqrt{\frac{\rho_V}{\rho_L}}\right] \tag{2-17}$$

The above correlation (2-17) was developed by Walker et al. [1] as early as 1937, on the basis of experiments using the test system air/water. Spirals and ceramic Lessing rings

Table 2-1. List of correlations for predicting the flooding point; A* ≡ valid for air/liquid systems, R* ≡ valid for destillation systems

No.	Author	Correlation	Remarks	Lit.
1	**Sherwood, Shipley and Holloway** (1938)	$u_{V,Fl}^2 \cdot \frac{\rho_V}{\rho_L \cdot g} \cdot \frac{a}{\varepsilon^3} \cdot \left(\frac{\eta_L}{\eta_{L,W}}\right)^{0.2} = f\left(\frac{L}{V} \cdot \sqrt{\frac{\rho_V}{\rho_L}}\right)_{Fl}$	A*, for ceramic Raschig rings	[4]
2	**Kirschbaum** (1969)	$u_{V,Fl} \cdot \sqrt{\frac{\rho_V}{\rho_L}} = \sqrt{\text{const.} \cdot d_T} \approx \text{const.}$	R*, for ceramic Raschig rings	[19]
3	**Lobo et al.** (1945)	$u_{V,Fl}^2 \cdot \frac{\rho_V}{\rho_L \cdot g} \cdot F_P \cdot \left(\frac{\eta_L}{\eta_{L,W}}\right)^{0.2} = f\left(\frac{L}{V} \cdot \sqrt{\frac{\rho_V}{\rho_L}}\right)_{Fl}$	A*	[6]
4	**Eckert** (1963 and 1970)	$u_{V,Fl}^2 \cdot \frac{\rho_V}{\rho_L \cdot g} \cdot (F_P)_{exp} \cdot \frac{\rho_W}{\rho_L} \cdot \left(\frac{\eta_L}{\eta_{L,W}}\right)^{0.2} =$ $f\left(\frac{L}{V} \cdot \sqrt{\frac{\rho_V}{\rho_L}}\right)_{Fl}$	A*, $F_P = f\left(\frac{L}{V} \cdot \sqrt{\frac{\rho_V}{\rho_L}}\right)$; $\delta\left(u_{V,Fl}\right) \leq \pm 50\%$ *for R**	[9, 8]
5	**Billet** (1968 and 1979)	$\frac{u_{V,FL}}{3.14} \cdot \sqrt{(F_P)_{exp} \cdot \frac{\rho_V}{\rho_L} \cdot \frac{\rho_W}{\rho_L} \cdot \eta_L^{0.12}} =$ $f\left(\frac{L}{V} \cdot \sqrt{\frac{\rho_V}{\rho_L}}\right)_{Fl}$	R*, $\delta\left(u_{V,Fl}\right) \leq \pm 15\%$, for metal Raschig and Pall rings	[5]
6	**Billet -Maćkowiak** (1985)	$F_{V,Fl}^* = \left(u_{V,Fl} \cdot \sqrt{\frac{\rho_V}{\rho_L - \rho_V}}\right)_{N=const.} =$ $f\left(\frac{L}{V} \cdot \sqrt{\frac{\rho_V}{\rho_L}}\right)_{Fl}$	A* and R*, $\delta\left(u_{V,Fl}\right) \leq \pm 10\%$	[40–48]
7	**Planowski – Kafarow** (1972)	$\log\left(u_{V,Fl}^2 \cdot \frac{\rho_V}{\rho_L \cdot g} \frac{a}{\varepsilon^3} \cdot \eta_L^{0.16}\right) =$ $C - 1.75 \cdot \left(\frac{L}{V} \cdot \sqrt{\frac{\rho_V}{\rho_L}}\right)^{1/4}$	A*: C = +0.022 R*: C = −0.125	[11, 50]
8	**Mersmann** (1965) **Vogt** (1985)	$\frac{\Delta p_0}{\rho_L \cdot g \cdot H} = f\left[\left(\frac{\eta_L}{\rho_L \cdot g^2}\right)^{1/3} \cdot \frac{u_L}{\varepsilon} \cdot \frac{1-\varepsilon}{\varepsilon \cdot d_P}\right] = f(B_L)$	$B_L \sim \delta_L/d_h$ A*, $\delta\left(u_{V,Fl}\right) \leq \pm 20\%$	[2, 3, 34]
9	**Reichelt** (1973)	$\frac{\Delta p_0}{\rho_L \cdot g \cdot H} =$ $f\left[\left(\frac{\eta_L}{\rho_L \cdot g^2}\right)^{1/3} \cdot \frac{u_L}{\varepsilon} \cdot \frac{1-\varepsilon}{\varepsilon \cdot d_P \cdot K} \cdot \left(\frac{\nu_V}{\nu_L}\right)^{0.1}\right]$ $\frac{\Delta p_0}{\rho_L \cdot g \cdot H} =$ $\left[50 \cdot \left(B_L^+\right)^{0.16} + 2.5 \cdot 10^6 \cdot \left(B_L^+\right)^{1.9}\right]^{-1}$	A*, $\delta\left(u_{V,Fl}\right) \leq \pm 20 - 50\%$ acc. to [37], valid for $d_S/d = 1.6-1.7$	[12]

Table 2-1. (continued)

No.	Author	Correlation	Remarks	Lit.
10	**Schumacher** (1976)	$B_L^* =$ dimensionless liquid load, C = gas flow parameter $B_L^* = \frac{1}{x_P} \cdot \sqrt{\frac{u_L^2}{d \cdot g} \cdot \left(\frac{x_P \cdot \nu_L}{d \cdot u_L}\right)^{0.1}}$; $C = \frac{1}{x_P} \cdot \sqrt{\frac{u_V^2}{d \cdot g} \cdot \frac{\rho_V}{\rho_L} \cdot \frac{\psi_V}{\psi_{Luft}}}$	A^* and R^* $x_P = f(d)$ $x_P =$ packing parameter, $C = f(B_L^*)$	[10]
11	**Blaß and Kurz** (1976)	$Re_{V,Fl} =$ $k_0 \cdot \left(\frac{\eta_L}{\eta_V}\right)^{0.85} \cdot \left[\frac{d_h}{\left(\frac{3 \cdot \nu_L^2}{g}\right)^{1/3}} - 1\right]^{m_1} \cdot Re_{L,Fl}^{m_2}$	$A^*, k_0, m_1, m_2 =$ const. $\delta(u_{V,Fl}) \leq \pm 50\%$	[37]
12	**Kister/Gill** (1992)	GRDC- graphical correlation, extended flood load factor $F_{V,Fl}^* = f(X_{Fl})$	flow parameter $X_{Fl} = \left(\frac{L}{V}\right)_{Fl} \cdot \sqrt{\frac{\rho_V}{\rho_L}}$	[88]
13	**Bornhüter/ Mersmann** (1992)	graphical correlation $\frac{\Delta p_0}{\rho_L \cdot g \cdot h} \cdot P = f(B_L) \qquad P \equiv$ $\frac{\varepsilon - h_{L,Fl}}{a \cdot d_T} \cdot (1 - h_{L,Fl})^{2.5}$	extended for lattice packings 25–90 mm	[66, 87]
14	**Billet, Schultes** (1995)	$u_{V,Fl} = \sqrt{\frac{2g}{\xi_{Fl}}} \cdot \frac{(\varepsilon - h_{L,Fl})^{1.5}}{\varepsilon^{0.5}} \cdot \sqrt{\frac{h_{L,Fl}}{a}} \cdot \sqrt{\frac{\rho_L}{\rho_V}}$ with $h_{L,Fl} = 0.3741\varepsilon \cdot \left(\frac{\eta_L}{\rho_L} \cdot \frac{\rho_w}{\eta_w}\right)^{0.05}_{\eta_L > 10^{-4}}$ and $h_{L,Fl} = f\left(\frac{\eta_L}{\rho_L}\right), \varepsilon/3 \leq h_{L,Fl} \leq \varepsilon$	$\xi_{Fl} =$ resistance coefficient $= f\left(C_{Fl} \cdot \left(\frac{L}{V} \cdot \sqrt{\frac{\rho_L}{\rho_V}} \cdot \left(\frac{\eta_L}{\eta_V}\right)^{0.2}\right)^{2 \cdot n_{Fl}}\right) n_{Fl} =$ $f(X_{Fl})$ $C_{Fl} =$ packing specific constant for $u_L <$ 80 $m^3(m^2\ h)^{-1}$, valid for 44 packing types	[89]
15	**Grabbert, Bonitz (1995)**	$u_{V,Fl} = u_{V,Fl,film} \cdot \gamma_{film} + u_{V,Fl,SBD} \cdot \gamma_{SBD}$ with $\gamma_{film} = \left(1 + 7.486 \cdot 10^{-3} \cdot \frac{\varepsilon}{\lambda_{0,Fl}}\right)^{-1}$	Compilation of two models: $u_{V,Fl,\ film}$ acc. to film model by Mersmann (1965) $u_{V,Fl,\ SBD}$ acc. to SBD model by Maćkowiak (1991)	[92]

Sy	System	p_T [mbar]	d_S [m]	H [m]	N [1/m³]	Ref.
⦶	chlorobenzene / ethyl benzene, $\dot{L}/\dot{V}=1$	33	0.22	1.45	250000	A
⊖		66.7				
⊘		133				
◒	toluene/n-octane	80	0.22	1.45	259820	A
◓		133				
●	ethyl benzene/styrene	133	0.5	1.5	236000	[5]
			0.8	2	224000	
○	air/water, 293 K	1000	0.15	1.25	229000	A
◆	methanol/ethanol, $\dot{L}/\dot{V}=1$	1000	0.5	1.5	230700	[5]

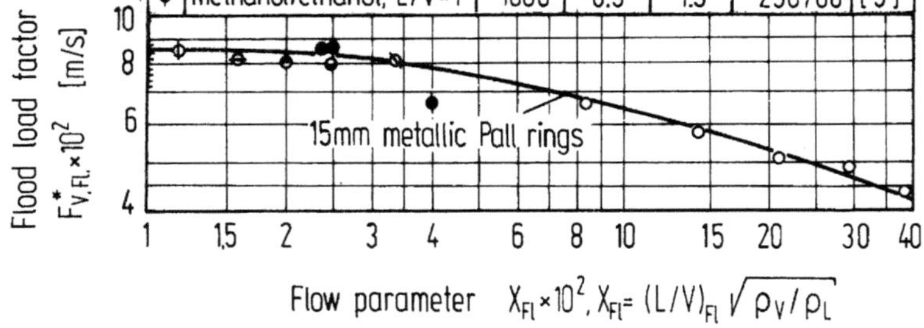

Figure 2-4a. Capacity diagram $F^*_{V,Fl} = f(X_{Fl})$ for random 15 mm metal Pall rings

were among the types of packings used. Sherwood, Shipley and Holloway [4] further modified the model developed by Walker et al. [1] by introducing the empirical factor $(\eta_L/\eta_{L,W})^{0.2}$ and the effective gas velocity $u_{V,eff.} = u_V/\varepsilon$ into Eq. (2-17). The numerical factor 1/2 in Eq. (2-17) was dropped. The result is as follows:

$$\frac{u_{V,Fl}^2 \cdot \rho_V}{g \cdot \rho_L} \cdot \frac{a}{\varepsilon^3} \cdot \left[\frac{\eta_L}{\eta_{L,W}}\right]^{1/5} = f\left[\left(\frac{L}{V}\right)_{Fl} \cdot \sqrt{\frac{\rho_V}{\rho_L}}\right] \tag{2-18}$$

This type of presentation of flooding point data is still used by packing manufacturers today, albeit with minor modifications [31, 88]. There have been a number of attempts over the years to improve the accuracy of the correlation expressed in Eq. (2-18). Lobo et al. [6], Eckert [9] introduced the quotient $(\rho_{L,W}/\rho_L)$ into the left-hand side of Eq. (2-18) and proposed to replace the packing factors $F_P = a/\varepsilon^3$ of the dry packing with experimentally determined variables $F_{p,exp}$, in order to show the flooding point for all types of packings in a single load curve. The numerical values for the experimentally derived packing factors $F_{P,exp}$, using the test system air/water, were listed the study [9] in relation to various operating ranges. In Eckert's correlation, the packing factor $F_{P,exp}$ is not a packing constant, due to the fact that it is also load-dependent.

Billet [5] found new numerical values for the packing factors $F_{P,exp}$, based on comprehensive rectification investigations for 15–50 mm metal Pall and Raschig rings. They are considerably higher for given packing sizes than those derived by Eckert [9] from experiments using the air/water system. The following numerical values are given as an

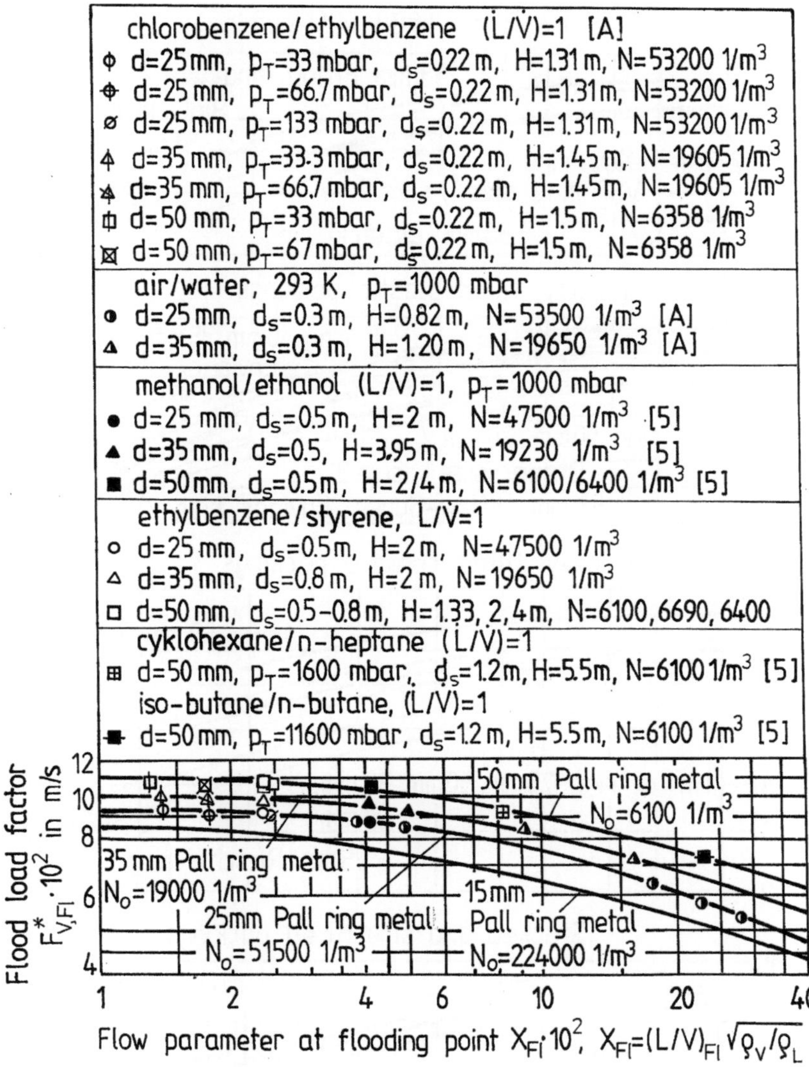

Figure 2-4b. Capacity diagram $F^*_{V,Fl} = f(X_{Fl})$ for random 15–50 mm metal Pall rings. The experimental values are based on the standard packing density N_0

example for 50 mm metal Pall rings (s = 1 mm):

$$F_{p,\text{Billet}}[5] = 107\,\text{m}^{-1} \qquad F_{p,\text{Eckert}}[9] = 65.6\,\text{m}^{-1}$$

and for 25 mm metal Pall rings:

$$F_{p,\text{Billet}}[5] = 202\,\text{m}^{-1} \qquad F_{p,\text{Eckert}}[9] = 157.4\,\text{m}^{-1}$$

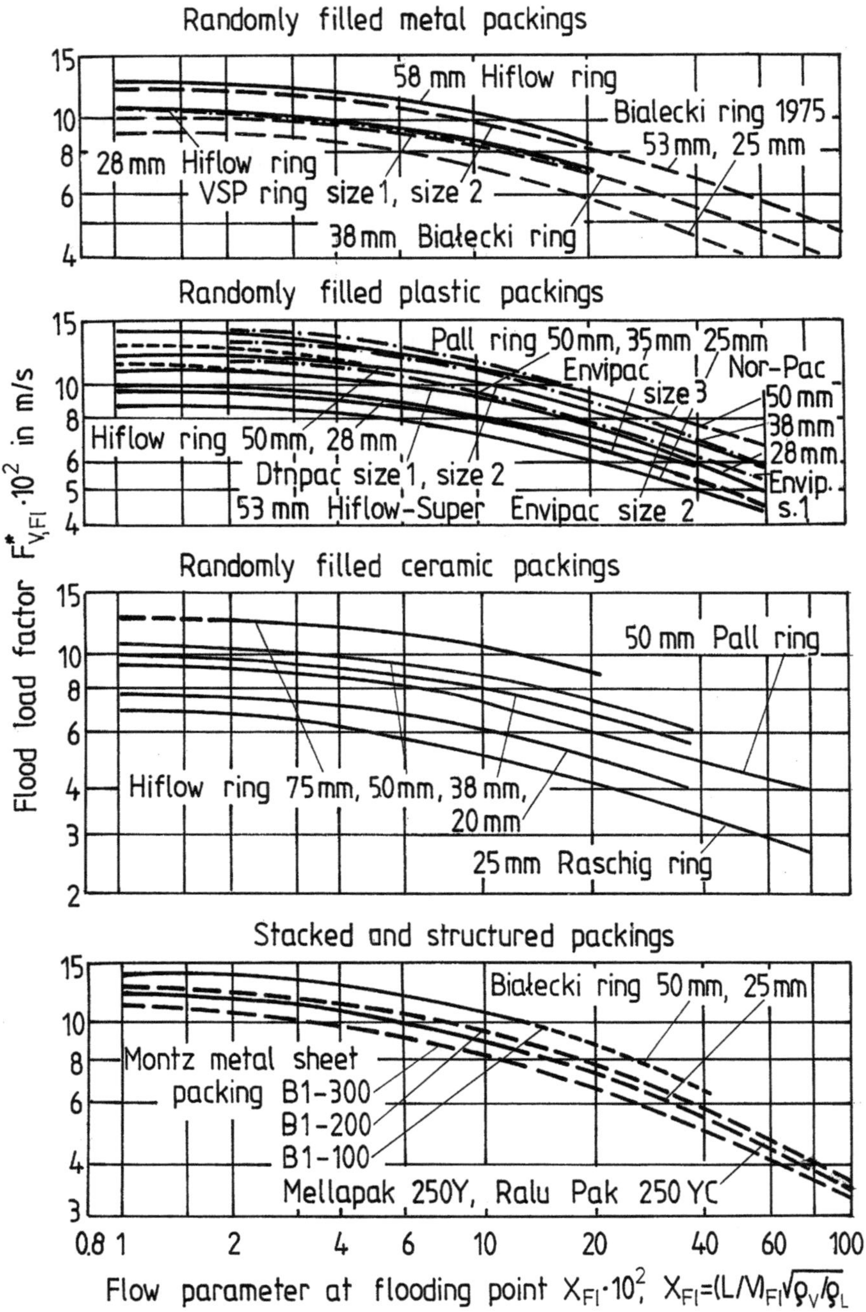

Figure 2-4c. Capacity diagram $F^*_{V,Fl} = f(X_{Fl})$ for various packing shapes. The experimental values are based on the standard packing density N_0. Experimental data – see figures in annex to Chap. 2

The empirical flooding point correlation developed by Billet [15] applies to metal Raschig and Pall rings and reflects the experimental data for the vapour/liquid systems much more accurately, $\delta(u_{V,Fl}) = \pm\,15{-}20\%$, than Eckert's correlation [9] does.

Depending on the packing size, acc. to Eq. (2-4), Table 2-1, the use of packing factors $F_{p,exp}$, based on Eckert [9] correlation, results in a deviation of approx. +29.1% when determining the vapour velocity $u_{V,Fl}$.

Bolles and Fair [8] compared data found in literature, in particular experimentally derived data provided by Billet [15], with the values calculated using Eckert's correlation [9]. The comparison also showed that the values for the vapour velocity $u_{V,Fl}$, which were calculated using Eckert's correlation, were 50% higher than the numerical values which were experimentally derived, particularly for systems in the vacuum range and for larger-diameter packing elements. Consequently, the correlation valid for air/water is not applicable to vapour/liquid systems.

Planovski and Kafarov [50], quoted by Weiß et al. [11], also differentiate between flooding in distillation columns and in absorption columns. This manifests itself quantitatively in the different numerical values for the constant C_i (see Table 2-1).

The other group of methods (no. 8–11 in Table 2-1), which apply to classic packing elements with smaller sizes of up to 0.025 m, are based on the so-called *film model* [3, 60] developed by Mersmann (1965). Here, the packing is assumed to be a bundle of flow channels with equal diameters d_h. Both phases pass evenly through each flow channel, which results in the formation of a downward flowing liquid film with a thickness of δ_L. The capacity diagram shows the correlation between the dimensionless pressure drop $\Delta p_0/(\rho_L \cdot g \cdot H)$ at the flooding point (proportional to $(F_{V,Fl})^2$) and the dimensionless film thickness δ_L/d_h, Fig. 2-5a. This film thickness δ_L determines the maximum capacity of the column. A rise in the liquid loads u_L leads to an increase in the film thickness δ_L, which, in turn, leads to a reduction of the free channel cross sections. This manifests itself in the increasing pressure drop $\Delta p/H$ of the gas, see Fig. 2-1b.

The measure used by Mersmann for the film thickness is the dimensionless liquid load B_L, which is based on the known proportionality $B_L \sim (\delta_L/d_h)^2$ [2, 3]. Further modifications by Reichelt [12], Gieseler [35], Vogt [49], Kleinhückenkotten [51], Bornhütter [66, 87], Bylica, Jaroszyński [81], Grabbert, Bonitz [92] et al. [20, 30, 78] have extended the range of validity of Mersmann's correlation [2, 3, 34], see Table 2-1. The correlation developed by Billet and Schultes [89] is also based on the film model.

Conclusions Paragraph 2.2.3 – Literature Overview

When the first correlations for determining the flooding point were developed, see Table 2-1, the only types of packing available were classic, ceramic packing elements such as Raschig rings, Lessing rings, Intalox and Berl saddles as well as spheres, with preferred diameters of up to 35 mm. It was only in the late 1960s that metal packings [5] became increasingly popular and the first plastic packings came on the market. The flooding point model developed by Mersmann (1965) [2, 3] therefore assumed a film flow in the packing. It is based on the assumption that the thrust force of the gas acts on the surface of the trickle film moving downwards in the packing. This is why the model is called *film gas thrust shear force model or film model* [60].

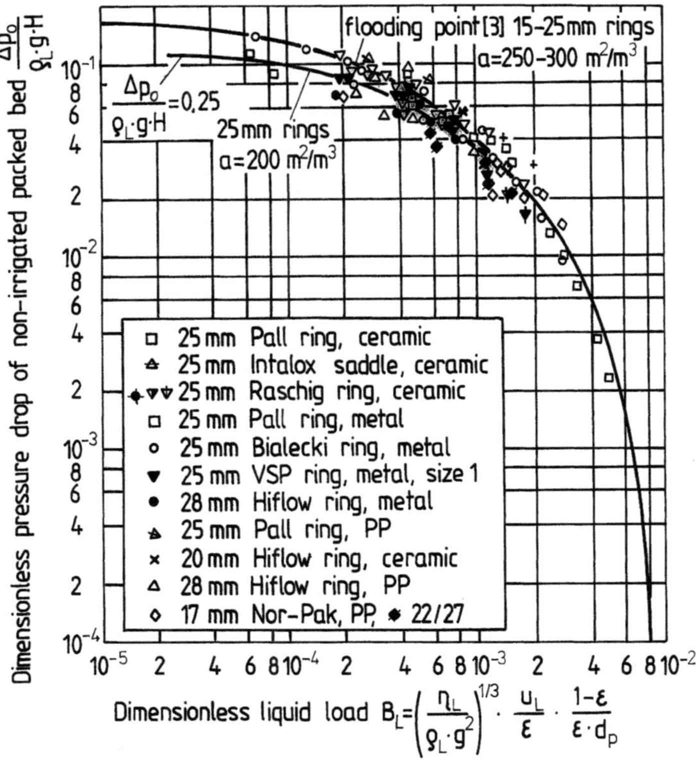

Figure 2-5a. Flooding line, dependent on the dimensionless liquid load acc. to Mersmann [3] with experimental data for various packings and test systems [20, 30, 44, 45, 46]

When, at a given liquid load u_L, certain gas velocities u_V are exceeded, the trickle film builds up in the void spaces of the packing elements, which are then gradually filled with liquid until flooding occurs. The film model is therefore a good method of describing the flooding point mechanism for operations with little droplet formation, i.e. for large-surface packing elements with low void fraction and *very high liquid loads*, e.g. for pressure rectification and pressure absorption.

This has also been confirmed by experimental data presented in this book. Figures 2-5a, 2-5b, 2-5c and 2-5d show the flooding point diagram developed by Mersmann. The diagram indicates that the experimental values for 25 mm Raschig rings, 25 mm metal Pall, Białecki and Hiflow rings, VSP rings etc., only deviate marginally from the flooding line plotted by Mersmann [2, 3]. This is based on a dimensionless dry pressure drop $\Delta p_0/\rho_L{\cdot}g{\cdot}H$ of 0.25. In the case of a low dimensionless liquid load B_L and increasing packing sizes, however, the deviation from the basic line plotted by Mersmann increases, see Figs. 2-5b, 2-5c and 2-5d. The position of the test points on the diagram reveals the following trend: A column filled with packing elements with a geometric surface of a $\approx$150 m^2 m^{-3}is flooded at a dimensionless dry pressure drop of $\Delta p_0/\rho_L{\cdot}g{\cdot}H$

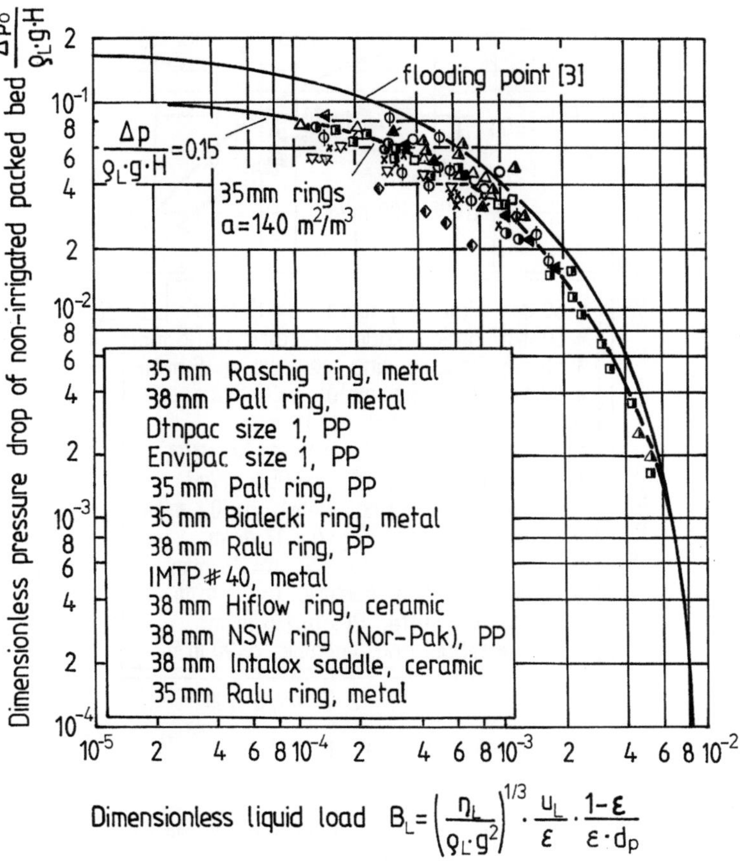

Figure 2-5b. Flooding line, dependent on the dimensionless liquid load acc. to Mersmann [3] with experimental data for various packings and test systems

≈ 0.15. If the surfaces are even larger, i.e. $a \cong 100\ m^2\ m^{-3}$ and $a \cong 80\ m^2\ m^{-3}$, flooding is triggered in the case of dimensionless dry pressure drops of $\Delta p_0/\rho_L{\cdot}g{\cdot}H = 0.1$ and 0.05, respectively. As the dimensionless liquid load B_L increases, the individual surface-related flood lines move closer to the basic line given by Mersmann. This indicates that Mersmann's *film model* [3] can be used to describe the flooding point mechanism even in the case of larger packing elements, if the dimensionless liquid loads B_L are extremely high. Under such operating conditions, there is little droplet formation, and the energy of the gas is not sufficient to keep the droplets suspended.

This clearly shows that the validity range of the *film model* covers operations running at very low gas velocities and using liquids with high surface tension. This applies, in particular, to large-surface, smooth and easily irrigatable packing materials, such as ceramic. In order to describe the fluid dynamics outside this validity range, it would be better to

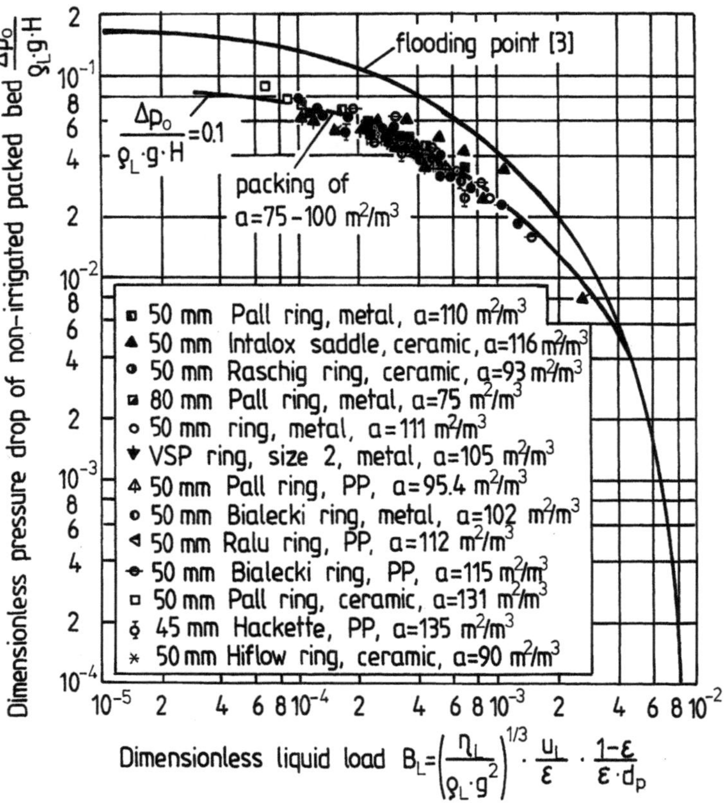

Figure 2-5c. Flooding line, dependent on the dimensionless liquid load acc. to Mersmann [3] with experimental data for various packings and test systems

develop a new model applicable to random, perforated packings with dimensions of d ≥ 0.015 m as well as to structured packings, which takes into account a considerable droplet entrainment by the gas flow.

2.2.4 New Model of Suspended Bed of Droplets (SBD) for Determining Gas Velocity $u_{V,Fl}$ at Flooding Point

One of the basic processes in technology occurs when a bed of particles is passed through by another phase. Examples of this process are: fixed bed flow, fluidised bed flow, pneumatic transport etc., all of which are qualitatively similar, as the flow occurs around single particles in the presence of surrounding particles. The basic processes mentioned above also include the flooding of packed columns under certain operating conditions, i.e. in

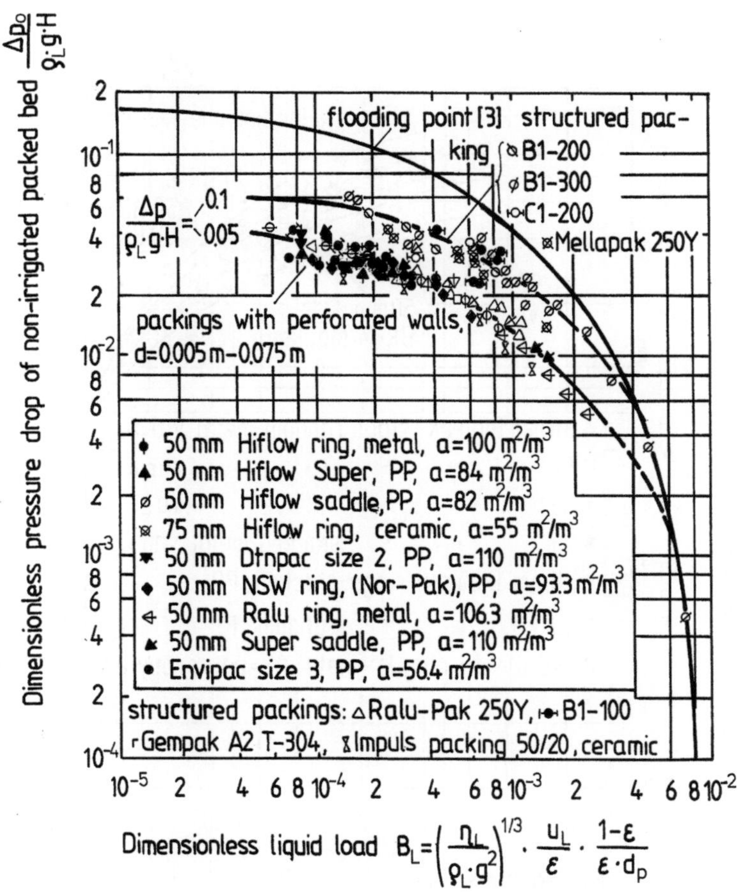

Figure 2-5d. Flooding line, dependent on the dimensionless liquid load acc. to Mersmann [3] with experimental data for various packings and test systems

the case of gas flowing through the suspended bed of droplets. This model is therefore known as the *suspended bed of droplets model*, see Fig. 2-6. First evaluated in 1986, it was presented at the GVC-VDI symposium in Strasburg (F) and published in the first German edition (1991) of this book.

Figure 2-6 shows a diagram of the terms which are useful for deriving the formula to calculate the flooding point. u_0 is used to express the superficial velocity of the gas which enables the suspension of a droplet with a diameter of d_T. $u_{V,Fl}$ describes the superficial velocity of the gas which enables the suspension of a droplet swarm. If the amount of droplets $h^0_{L,Fl}$ in the gas increases, flooding occurs at lower gas velocities $u_{V,Fl}$ in the column. The liquid hold-up $h^0_{L,Fl}$ is therefore the most important factor which influences the ratio of both gas velocities $u_{V,Fl}/u_0$. Figure 2-7 shows an example of the experimental flooding point data, acc. to Fig. 2-2a, using the function $u_{V,Fl} = f_1\,(1\text{-}h^0_{L,Fl})$. This confirms

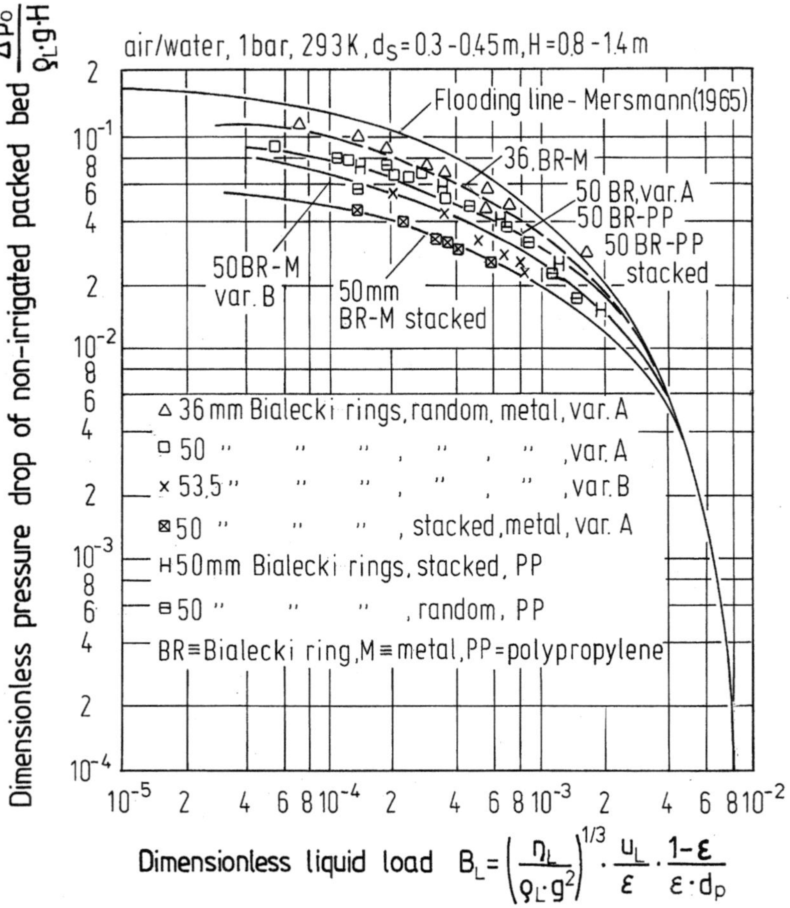

Figure 2-5e. Flooding line, dependent on the dimensionless liquid load acc. to Mersmann [3] with experimental data for random and stacked Białecki rings

the applicability of the following model:

$$\frac{u_{V,Fl}}{u_0} = \left(1 - h^0_{L,Fl}\right)^n \tag{2-19}$$

where $h^0_{L,Fl}$ is the liquid hold-up $h^0_{L,Fl}$ of the free packing volume at the flooding point, acc. to Eq. (2-19a),

$$h^0_{L,Fl} = \frac{V_{L,Fl}}{V_S \cdot \varepsilon} = \frac{h_{L,Fl}}{\varepsilon} \quad \left[m^3 m^{-3}\right] \tag{2-20}$$

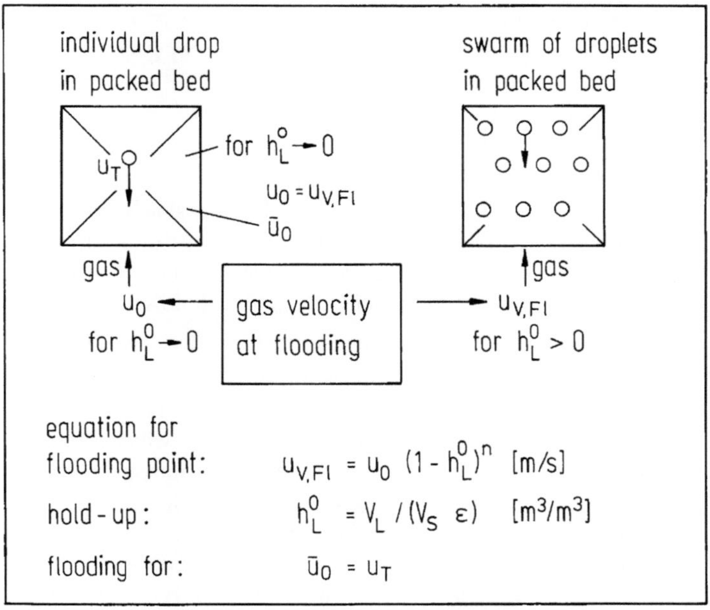

Figure 2-6. Model of suspended bed of droplets (SBD) for describing the flooding point in packed columns for gas (or vapour)/liquid systems

Figure 2-7. Dependence $u_{V,Fl} = f(1-h^0_{L,Fl})$ for random metal Białecki rings. Experimental data see Fig. 2-3

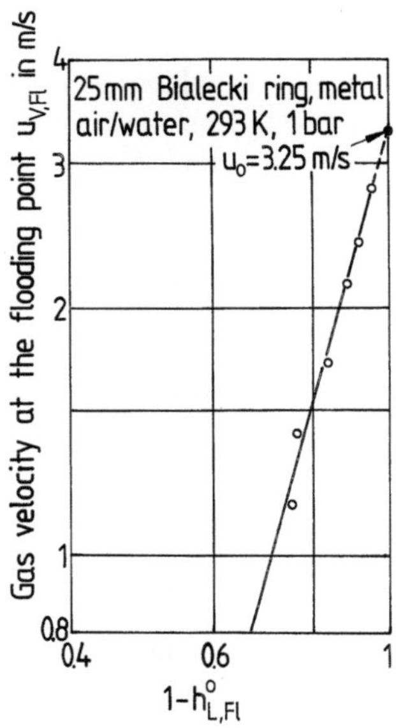

Based on this model, flooding occurs when the effective gas velocity in the packing $\bar{u}_0$ has reached the value of the effective droplet velocity u_T in the packing, i.e.:

$$h^0_{L,Fl} \to 0 \quad \Rightarrow \quad \bar{u}_0 = u_T \tag{2-21}$$

An analogous condition applies to droplet swarms.

Acc. to Eqs. (2-19) and (2-21), a droplet swarm falls more slowly than a single droplet. This is due to the increased relative velocity of both phases under the influence of a modified lifting force [55], which changes as a result of the mean density change of the two-phase mixture, similarly to the sedimentation and/or fluidisation process.

The function $f_1\left(1 - h^0_{L,Fl}\right)$ describes the contraction effect, which occurs as the gas flows through the droplet swarm. The application of Eq. (2-19) for calculating the gas velocity at the flooding point $u_{V,Fl}$ requires the knowledge of the velocity u_0, the exponent n as well as the liquid hold-up at the flooding point $h^0_{L,Fl}$, which is dependent on the phase flow ratio λ_0 at the flooding point [13, 17, 18, 38], in analogy to liquid/liquid systems. Section 2.2.4.4 takes a closer look at the derivation of the equation for calculating the liquid hold-up $h^0_{L,Fl}$ for gas/liquid systems in packed columns. Tables 2-2a and 2-2b contains a list of mixtures and their properties at different top pressures p_T up to 100 bar, based on the author's own experimental flooding point data as well as on literature data.

2.2.4.1 Effective Falling Velocity of a Single Droplet in the Packing u_T

The effective falling velocity u_T of droplets in the packing is determined on the basis of Eq. (2-12). Analogy to Eq. (2-13), the balance of forces acting on a droplet leads to the following formula (2-22):

$$u_T = \sqrt{\frac{4}{3}} \cdot \sqrt{\frac{d_T \cdot g}{\psi_0}} \cdot \sqrt{\frac{\rho_L - \rho_V}{\rho_V}} \tag{2-22}$$

where ψ_0 is the resistance coefficient of a single droplet falling through the packing.

The term

$$\sqrt{\frac{4}{3}} \cdot \sqrt{\frac{d_T \cdot g}{\psi_0}} \equiv \bar{u}_T \quad \Rightarrow \quad \bar{u}_T = u_T \cdot \sqrt{\frac{\rho_V}{\rho_L - \rho_V}} \tag{2-23}$$

represents the reduced falling velocity of a single droplet $\bar{u}_T$ in the packing.

Figures 2-4a, 2-4b and 2-4c as well as the analysis of the models described in paragraph 2.2.3 provide important information on the parameters which influence the effective droplet velocity u_T. In the case of decreasing flow parameters $X_{Fl} < 10^{-2}$, the flood load factor $F^*_{V,Fl}$ remains virtually constant and is proportional to the reduced droplet

Table 2-2a. Physical properties of the test systems under vacuum and in the normal pressure range

No.	System	p_T [mbar]	ρ_V [kg/m^3]	ρ_L [kg/m^3]	$\eta_L \cdot 10^3$ [kg/(m.s)]	$\eta_V \cdot 10^6$ [kg/(m.s)]	$\sigma_L \cdot 10^3$ [N/m]	Literature
1	**air/ water**	1000	1.180	998.2	1.0	18.2	72.5	[A] et al.
2	**ethylbenzene / styrene**	133.0	0.486	826.0	0.38	7.57	23.0	[5]
	$y_T \approx 0.8$ mol/mol	66.7	0.256	837.0	0.44		25.0	
3	**methanol / ethanol**	1000	1.250	742.0	0.355	11.2	17.7	
	$y_T \approx 0.8$ mol/mol							
4	**cyclohexane / n-heptane**	344	1.140	727.0	0.505	7.06	20.6	
	$y_T \approx 0.8$ mol/mol							
5	**1,2-propylene glycol/ ethylene glycol**	13	0.032	1018.0	4.0	9.5	35.0	
6	**chlorbenzene / ethylbenzene**	33.0	0.142	954.0	0.58	7.53	30.0	[A] et al.
	$y_T \approx 0.65–0.8$ mol/mol	66.7	0.270	992.0	0.50	7.90	27.3	
		66.7	0.268	941.0	0.43	7.68	25.0	
		133.0	0.517	974.0	0.38	8.27	23.2	
		266.0	0.963	911.0	0.32	9.76	22.4	
		532.0	1.863	957.8	0.27		19.5	
		1000.0	3.334	995.0				
7	**toluene / n-octane**	133	0.473	792.0	0.39	7.23	23.0	
	$y_T \approx 0.85$ mol/mol	80	0.300	829.9	0.42	6.83	24.7	
8	**ethanol / water**	1000	1.210	786.0	0.495	10.5	25.0	
	$y_T \approx 0.65$ mol/mol							
9	**iso-octane / toluene**	133	0.545	713.3	0.40	6.71	22.5	[8, 16, 25]
		1000	3.460	660.9	0.23	8.00	14.4	
		400	1.170	831.0	0.37	8.3	23.0	[7, 26]
10	**benzene / toluene**	1000	2.740	803.0	0.28	9.0	20.0	
11	**trans-decaline / cis-decaline**	13	0.066	850.0	1.27	7.0	26.3	[5]

Table 2-2a. (continued)

No.	System	p_T [mbar]	ρ_V [kg/m^3]	ρ_L [kg/m^3]	$\eta_L \cdot 10^3$ [kg/(m.s)]	$\eta_V \cdot 10^6$ [kg/(m.s)]	$\sigma_L \cdot 10^3$ [N/m]	Literature
12	**p-xylene / m-xylene**	66.7	0.250	830.0	0.42	7.28	24.2	[A] et al.
13	**benzene / ethylene chloride**	1000	3.0	816.0	0.45	8.9	20.0	[5, 19]
14	**ethylene chloride / toluene**	20	0.088	1127.0	1.11	10.9	36.0	
		1000	3.300	1158.0	0.40		24.0	
15+	**methanol / water**	466	0.54	778.0	0.41	11.81	21.7	[A]
	$y_T \approx 0.9$ mol/mol	1000	1.10	760.0	0.33	11.60	19.8	
16	**water vapour / water**	1000	0.598	958.3	0.282	15.0	58.8	[19]
17	**air / ethylene glycol** 305 K	1000	1.160	1110.0	18.50	18.2	48.0	[5, 36, 37]
18	**air / engine oil** 305 K	1000	1.160	850.0	90.0	18.2	25.6	[5]
19	**air / propylene carbonate** 293 K	1000	1.160	1259.0	2.70	18.2	44.5	[39]
20	**air / silicone oil**	1000	1.160	932.0	9.60	18.2	21.3	[36, 37]
21	**air / 98% sulphuric acid** 293 K	1000	1.160	1830.0	29.0	18.2	80.0	[29]
22	**air / 4% NaOH** 293 K	1000	1.160	1040.0	1.50	18.2	27.0	[A]
23+	**1,2-dichloroethane / toluene** $y_T \approx 0.7$ mol/mol	1000	3.260	1020.0	0.36	0.45	22.0	
31	**benzene / n-heptane**	1000	2.84	765.9	0.296	8.55	19.5	[A]
32	**ethanol / benzene**	1000	2.55	787.7	0.375	8.95	21.7	[A]

A ≡ Author
+ ≡ there is no flooding point data for these systems in the flooding point database; only pressure drop data $\Delta p/H$ is available

Table 2-2b. Physical properties of test systems for higher pressures

No.	System	p_T [bar]	ρ_V [kg/m^3]	ρ_L [kg/m^3]	$\eta_L \cdot 10^3$ [kg/(m.s)]	$\eta_V \cdot 10^6$ [kg/(m.s)]	$\sigma_L \cdot 10^3$ [N/m]	Literature
24	**air / propylene carbonate**	5	5.92	1204	2.69	18.2	43.7	[39]
		10	11.61				42.2	
		15	17.40				41.4	
25	**cyclohexane / n-heptane**	1.69	4.74	682.9	0.285		15.7	[5]
26	**iso-butane / n-butane**	11.6	32.2	477	0.083		6.15	[5]
27	**methanol / nitrogen (N_2)**	5	6.7	830	1.32	15.6	26.3	[75]
		10	12.5	813	0.87	16.8	23.8	
		15	20.9	833	1.27	15.9	25.4	
		20	26.6	826	1.10	16.5	24.4	
		30	41.6	834	1.32	16.0	24.5	
		40	53.8	826	1.11	16.7	23.0	
		65	86.5	822	1.03	17.4	21.5	
		80	110.4	826	0.93	18.2	20.7	
		90	127.0	833	1.27	17.9	20.7	
		100	128.8	816	0.91	18.8	18.6	
28	**ethane diol / nitrogen (N_2)**	20	23/25	1111/1128	19.1	18.6	49.7	[75]
29	**water / nitrogen (N_2)**	12.8	15.2	1000	1.30	17.5	73.8	
		20	24.0		1.27	17.7		
30	**methane / ethane**	30	51/63	393	0.08	9.0	1.8/2.5	[72]

velocity u_T, which for $h^0_{L,Fl} \rightarrow 0$ tends to the value $F^*_{V,Fl}$, i.e. $u_T \rightarrow F^*_{V,Fl}$. This applies to a given mixture and packing size.

In addition, Figs. 2-4a, 2-4b and 2-4c indicate that the reduced droplet velocity u_T increases with the packing size d. The material of the packing elements also has an influence on the parameter u_T. It follows from this that the resistance coefficient ψ_0 in Eq. (2-22) is a function of the size and the surface properties of the packing. The first influencing factor can be expressed dimensionless by the quotient $f_2(d_h/d_T)$. The second factor is linked to the resistance coefficient ψ of the dry packing. These two effects are reflected in the general correlation (2-24):

$$u_T = C_1 \cdot \sqrt{\frac{d_T \cdot \Delta\rho \cdot g}{\rho_V}} \cdot f_2\left(\frac{d_h}{d_T}\right) \cdot f_3(\psi) \tag{2-24}$$

where C_1 is a dimensionless constant. Sections 2.2.4.5 and 2.2.4.6 take a closer look at the derivation of the functions $f_2\left(d_h/d_T\right)$ and $f_3\ (\psi)$.

2.2.4.2 Droplet Size and Range of Droplet Movement

Figure 2-8 [15] shows that, in the case of droplet Reynolds numbers $Re_T > 400$, the resistance coefficient ψ_0 increases with the Reynolds number Re_T, contrary to expectations. Acc. to the research of numerous authors, such as Mersmann [33, 38], Reinhart [63] and Kovalenko [64], this is applicable in the range of $Re_T > 400$ only to large, deformed droplets in excess of 1 mm, falling in liquids and gases, and for $\eta_V/\eta_L \rightarrow 0$ and $\rho_V/\rho_L \ll 1$. Mersmann [2, 38, 60] found that such types of droplets are stable in the following range:

$$\frac{d_T^2 \cdot \Delta\rho \cdot g}{\sigma} < 9 \tag{2-25}$$

To calculate the droplet Reynolds number in the packing, it is necessary to ascertain the droplet velocity u_T as well as the droplet size d_T and the kinematic viscosity of the surrounding phase ν_V.

There are equations available to calculate the size of deformed droplets generated in liquid/liquid systems in packed columns. These equations can be found in the work of Mersmann [38] as well as Maćkowiak and Billet [17, 18].

For simplification purposes, the fluid dynamics model described here is based on a mean droplet diameter of:

$$d_T = C_T\sqrt{\frac{\sigma_L}{(\rho_L - \rho_V)g}} \quad \text{where } C_T = 1 \quad [17, 18] \tag{2-26}$$

Based on this, it is possible to estimate the Reynolds numbers Re_T, acc. to Eq. (2-9), with $u_V = u_T$ [17, 18, 38].

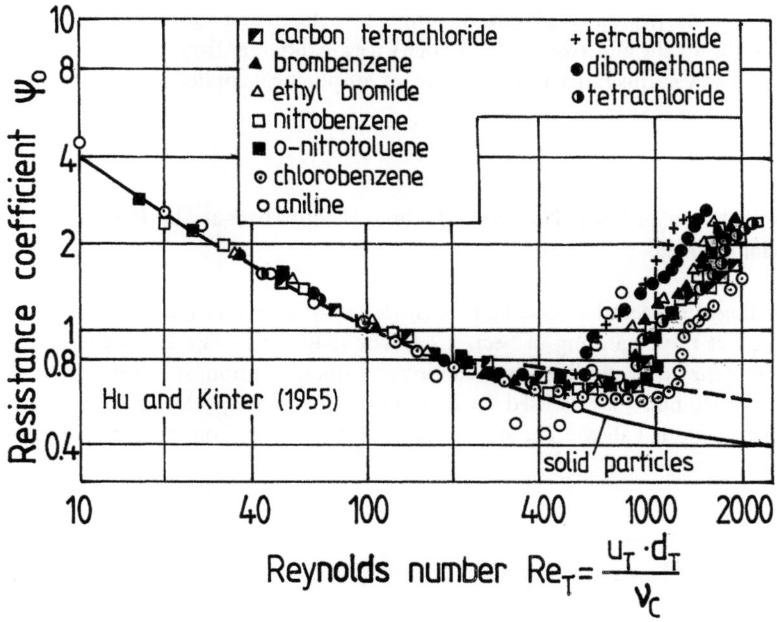

Figure 2-8. Dependence of the resistance coefficient of individual droplets ψ_0 on the Reynolds number Re_T acc. to Hu and Kinter [15]. ν_C = kinematic viscosity of the surrounding liquid. $\nu_C = \nu_V$. applies to droplet fall in gases

$$Re_T = \frac{u_T \cdot d_T}{\nu_V} \tag{2-9}$$

For the maximum stable droplet size, $C_T = 2.44$ is substituted into Eq. (2-26) [38, 57].

Bornhütter [66] has published new experimental data on the prediction of droplet size in gas/liquid systems with relation to dripping processes as well as the breakdown of sprays and threads, using water, ethylene glycol and methanol. The evaluation of the experiments shows that the droplet diameter, acc. to Eq. (2-26), is independent of the specific liquid load and of the size and type of the packing. It does, however, depend on the wetting properties and the physical properties of the liquid. In the case of ceramic, the droplets are larger than for other materials, such as PP, PTFE and stainless steel. In terms of the adhesion work $(1 + cos\theta) \cdot 10^3 = 80 - 120$, the constant C_T is approx. 1 ± 0.15.

Based on Eq. (2-26), with $C_T = 1$, the droplet sizes for the systems listed in Table 2-2 are approx. $1.5 \cdot 10^{-3}$ to $2.7 \cdot 10^{-3}$ m. The corresponding Reynolds numbers, acc. to Eq. (2-9), are in the range of $400 < Re_T < 1400$. Based on Fig. 2-8, this is the range in which the resistance coefficient ψ_0 increases as the Reynolds number Re_T increases, see Fig. 2-6.

Systems with very low surface tensions σ_L are characterised by smaller Reynolds numbers Re_T in the transition and/or laminar range, and the resistance coefficient ψ_0 in Fig. 2-8 decreases as the Reynolds number increases.

As a result, C_l in Eq. (2-24) can no longer be defined as a constant. In this range, the correlations for the effective droplet velocity u_T are different from those expressed in Eq. (2-24). However, this range is less significant for practical applications.

2.2.4.3 Analogy Between the Falling Process of Particles in Fluidised Beds and the Droplet Fall in Random Packings

The analogy between the flooding process in packed columns and in fluidised beds was discussed at the beginning of Sect. 2.2.4. The diagram in Fig. 2-9, which was developed by Zenz [61], modified by Wunder and quoted, amongst others, by Mersmann [2] and Stichlmair [55], is used for determining the falling velocity of particles in the systems mentioned above. The superficial velocity is plotted on the ordinate axis of this diagram,

$$u_0 \cdot \sqrt[3]{\frac{\rho_V^2}{\Delta\rho \cdot g \cdot \eta_V}}$$

whereas the dimensionless particle diameter $d_T{}^*$, acc. to Eq. (2-27)

$$d_T^* = Ar^{1/3} \qquad (2\text{-}27)$$

is plotted on the abscissa axis. Ar stands for the Archimedes number, described as:

$$Ar = \frac{d_T^3 \cdot \rho_V}{\eta_V^2} \cdot \Delta\rho \cdot g. \qquad (2\text{-}28)$$

Based on Eq. (2-27), the dimensionless particle diameters $d_T{}^*$ for the systems listed in Table 2-2 range between 40 and 170. Acc. to Fig. 2-9, the dimensionless superficial velocity in this range is approximately dependent on the root of the dimensionless particle diameter $d_T{}^*$. The following correlation applies:

$$u_0 \cdot \sqrt[3]{\frac{\rho_V^2}{\Delta\rho \cdot g \cdot \eta_V}} = A_i \cdot d_T^{*1/2}. \qquad (2\text{-}29)$$

where A_i is the packing constant.

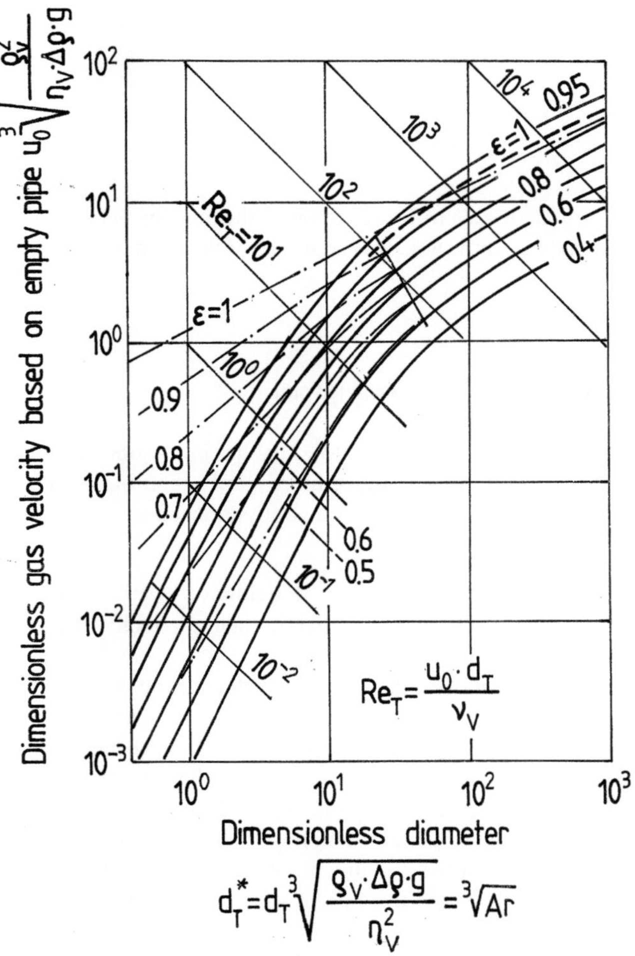

Figure 2-9. Diagram for determining the velocity, at which particles are kept in suspension [55]. ________ homogenous two-phase bed (particulate fluidisation); -.-.-.-.-.- inhomogenous two-phase bed (aggregative fluidisation)

By substituting Eqs. (2-27) and (2-28) into Eq. (2-29), it possible to find the correlation (2-30) for the superficial velocity u_0, in which droplets are suspended in a certain type of packing.

$$u_0 = A_i \cdot \sqrt[3]{\frac{\Delta\rho \cdot \eta_V \cdot g}{\rho_V^2}} \cdot (\sqrt[3]{Ar})^{1/2} = A_i \cdot \sqrt{\frac{\Delta\rho \cdot d_T \cdot g}{\rho_V}} \qquad A_i = \text{packing constant} \qquad (2\text{-}30)$$

This functional correlation is comparable with Eq. (2-22).

It follows from this that both velocities – u_0 acc. to Eq. (2-30) and u_T acc. to Eq. (2-22) – are dependent on the same term:

$$u_0,u_T \approx \sqrt{\frac{d_T \cdot \Delta\rho \cdot g}{\rho_V}} \quad \text{for } d_T^* \in (40 - 170) \tag{2-31}$$

Deriving the formula, (2-27) to (2-31), showed the analogy between the model of *suspended bed of droplets* and the well-known model of fluidised beds, Fig. 2-9, in the range of $d_T^* \geq 40$.

2.2.4.4
Determining Liquid Hold-Up $h_{L,Fl}^0$ at Flooding Point

The free cross section available for gas flow changes in the presence of several droplets (droplet swarm). Droplet swarms move more slowly than individual droplets, which is due to the different lifting forces. The contraction effect is given by the function $f_1\left(h_{L,Fl}^0\right) = \left(1 - h_{L,Fl}^0\right)^n$, see Eq. (2-19).

Acc. to the SBD model described above, flooding is triggered for $h_{L,Fl}^0 \rightarrow 0$, when the effective gas velocity $\bar{u}_0$ has reached the effective falling velocity of a droplet u_T, see Eq. (2-21).

It is possible, for a given packing element, to calculate the gas velocity u_0 of an empty column as well as the exponent n, using Eq. (2-19), provided there is sufficient experimental data available on the liquid hold-up $h_{L,Fl}^0$ and the corresponding gas velocities at the flooding point $u_{V,Fl}$. The variables u_0 and n for Eq. (2-19) were determined using the data for 25 mm Białecki rings shown in Figs. 2-2a and 2-2b. The results are listed in Table 2-3, see Fig. 2-7. The variables $u_0$and n were determined using the experimentally derived gas velocities $u_{V,Fl}$ at the flooding point, the corresponding specific liquid loads u_L and the respective liquid hold-ups $h_{L,Fl}^0$ as well as the properties ρ_V, ρ_L and σ_L of the test system air/water under ambient conditions. The calculation was performed using the optimisation method, based on the simplex method:

$$u_0 \cong 3.25\ ms^{-1} \pm 3\% \quad \text{and } n = 3.5 = const. \tag{2-32}$$

Table 2-3. Flooding point data acc. to Figs. 2-2a and 2-2b, valid for 25 mm random Białecki rings made of metal. Sytstem: air/water,1 bar, 293 K, $d_S = 0.15$ m, H = 1.4 m, $\varepsilon = 0.94\ m^3\ m^{-3}$, $a = 238\ m^2\ m^{-3}$, $Re_L \geq 2$

TP	$u_{V,Fl}$	$u_L \cdot 10^3$	$h_{L,Fl} \cdot 10^2$	$\lambda_0 \cdot 10^3$	$h_{L,Fl}^0 \cdot 10^2$	$\frac{u_{V,Fl}}{(1-h_{L,Fl}^0)^{3.5}}$	u_0
–	ms^{-1}	ms^{-1}	$m^3\ m^{-3}$	[–]	$m^3\ m^{-3}$	ms^{-1}	ms^{-1}
1	2.80	1.39	4.26	0.496	4.84	3.330	3.331
2	2.45	2.78	6.38	1.135	7.20	3.183	3.184
3	2.15	5.55	9.79	2.58	10.60	3.186	3.187
4	1.75	11.1	16.0	6.53	16.2	3.164	3.249
5	1.40	16.7	21.3	11.93	21.2	3.225	3.227
6	1.15	22.2	23.4	19.3	26.0	3.302	3.305

The simplex method is used to minimise the mean square error in the calculation of the gas velocity at the flooding point $u_{V,Fl}$, acc. to the formula:

$$\bar{\delta}(u_{V,Fl}) = \frac{1}{n_i}\sqrt{\sum_{i=1}^{n_i}\left[\frac{u_{V,Fl,exp} - u_{V,Fl,calc}}{u_{V,Fl,exp}}\right]_i^2} \tag{2-33}$$

where n_i is the number of test points. $\bar{\delta}\left(u_{V,Fl}\right)$ is the mean error in the determination of the gas velocity at the flooding point.

The constant numerical value of the exponent $n = 3.5$ also applies to other packing elements, for which the experimental flooding point data $u_{V,Fl}$ and u_L as well as the liquid loads $h^0_{L,Fl}$ are available, e.g. [3, 12, 22, 36, 37].

The validity verification of the method based on Eq. (2-19), for any type of system is only possible if the specific liquid hold-up $h^0_{L,Fl}$ at the flooding point is known or can be calculated. There is no such correlation available in literature for gas/liquid systems. However, the velocities $u_{V,Fl}$ and u_L at the flooding point can generally be ascertained from experiment data. In order to evaluate Eq. (2-19), it is therefore necessary to derive a correlation for determining the liquid hold-up $h^0_{L,Fl}$, which is valid for all systems.

Deriving the Correlation for Determining the Liquid Hold-Up at the Flooding Point $h^0_{L,Fl}$

Based on the two layer model for counter-current processes [13, 14, 38], the relative velocity u_R is given by the sum of the effective gas velocity of both phases $u_{V,eff}$ and $u_{L,eff}$

$$u_R = u_{V,eff} + u_{L,eff} = \frac{u_L}{\varepsilon \cdot h^0_L} + \frac{u_V}{\varepsilon \cdot \left(1 - h^0_L\right)} = f_1\left(1 - h^0_L\right). \tag{2-34}$$

The mean relative velocity u_R is usually referred to as 'slip velocity'. It is dependent on the gas content $(1–h^0_L)$ and the falling velocity of a droplet in an infinitely extended continuous phase. The falling velocity of a droplet, acc. to Eq. (2-31), only takes into account the physical properties of the system and the droplet size, whereas the gas content $(1–h^0_L)$ also includes the mutual interaction of the individual droplets. There are numerous formulas for calculating the function $f_1(1-h^0_L)$ in Eq. (2-34) for bubble columns, fluidised beds, liquid/liquid extractors, and for fluidisation and sedimentation [14, 38, 52]. These can be found in literature. However, there are no such methods available for systems, in which gas, as a continuous phase, flows through the packing counter-current to the downward moving liquid phase.

Figure 2-10 shows the slip velocity u_R plotted against the liquid hold-up h^0_L for randomly filled Białecki rings. The parameter here is the liquid load u_L.

A set of curves is generated by different liquid loads $u_{L,i}$, which can be expressed near the flooding point by the following equation:

$$u_{R,i} = P_i \cdot \left(1 - h^0_L\right)^m \tag{2-35}$$

Figure 2-10. Experimentally derived relative velocity u_R as a function of the liquid hold-up h_L^0, valid for random 25 mm metal Białecki rings, based on data shown in Fig. 2-2b

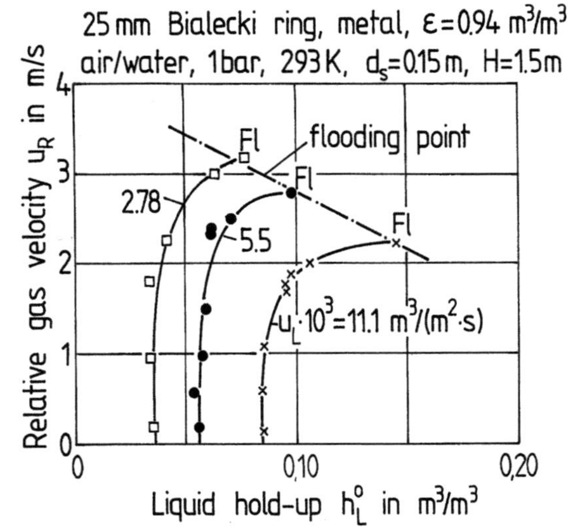

in analogy to the known models developed, e.g., by Mersmann [38]. Figure 2-10 indicates that the characteristic gas velocity P_i is a function of the specific liquid load, i.e. $f(u_L)$, which means $P_i$is only constant for a specific liquid load $u_{L,i}$, and the curve $u_R = f(h_L^0)$ tends to zero near the flooding point. The two layer model is therefore suitable for determining the liquid hold-up at the flooding point $h^0_{L,Fl}$. However, it does not lead to a general method for determining the gas velocity at the flooding point. This is why the method for calculating the gas velocity $u_{V,Fl}$ was derived using Eq. (2-19).

Combining Eqs. (2-34) and (2-35) gives the following equation:

$$\frac{u_L}{\varepsilon \cdot h_L^0} + \frac{u_V}{\varepsilon(1 - h_L^0)} = P_i(1 - h_L^0)^m \tag{2-36}$$

Figure 2-11 shows a diagram of the correlation between the liquid hold-up h_L^0 and the gas velocity u_V. The correlation for $h^0_{L,Fl}$ was derived, based on the following assumptions: $\left(\partial u_V / \partial h_L^0\right) = 0$ at the flooding point for $u_L = \text{const}$.

In the operating range above 65% of the flooding point, the differentiation of the transposed Eq. (2-36)

$$u_V = \varepsilon \cdot \left[P_i \cdot \left(1 - h_L^0\right)^{m+1} - u_L \cdot \left(h_L^0 - 1\right)^{-1}\right] \tag{2-37}$$

for $\left(\partial u_V / \partial h_L^0\right) = 0$ and for selected specific liquid loads $u_{L,i} = \text{const}$.

$$\left[\frac{\partial u_V}{\partial h_L^0}\right]_{uL,i=const} = (P_i \cdot \varepsilon) \cdot (m+1) \cdot \left(1 - h_L^0\right)^m \cdot (-1) - u_L \cdot \left(-\frac{1}{\left(h_L^0\right)^2}\right) = 0 \tag{2-38}$$

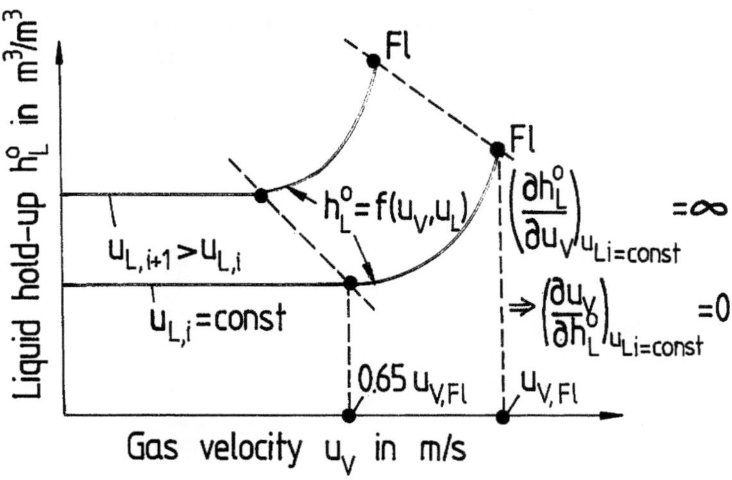

Figure 2-11. Schematic representation of the dependence of h_L^0 on the gas velocity u_V in packed columns

leads to the following solution:

$$(1-h_L^0)_{Fl}^m = \frac{u_{L,Fl}}{(P_i \cdot \varepsilon)\cdot\left(h_{L,Fl}^0\right)^2\cdot(m+1)} \tag{2-39}$$

Substituting Eqs. (2-1) and (2-39) into Eq. (2-36) gives the relationship between the phase flow ratio and the liquid hold-up $h_{L,Fl}^0$ at the flooding point:

$$\lambda_0 = \frac{(h_{L,Fl}^0)^2(m+1)}{[1-h_{L,Fl}^0][1-h_{L,Fl}^0(m+1)]} \tag{2-40}$$

The differentiation of Eq. (2-36) leads to the same Eq. (2-40), under the following conditions:

for u_V = const. $\rightarrow$ $(\partial u_V/\partial h_L^0)=0$ and
for u_L/u_V = const. $\rightarrow$ $(\partial u_V/\partial h_L^0)=0$

For reasons of brevity, the analogous derivation process, based on Eqs. (2-36) and (2-40), is not discussed here.

The derived Eq. (2-40) shows that the liquid hold-up at the flooding point $h_{L,Fl}^0$ is only dependent on the phase flow ration λ_0 at the flooding point and on the parameter *m*. The exponent *m* can be determined from experiments with known value pairs $\lambda_0 = (u_L/u_V)_{Fl}$, acc. to Eq. (2-1), and $h_{L,Fl}^0$ by transposing Eq. (2-30), using the following formula:

$$m = \left[h_{L,Fl}^0\left(\frac{h_{L,Fl}^0}{\lambda_0\cdot(1-h_{L,Fl}^0)}+1\right)\right]^{-1}-1 \tag{2-41}$$

Evaluation of Experimental Results for the Range of Low and Moderate Phase Flow Ratios λ_0 at Flooding Point

The experimental data available for the evaluation of Eq. (2-41) gives a constant numerical value for the exponent m, based on $\lambda_0 \leq 0.025$ and on the Reynolds numbers of the liquid $Re_L = u_L/a \cdot \nu_L \geq 2$, Eq. (2-42)

$$m \approx -0.8 \pm 12\ \% \tag{2-42}$$

The numerical value of the parameter m was determined for various systems, namely air/water, ethanol/water, steam/water [7], air/silicone oil [36, 37], using various random and structured packings made of different materials. In the range of low and moderate λ_0 numbers below $\lambda_0 < 0.025$, the parameter m is not dependent on the phase flow ratio λ_0 at the flooding point, see Fig. 2-12a.

For $h^0_{L,Fl} \leq 0.3$, the Archimedes number was found to have no impact. The same applies to fluidised beds [2, 55, 38]. Figure 2-12b is based on experimental data taken by Pliss, Ender [67] and Bornhütter [66] at an industrial pilot plant at the Technical University of Munich, using modern 50–90 mm lattice packings produced by Envipac (size 3), Hiflow rings, VSP rings and 50 mm Pall rings made of metal and plastic, with $d_S = 1$ m and a packing height of approx. 3.5 m, as well as experimental data for structured packings (Mellapak 250Y, 250X and 500Y), taken at an industrial pilot plant, operated by Sulzer, with $d_S = 1$ m and $H = 3.7$ m [67–69], plus additional data [79, 71]. The diagram

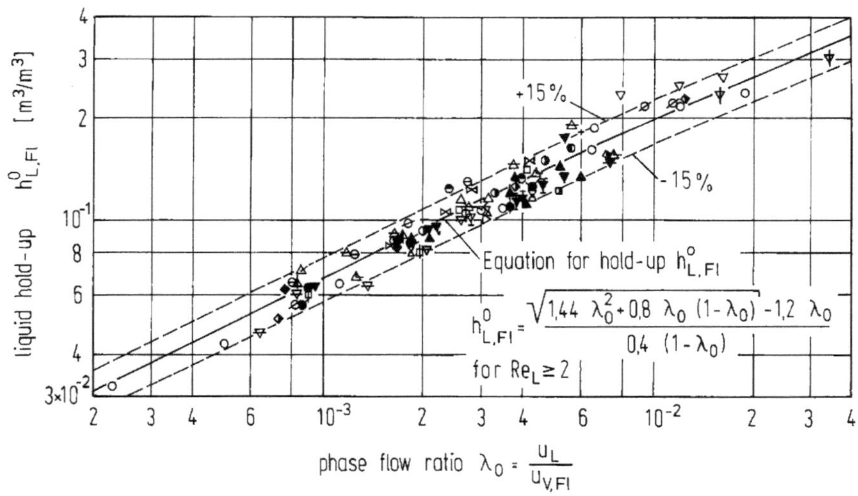

Figure 2-12a. Dependence of the liquid hold-up at the flooding point $h^0_{L,Fl}$ on the phase flow ratio λ_0, valid for various systems and packings made of metal, ceramic and plastic for $Re_L \geq 2$ – experimental data taken by the author in comparison to calculated values based on Eq. (2-47)

Table Relating to Fig. 2-12 a Experimental hold-up data at the flooding point $h^0_{L,Fl}$

TP	d.10³ [m]	Packing	Material	System	ε [m³/m³]	d_S [m]	H [m]	Lit.
○	25	**Bialecki rings** random	metal	air/water 1bar, 293 K	0.94	0.154	1.5	[43]
▽	15	**Raschig rings** random	ceramic		0.676	0.226	0.63	[22]
▼	25	**Raschig rings** random	glass		0.82	0.15	1.0	[36,37]
▽̸	25			air/silicone oil 1 bar, 293 K				
⊽	25		ceramic	C_2H_5OH/H_2O	0.667	0.3	2.0	[7]
▼̶				H_2O/H_2O				
+	20	**Hiflow ring** random		air/water 1 bar, 293 K	0.762	0.3	1.2	[A]
⊕	75				0.865	0.45	2.0	
●	28	**Hiflow ring** random	metal		0.962	0.3	1.46	
●	58				0.97	0.45	2.0	
□	25	**Pall ring** random			0.95	0.3	1.46	
◨	35				0.95	0.3	1.46	
◧	58				0.97	0.45	2.0	
▷	32 (Gr.1)	**VSPring** random			0.976	0.30/0.45	1.46/2.0	
▶	50 (Gr.2)				0.98	0.45	2.0	
▲	30 (Gr.1A)	**Envipac** random	plastic (PP)		0.932	0.3	1.96	
△	35	**Intalox saddle** random	plastic		0.90	0.3	1.36	
▲	50	**Hiflow ring** random			0.94	0.45	2.0	
⌧		**Pall ring**			0.926			
△	25	**Intalox saddle** random	ceramic		0.73	0.30	0.9	
△	38				0.757		1.4	
◆	50	**NSW ring**	plastic		0.950	0.45	2.0	
△		**Ralu ring**			0.940			
◆	25	**NSW ring, Typ C** random			0.92	0.15	1.3	[40]
Ø	25	**Bialecki ring stacked**	metal		0.928	0.15	1.5	[A]
◐	–	**Montz packing B1-300**	sheet metal		0.972	0.3	1.4	
◑	–	**Montz packing B1-200**			0.978			
⦶	–	**Montz packing C1-200**	plastic		0.954			
⊖	–	**Mellapak 250Y**	sheet metal		0.96	0.22	1.25	
⋈	70 (Gr.2)	**Dtnpac**	plastic		0.938	0.3	1.4	
⧖	45 (Gr.1)				0.92	0.45	2.0	

shows the hold-up at the flooding point $h^0_{L,Fl}$ as a function of the phase flow ratio at the flooding point λ_0 in the range up to $\lambda_0 \to 1$.

The parameter m, calculated by means of the simplex method, can be described by the new model for phase flow ratios ranging from $\lambda_0 = 0.001$ to 1.0:

$$m = -0.82 + \frac{\lambda_0}{\lambda_0 + 0.5} \tag{2-43}$$

This changes to $m \approx -0.80$ in Eq. (2-42) for lower phase flow ratios λ_0 in the range shown in Fig. 2-12a. Equation (2-43) leads to a correlation for $m = (\lambda_0)$, which covers practically all areas of application of packed columns for gas/liquid systems, ranging from vacuum rectification to pressure rectification and pressure absorption.

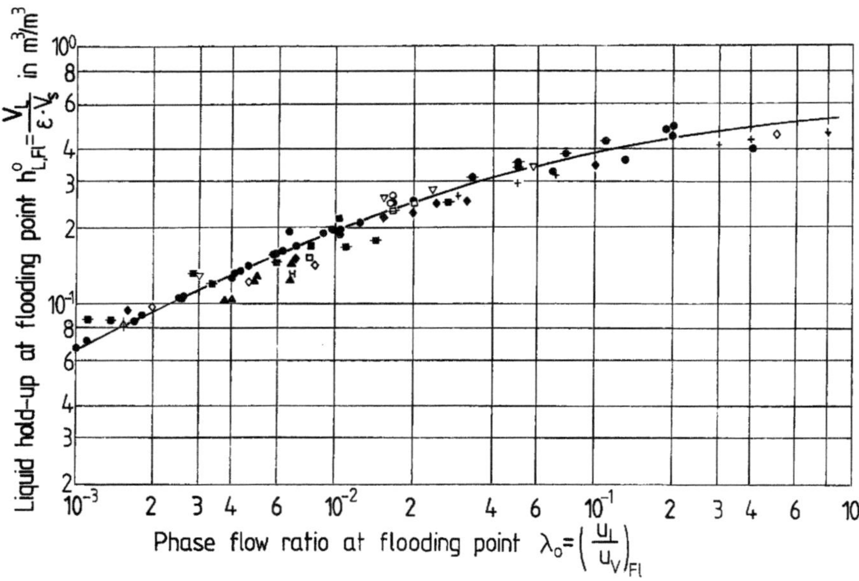

Figure 2-12b. Dependence of the liquid hold-up at the flooding point $h^0_{L,Fl}$ on the phase flow ratio λ_0, valid for various systems and packings made of metal, ceramic and plastic for $Re_L \geq 2$ – experimental data [67–71] for high phase flow ratios at the flooding point λ_0 and large column diameters in comparison to calculated values based on Eq. (2-47)

Table Relating to Fig 2-12b Experimental hold-up data at the flooding point $h^0_{L,Fl}$. System: air / water, 1 bar, 293 K

TP	$d \cdot 10^3$ [m]	Packing	Material	a [m²/m³]	ε [m³/m³]	d_S [m]	H [m]	Literature
▲	50	**Hiflow ring** 6874 m⁻³	PP	90.7	0.926	1.0	3.70	1991 [66]
⚠	58	**Hiflow ring** 5060 m⁻³	metal (1.4301)	93.0	0.979	1.0	3.15	1991 [66]
H	90	**Hiflow ring** 1415 m⁻³	PP	61.0	0.954	1.0	3.51	1991 [60]
○	–	**K packing** (25 x 12 x 2 mm)	ceramic	192	0.810	0.2	1.00	[70]
⊠	–	**Mellapak 250 X** φ = 30°	sheet metal	250	0.980	1.0	3.50	1995 [68,69]
⊟	–	**Mellapak 250 Y** φ = 45°	sheet metal	256	0.975	1.0	3.50	1995 [68,69]
▽	–	**Mellapak 500 Y** φ = 45°	sheet metal	500	0.975	1.0	3.50	1995 [68,69]
+	50	**Pall ring** 6846 m⁻³	PP	112	0.929	1.0	3.43	1991 [66]
●	15	**Raschig ring**	ceramic	300	0.700	0.2	1.00	[71]
△	50	**VSP ring** Gr.2	metal (1.4571)	100	0.98	1.0	3.50 6.00	1991 [67]

For Reynolds numbers $Re_L < 2$ (Chap. 4) in laminar liquid flow, the numerical value is as follows:

$$m = -0.9 + \frac{\lambda_0}{\lambda_0 + 0.5} \qquad (2\text{-}44)$$

which changes to

$$m \approx -0.88 \pm 12\ \% \qquad (2\text{-}45)$$

for $h^0_{L,Fl} < 0.20$. The parameter m in Eq. (2-41) for gas/liquid systems is therefore dependent on the liquid flow in the packing as well as on the phase flow ratio at the flooding point λ_0, in the case of higher liquid hold-ups $h^0_{L,Fl} > 0.2$.

Now that the phase flow ratio λ_0 and the exponent m are known, it is possible to solve Eq. (2-40) for $h^0_{L,Fl}$. The general, real solution of the quadratic Eq. (2-40) $h^0_{L,Fl} = f(\lambda_0)$ is as follows:

$$h^0_{L,Fl} = \frac{\sqrt{\lambda_0^2(m+2)^2 + 4\lambda_0(m+1)(1-\lambda_0)} - (m+2)\lambda_0}{2 \cdot (m+1)(1-\lambda_0)} \quad [m^3 m^{-3}] \qquad (2\text{-}46)$$

which, for $\lambda_0 < 0.025$ with $m = -0.8$ and for moderate phase flow ratios $\lambda_0 < 0.025$, acc. to Fig. 2-12a, leads to Eq. (2-47) describing the liquid hold-up $h^0_{L,Fl}$ for $Re_L \geq 2$ at the flooding point.

$$h^0_{L,Fl} = \frac{\sqrt{1.44\lambda_0^2 + 0.8\lambda_0(1-\lambda_0)} - 1.2\lambda_0}{0.4 \cdot (1-\lambda_0)} \quad [\mathrm{m^3 m^{-3}}] \qquad (2\text{-}47)$$

This equation is applicable to vacuum rectification and absorption, operated under low and/or moderate liquid loads u_L.

Figures 2-12a and 2-12b show the comparison between the calculated $h^0_{L,Fl}$ values and the experimental data. The evaluation took into account experimental data for randomly filled, stacked and structured packings with void fractions of $0.65 \leq \varepsilon \leq 0.98$ $\mathrm{m^3\ m^{-3}}$ as well as for packing elements with diameters from $d = 0.015$ to $d = 0.090$ m. The deviation of the experimental values from the curve, which was calculated using Eq. (2-47), is $\delta(h^0_{L,Fl}) \leq \pm 15\%$. Figures 2-12a and 2-12b show that the liquid hold-up $h^0_{L,Fl}$ at the flooding point is dependent on the phase flow ratio λ_0 at the flooding point and can therefore be determined for any type of system, acc. to the single-parameter Eq. (2-47).

The properties and constructive parameters of the test systems are varied in the following ranges:

$$\left.\begin{array}{l} \sigma_L = 26.............72\ mNm^{-1} \\ \eta_L = 0.35...........10\ mPas \\ \rho_V = 0.09..........1.2\ kgm^{-3} \\ \rho_L = 932...........1000\ kgm^{-3} \\ d \cdot 10^3 = 15........90\ m \\ d_S = 0.15...........1.0\ m \\ H = 0.7..............7.0\ m \\ \lambda_0 \cdot 10^3 = 0.2......1000\ [-] \end{array}\right\} \quad for\ \mathrm{Re}_L \geq 2 \qquad (2\text{-}48)$$

There is little data available for laminar liquid flow, with $\mathrm{Re}_L < 2$, for the test system air/water [A] for 12 mm metal Białecki rings, 15 mm Pall rings, 17 mm Nor-Pac rings and 18 mm Hiflow rings made of PP and/or for air/silicone oil [36, 37]. Figure 2-13 shows the comparison between the liquid hold-up $h^0_{L,Fl}$, which was calculated using Eq. (2-46), with the parameter m, based on Eq. (2-44), and the experimental data $(h^0_{L,Fl})_{exp}$. The deviations from the curve, reflecting Eq. (2-49), with $m = -0.88$, acc. to Eq. (2-45), are approx. $\pm 10\%$ in the range of the phase flow ratios $\lambda_0 = (0.2-6) \cdot 10^{-3}$.

$$h^0_{L.Fl} = \frac{\sqrt{1.254 \cdot \lambda_0^2 + 0.48 \cdot \lambda_0 \cdot (1 - \lambda_0)} - 1.12\lambda_0}{0.24 \cdot (1 - \lambda_0)} \quad \left[m^3 m^{-3}\right] \qquad (2\text{-}49)$$

The applicability of the derived Eqs. (2-47) and (2-49) for determining the gas velocity at the flooding point for low-viscosity and viscous mixtures is shown in Table 2-4, based on the example of 25 mm metal Pall rings.

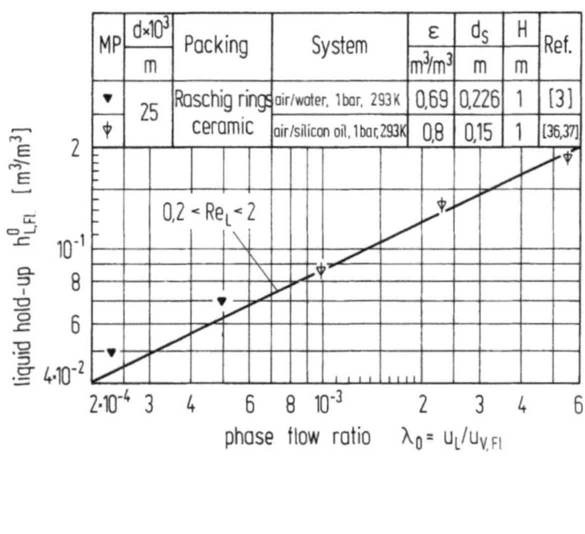

Figure 2-13. Dependence of the liquid hold-up at the flooding point $h^0_{L,Fl}$ on the phase flow ratio λ_0 for $\mathrm{Re}_L < 2$ – experimental data in comparison to calculated values based on Eq. (2-49)

Table 2-4. List of experimental data on the flooding point for random 25 mm metal Pall rings, valid for various systems

–	(a) valid for $Re_{L,Fl} \geq 2$, $h^0_{L,Fl}$ calculated acc. to Eq. (2-47), $u_{V,Fl}$ acc. to Eq. (2-69), $C_{Fl,0} = 0.566$					(b) valid for $Re_{L,Fl} \geq 2$, $h^0_{L,Fl}$ calculated acc. to Eq. (2-49), $u_{V,Fl}$, $C_{Fl,0}$ as (a)	
System	ethylbenzene/ styrol $\dot{L}/\dot{V} = 1$ $N = 47500\ m^{-3}$	air/water 293 K $N=51500\ m^{-3}$	air/water 293 K $N = 52388\ m^{-3}$	chlorobenzene/ ethylbenzene $\dot{L}/\dot{V} = 1$ $N=53200\ m^{-3}$	air/glycol 305 K $N=51500\ m^{-3}$	ethyleneglycol/ propylene-glycol $N=47500\ m^{-3}$	air/ engine oil 305 K
p_T [mbar]	133	1000	1000	33 66.7 133	1000	13.3	1000
d_S/H [mm^{-1}]	0.8/2	0.435/ 1.65	0.3/0.8	0.22/1.31 0.22/1.31 0.22/1.31	0.435/1.65	0.5/2	0.435/1.65
ρ_V [kgm^{-3}]	0.482	1.165	1.165	0.212 0.327 0.578	1.135	0.036	1.135
ρ_L [kgm^{-3}]	854.0	998.2	998.2	1026.0 995.0 974.7	1110	1017	850
a [$m^2\ m^{-3}$]	198.7	215.0	218.7	222.1 222.1 222.1	215.0	198.7	215
σ_L [mNm^{-1}]	21.7	72.74	72.74	30.0 28.0 25.6	48.0	30.0	25.6
ε [$m^3\ m^{-3}$]	0.947	0.942	0.941	0.94 0.94 0.94	0.942	0.947	0.942 0.942 0.940 0.940
$\eta_L.10^3$ [$kgm^{-1}\ s^{-1}$]	0.32	1.0	1.0	0.62 0.50 0.41	18.5	4.0	85.5 89.5 91.2 92.6

Table 2-4. (continued)

–	(a) valid for $Re_{L,Fl} \geq 2$, $h^0_{L,Fl}$ calculated acc. to Eq. (2-47), $u_{V,Fl}$ acc. to Eq. (2-69), $C_{Fl,0} = 0.566$					(b) valid for $Re_{L,Fl} \geq 2$, $h^0_{L,Fl}$ calculated acc. to Eq. (2-49), $u_{V,Fl}$, $C_{Fl,0}$ as (a)	
System	ethylbenzene/ styrol $\dot{L}/\dot{V} = 1$ $N = 47500\ m^{-3}$	air/water 293 K $N=51500\ m^{-3}$	air/water 293 K $N = 52388\ m^{-3}$	chlorobenzene/ ethylbenzene $\dot{L}/\dot{V} = 1$ $N=53200\ m^{-3}$	air/glycol 305 K $N=51500\ m^{-3}$	ethyleneglycol/ propylene-glycol $N=47500\ m^{-3}$	air/ engine oil 305 K
$(u_{V,Fl})_{exp}$ $[ms^{-1}]$	3.90	1.550 1.643 1.826	2.51 2.17 2.05 1.91	6.26 5.04 3.705	1.89 1.78 1.72	15.6	1.76 1.57 1.36 1.34
$(u_L.10^3)_{exp}$ $[ms^{-1}]$	2.63	14.92 13.05 10.83	2.78 5.54 8.40 11.10	1.30 1.66 2.20	9.27 11.08 12.90	0.55	5.27 6.95 8.60 10.30
$\lambda_0.10^3$ [-]	0.670	9.63 7.94 5.93	1.11 2.55 4.09 5.81	0.21 0.33 0.60	5.08 6.40 8.11	0.03	3.00 4.43 6.32 7.68
Re_L [-]	27.33	69.23 60.63 50.30	12.70 25.30 38.30 50.65	9.7 14.9 23.5	2.59 3.09 3.60	0.75	0.224 0.300 0.365 0.440
$(u_{V,Fl})_{calc}$ $[ms^{-1}]$	3.68	1.66 1.767 1.907	2.657 2.333 2.083 1.89	6.12 4.71 3.30	2.023 1.90 1.782	16.44	1.56 1.38 1.223 1.09
$\delta(u_{V,Fl})$ [%]	5.6	−8.38 −7.5 −4.4	−5.85 −7.5 −1.6 +0.5	+2.2 +6.6 +10.8	−7.04 −7.9 −3.6	−5.40	+11.4 +12.0 +10.04 +18.6
Literature	Billet [5]	Billet [5]	[A]	[A]	Billet [5]	Billet [5]	Billet [5]

2.2.4.5
Influence of Packing Size on Droplet Velocity u_0

The prediction of the effective individual droplet velocity u_T for packings of any given size, acc. to Eq. (2-24), is only possible, once the correlations for functions $f_2(d_h/d_T)$ and $f_3(\psi)$ have been found. Table 2-5 contains the numerical values of the reduced droplet velocity $\bar{u}_T$ for various types of packing elements. It shows that $\bar{u}_T$ is dependent on the type and size of the packing element and therefore on the packing-specific variables a and ε. Physically, this can be compared to the fall of an individual droplet in a tube near a circular wall. It was referred to as a *wall effect* by Strom and Kinter [62], Reinhart [63] and Clift et al. [65]. Figure 2-14 is a schematic representation of this effect.

If the droplet diameter d_T approximates to the tube diameter d_R, i.e. $d_T \cong d_R$, the ratio of the falling velocity u_T of the droplet to the maximum velocity $u_{T,max}$ tends to zero, see Fig. 2-14. The droplet only reaches its maximum falling velocity $u_{T,max}$ when the diameter ratio d_R/d_T is sufficiently high.

The wall effect in packed columns is analogous to the fall of droplets in tubes, acc. to [62, 63, 65], and can be expressed by the following functions:

$$\frac{u_T}{u_{T,\max}} = \left[1 - \left(\frac{d_T}{d_h}\right)^b\right]^c \qquad \text{b,c} = \text{exponent} \tag{2-50}$$

and/or based on the equation developed by Wallis [65]:

$$\frac{u_T}{u_{T,\max}} = d \cdot \left[\frac{d_T}{d_h}\right]^e \qquad \text{d} = \text{const.,} \quad \text{e} = \text{exponent} \tag{2-51}$$

where the hydraulic diameter of the packing d_h is introduced instead of d_p. For simplification purposes, the wall effect in packed columns is expressed by the correlation in Eq. (2-51), as this requires one less constant to be determined when deriving the final equation for the gas velocity at the flooding point $u_{V,Fl}$.

The proximity of the packing wall also has an influence on the droplet fall in the packing. In the channel model, the tube diameter depends on the channel diameter, acc. to Eq. (2-52), see Chap. 3.

$$d_h = 4 \cdot \frac{\varepsilon}{a} \tag{2-52}$$

When the droplet size is equal to the hydraulic diameter d_h of the packing, acc. to Eq. (2-52), it is not possible for the droplet to fall in the packing, i.e. $u_T \to 0$. It can therefore be assumed that the reduced droplet velocity $\bar{u}_T$ changes with the diameter ratio d_h/d_T, if d_h becomes larger than the droplet diameter d_T, acc. to Eq. (2-26).

Figure 2-15 leads to the following correlation between the reduced droplet velocity $\bar{u}_T$ and the ratio d_h/d_T and therefore between the effective droplet velocity $u_T$and the quotient (d_h/d_T), see Eq. (2-53):

Table 2-5. List of experimental data of the reduced gas velocity of an individual droplet $\bar{u}_T$ for various types of packings, used for determining the gas or vapour velocity $u_{V,Fl}$ – data relating to Fig. 2-15

No.	TP	Number of TP	Packing	Material	$d \cdot 10^3$ [m]	a [m^2/m^3]	ε [m^3/m^3]	$\bar{u}_T \cdot 10^2$ [m/s]	d_S [m]	H [m]	Literature
1	□	4	**Pall ring**	metal	15	376.8	0.93	9.63	0.22-0.3	1.4	[A], [5,28]
		5			15	368.4	0.933	10.4	0.15-0.8	2	
2	■	24			25	215	0.950	11.8	0.22-0.8	2	[A],[5,16,25]
		4			25	217	0.950	12.6	0.3	0.8	[A]
		5			25	292	0.970	12.0	0.15	1.3	[A]
3	■	12			35	150	0.946	12.8	0.3-0.8	1.2-4	[A],[5]
		5			38	149.6	0.952	13.3	0.22-0.3	1.45	[A]
4	□	8			50	110	0.952	13.5	0.5-1.2	2-5.5	[5]
		4			50	110	0.952	14.1	0.305-0.75	3.85	[16,25,28]
		5			58	112	0.979	15.5	0.45	2	[A]
5	■	4			80	78	0.96	18.0	0.75	3	[28]
6	■	12		plastic (PP,PVDF)	25	239.7	0.88	10.3	0.22-0.5	1.4-2	[A],[23]
7	■	5			35	142	0.922	13.0	0.3-0.45	1.4	[A]
8	■	2			50	110	0.92	12.9	0.75	3	[28]
		2			50	111	0.919	13.2	0.3	1.4	[A]
9	●	8	**Bialecki ring**	metal	25	238	0.94	11.3	0.3	1.4	[20]
		7			25	225	0.945	11.3	0.15	1.4	[A]
10	●	7			35	155	0.95	12.8	0.22-0.3	1.45	[A],[20]
11	●	7			53.5	110	0.973	14.8	0.3	1.4	[20]
12	▲	9	**VSP ring**	metal	32 (Gr.1)	200.8	0.972	13	0.22-0.45	1.4-2	[A]
13	▲	5			50 (Gr.2)	104.9	0.98	15.6	0.22-0.45	1.4-2	[A]
14	✱	2	**Intalox-saddle**	ceramic	25	255	0.73	8.1	0.3	2	[19]
		5			25	194	0.79	8.35	0.3	0.8	[A]
15	○	2	**Top-Pak**	metal	45 (Gr.1)	105	0.975	15.0	0.4	2	[27]
16	●	4			80 (Gr.2)	75	0.98	16.2	0.45	2	[A]
17	+	8	**Dtnpac**	plastic (PP,PVDF)	50	112.7	0.937	15.4	0.3	1.4	[A]
18	◆	4	**NSW ring (Nor-Pac)**		15	309	0.92	12.0	0.30	1.4	[40]
19	◆	6			28	193	0.922	12.8	0.15	1.4	
20	◆	7			50	90.3	0.952	17.2	0.3-0.45	2	
21	▲	3	**Hiflow ring**	metal	28	198.5	0.962	13.4	0.3	1.4	[A]
		4			28	182	0.965	13.7	0.45	2	[A]
22	▲	8			58	100.6	0.976	15.9	0.45	2	[45]
23	△	4		plastic (PP,PVDF)	28	190	0.920	12.3	0.30	1.4	[44]
24	△	5			50	108	0.935	15.1	0.30	1.4	[44]
25	△	3			90	69.5	0.965	18.8	0.45	2	[44]
26	▲	1		ceramic	20	272	0.763	8.3	0.22	1.4	[46]
		4			20	280	0.76	8.8	0.3	1.4	
27	▲	3			35	111.8	0.824	12.2	0.3	1.4	[A]
		5			35	110	0.926	12.1	0.3	1.4	
28	▲	4			50	89.8	0.809	12.5	0.3	1.2	[A]
		4			50	85.6	0.817	13.0	0.3	0.9	
29	+	4	**Hiflow saddle**	plastic	50	82.6	0.94	16.2	0.45	2	[A]
30	✳	3	**Ralu ring**	(PP)	50	112.7	0.938	14.0	0.45	2	[A]
31	○	10	**Mellapak**	sheet metal	-	250	0.96	13.2	0.22-1	1.2-4	[A],[24,26]
32	✳	7	**Ralu Pak**		-	250	0.963	13.45	0.45	2	[48]
33	◇	4	**Montz packing**		B1-100	100	0.987	18.2	0.3	1.4	[A]
34	◇	9			B1-200	200	0.978	14.36	0.3	1.4	[A]
35	◇	12			B1-300	300	0.972	12.44	0.22-0.45	1.4-2	[A]
36	◆	3		PP	C1-200	200	0.96	13.7	0.3	1.4	[A]
37	◆	3		sheet metal	B2-500	500	0.95	10.1	0.22	1.5	[A]

A ≡ Author

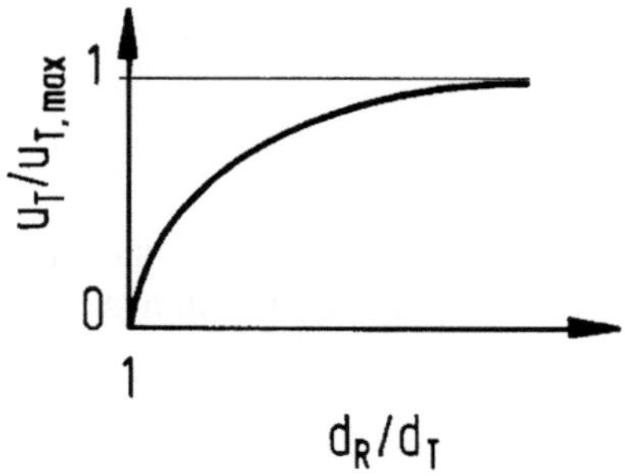

Figure 2-14. Schematic representation showing the influence of the column diameter ratio d_R/d_T on the falling velocity of droplets

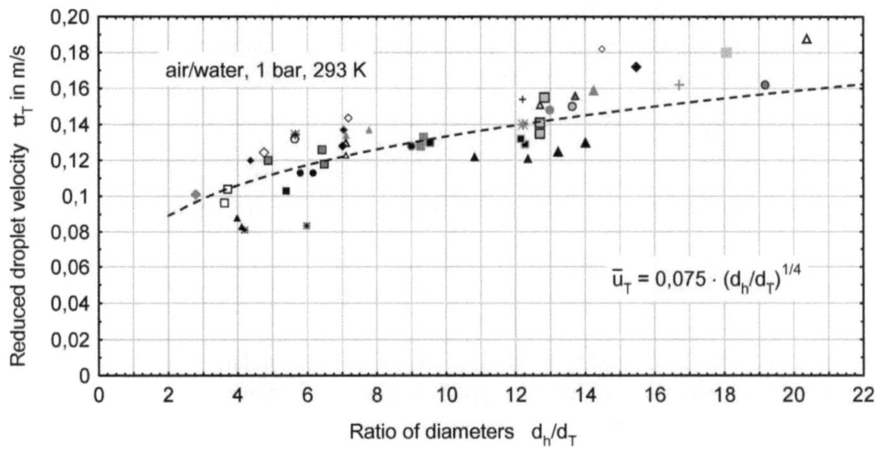

Figure 2-15. Reduced gas velocity of an individual droplet u_T as a function of the diameter ratio d_h/d_T, valid for Pall, Białecki and VSP rings, d = 15–80 mm, and the air/water system under normal conditions

$$\text{for } \bar{u}_T \approx \left[\frac{d_h}{d_T}\right]^{1/4} \quad \rightarrow \quad u_T \approx \left[\frac{d_h}{d_T}\right]^{1/4} \tag{2-53}$$

Hence, $f_2(d_h/d_T)$ in Eq. (2-24) takes the form:

$$f_2\left(\frac{d_h}{d_T}\right) = C_2 \cdot \left(\frac{d_h}{d_T}\right)^{1/4} \quad \text{for } \frac{d_h}{d_T} > 3 \tag{2-54}$$

2.2.4.6
Deriving the Final Equation for Gas Velocity at Flooding Point $u_{V,Fl}$

Stichlmair [53] found that the effective velocity $\bar{u}_0$ of droplets flowing through individual packing elements increases to a value higher than u_0/ε, which is due to the deflection of the fluid between the particles and the vortex shedding. As a result, the effective gas velocity $\bar{u}_0$ can be expressed as:

$$\bar{u}_0 = \frac{u_0}{f_4(\varepsilon)} \tag{2-55}$$

Assuming that $f_4(\varepsilon) = \varepsilon^q$, the correlation for $\bar{u}_0$ is as shown in Eq. (2-56), and the exponent q must be determined experimentally:

$$\bar{u}_0 = \frac{u_0}{\varepsilon^q} \tag{2-56}$$

Flooding occurs for $h^0_{L,Fl} \rightarrow 0$, acc. to the model shown in Fig. 2-6 and Eq. (2-21), when the effective falling velocity u_T of an individual droplet is equated with the effective gas velocity $\bar{u}_0$, which keeps the droplet suspended, acc. to (2-21), i.e.

$$\bar{u}_0 = u_T \Rightarrow \frac{u_0}{\varepsilon^q} = u_T \tag{2-57}$$

Equations (2-57), (2-22), (2-53), (2-54) and (2-56) result in the following equation:

$$u_0 = C_1 \cdot C_2 \left[\frac{d_T.\Delta\rho \cdot g}{\rho_V}\right]^{1/2} \cdot \left[\frac{d_h}{d_T}\right]^{1/4} \cdot \varepsilon^q \cdot f_3(\psi) \tag{2-58}$$

Equations (2-19), (2-32) and (2-58) lead to the following correlation for the gas velocity at the flooding point $u_{V,Fl}$, which also applies to packing elements with similar resistance coefficients ψ_{Fl}, see Table 6-1a:

$$u_{V,Fl} = u_0 \cdot \left(1 - h^0_{L,Fl}\right)^n \quad \Rightarrow$$
$$C_{Fl} \cdot \left[\frac{d_T \cdot \Delta\rho \cdot g}{\rho_V}\right]^{1/2} \left[\frac{d_h}{d_T}\right]^{1/4} \cdot \varepsilon^q \cdot \left(1 - h^0_{L,Fl}\right)^{7/2} \tag{2-59}$$

where $C_{Fl} = C_1 \cdot C_2 \cdot f_3(\psi_{Fl})$.

Using the minimisation procedure, the evaluation of the experimental data listed in Tables 2-6, 2-7, 2-8 and 2-9 for various types of packings leads to the following model:

$$u_{V,Fl} = C_{Fl} \cdot \varepsilon^{6/5} \cdot \left[\frac{d_h}{d_T}\right]^{1/4} \left[\frac{d_T \cdot \Delta\rho \cdot g}{\rho_V}\right]^{1/2} \cdot \left(1 - h^0_{L,Fl}\right)^{7/2} \tag{2-60}$$

where C_{Fl} is a dimensionless constant, which has been calculated as:

$$C_{Fl} \approx 0.50 \tag{2-61}$$

for classic, randomly filled packing elements.

Table 2-6. Data relating to the experimental flooding point values, diagrammed in Fig. 2-17a, in columns with randomly filled metal packing elements. No. of test system acc. to Table 2-2

TP	$d \cdot 10^3$ [m]	Packing	Test system	d_S [m]	H [m]	Literature
PART 1						
●	12	**Bialecki rings**	1,6	0.22-0.3	0.8-1.4	[A]
●	25		1,6,10	0.15-0.30	0.7-1.4	[A],[20]
●	35		1,6,10	0.15-0.50	0.7-2.0	[A],[20,39]
○	50		1,6,8	0.15-0.80	1.0-2.5	[A],[20,39]
◇	15	**Pall rings**	1,2,3,6,7	0.30-0.788	1.4-2.0	[A],[5]
◆	25		1,2,6,16	0.15-0.75	1.4-4.0	[A],[5]
◇	30-Glitsch		12	0.22	1.5	[A]
◆	35		1,2,3,6	0.22-0.80	1.4-4.0	[A],[5]
◆	38-VFF		1,6	0.22-0.30	1.46	[A]
◆	50		1,2,3,4,8,11	0.50-1.20	1.4-5.5	[A],[5,7,66]
◆	58		1	0.45-1.00	2-3.5	[A],[67]
◆	80		2	0.80	2.0	[5]
+	35	**Ralu rings**	1	0.75	3.0	[28]
+	50					
▲	15	**Raschig rings**	2	0.5	2.0	[5]
▲	25		2			
△	35		2			
▲	50		2,3			
□	0.3	**Raschig Super rings**	1	0.288	1.0	[94]
■	0.5			0.288	1.0	[94]
■	0.7			0.28	2.0	[94]
□	1.0			0.30-0.40	1.4-3.0	[86, 94]
■	1.5			0.28	2.0	[94]
■	2.0			0.30-0.75	1.4-3.0	[86, 94]
■	3.0			0.44	2.0	[94]
PART 2						
●	0.5A	**Glitsch CMR 304 rings**	1,12	0.22-0.30	1.4-1.54	[A]
●	1A		1,6,12	0.22-0.30	1.4-1.54	
●	1.5A		1,6,12	0.22-0.30	1.4-1.54	
○	1.5A Turbo		6	0.22	1.54	
○	2A		1	0.45	2.0	
●	3A		1	0.45	2.0	
+	30P	**Glitsch rings**	1	0.30	1.4	[A]
■	28	**Hiflow rings**	1,6	0.22-0.60	1.4-2.0	[A]
□	58		11	0.45-1.00	2.0-3.5	[A],[45,60]
+	25	**I-13 rings**	1	0.30	0.7-1.4	[20]
+	40	**IMPT rings**	1	0.45	2	[A]
✳	10/15/20	**Interpack**	6	0.218	1.5	[5]
▲	0.7/1.0/1.5	**Nutter rings**	1	0.3	2.25	[95]
△	2.0/2.5/3.0					
■	size 1	**Mc-Pac**	1,8	0.18-0.32	2.8-3.0	[A],[79,80]
■	size 2		1,6	0.32-0.60	1.5-3.0	
✳	50	**PSL ring**	1,6	0.22-0.3	1.4	[A]
△	25	**Rauschert metal saddle rings (RMSR)**	1	0.8	–	[93]
▲	40					
▲	50					
▲	60					
▲	70					
◆	no.1	**Top-Pak**	1	0.30-0.45	1.45-2.0	[A]
◆	no.2			0.45	2.0	
◆	no.1	**VSP rings**	1,6	0.218-0.45	1.45-2.0	[A],[27,93]
◆	no.2			0.218-1.00	1.45-6.0	[A],[67,93]

Number of test points: 435, mean relative deviation $\overline{\delta}(u_{V,Fl}) = 6{,}15\ \%$

A ≡ Author

Table 2-7. Data relating to the experimental flooding point values, diagrammed in Fig. 2-17b, in columns randomly filled with plastic (PP, PVDF) packing elements. No. of test system acc. to Table 2-2

TP	$d \cdot 10^3$ [m]	Packing	Material	Test system	d_S [m]	H [m]	Literature
PART 1							
+	50	**Białecki rings**	PP	1	0.30	1.4	[20]
+ +	size 1 size 2	**Dtnpac**	PP	1	0.45 0.30	2.0	[A]
◆ ◇ ◆	size 1 size 2 size 3	**Envipac**	PP	1	0.30 0.30-0.45 0.45	1.4 1.4-2.0 2.0	[A]
✱ ✱	no.1 no.2	**Glitsch CMR**	PP	1	0.30-0.45 0.45	1.4-2.0 2.0	[A]
●	45	**Hackette**	PP	1	0.45	2.0	[A]
■	25	**Pall rings**	PVDF	6,7	0.218-0.22	1.4	[A]
■ □ ■ □	25 35 50 90	**Pall rings**	PP	1	0.30 0.30-0.45 0.30-1.00 0.75	1.4 1.4 1.4-3.5 3.0	[A], [44] [A] [A], [44,66] [96]
▲ △	38 50	**Ralu rings**	PP	1	0.75 0.30-0.75	3.0 1.4-3.0	[28] [A], [28]
✱	2"	**Super saddles PP**	PP	1	0.75	3.0	[28]
● ○	no.1 no.2	**Telerette**	PP	1	0.15 0.45	1.3 2.0	[A], [40] [A], [54]
PART 2							
+ +	28 50	**Hiflow rings**	PVDF	1,6	0.218-0.30	1.4	[A],[44]
□ ○ △	15 38 90	**Hiflow rings**	PP	1	0.30 0.8 0.45-1.00	1.4 – 2.0-3.5	[A] [97] [66],[A]
□	50	**Hiflow Super**	PP	1	0.30-0.45	1.4-2.0	[A]
○	50	**Hiflow saddles**	PP	1	0.45	2.0	[A]
■ ●	35 50	**Intalox saddles**	PP	1	0.30 0.45	1.4 2.0	[A]
■ ● ▲	17 22 x 27 28	**Nor-Pac**	PVDF	7 6,7 6	0.218	1.4	[A]
■ ● ▲ ◆ ✱	17 22 x 27 28 38 50	**Nor-Pac**	PP	1 1 1,21b 1 1,8,22	0.30 0.30 0.15-1.40 0.30 0.30-1.40	0.9-1.4 1.4 1.0-2.0 1.4 0.9-2.0	[A] [A] [A], [40] [A] [A], [Fi/Lu]
◆	90	**Raflux rings**	PP	1	0.8	–	[97]
▲	50	**VSP rings**	PP	1	0.45	2.0	[A]

Number of test points: 260, mean relative deviation $\overline{\delta}(u_{V,Fl}) = 4.10\,\%$

A ≡ Author

This packing group includes: 15 to 80 mm Pall rings made of metal, plastic and ceramic, 25 to 50 mm metal Białecki rings, metal VSP rings (sizes 1 and 2), metal Top-Pak packings (sizes 1 and 2), 10 to 50 mm Intalox saddles made of plastic and ceramic, 8 to 50 mm ceramic Raschig rings, Glitsch rings made of metal and plastic (sizes 0.5–3), I-13 rings and PSL rings.

The flooding point constant C_{Fl}

$$C_{Fl} \approx 0.55 \tag{2-62}$$

Table 2-8. Data relating to the experimental flooding point values, diagrammed in Fig. 2-17c, in columns randomly filled with ceramic packing elements. No. of test system acc. to Table 2-2

TP	$d \cdot 10^3$ [m]	Packing	Test system	d_S [m]	H [m]	Literature
●	20	**Hiflow rings**	1,6	0.218-0.30	1.25-1.4	[A]
○	38		1,6	0.218-0.45	1.4-2.0	
●	50		1,6	0.218-0.30	0.9-1.4	
○	75		1	0.45	2.0	
▲	25	**Intalox saddles**	1,6	0.220-0.30	1.4	[A]
△	38		1,6,21a	0.220-0.50	1.4-2.0	[A], [29]
▲	50		1,6	0.220-0.75	1.4-3.0	[A], [28]
✱	25	**Pall rings**	3,6,11	0.218-0.50	1.0-1.4	[A],[23]
✳	50		6	0.220		[A],[21]
■	8	**Raschig rings**	8,10,14	0.10	1.0	[A],[5]
■	15		1,6	0.20-0.220	1.0	[A],[3,5,71]
□	19		4	1.2	5.5	[5]
■	25		1,3,6,14	0.220-0.50	1.0-2.0	[A],[5,32]
□	25 Glas		20	0.150	1.6	[36,37]
◆	35		1	0.30	0.7	[20]
◇	50		1,6,8	0.30-0.60	0.7-1.4	[A],[Fi],[20,82]
◆	100		21a	1.20	2.0	[84,85]
+	size 1	**R-Pac**	1	0.316	1.0-3.0	[82,83]
+	size 2		1,6	0.320-0.60		
+	size 2	**SR-Pac**	1,6	0.320-0.60	1.0-3.0	[82,83]

Number of test points: 181, mean relative deviation $\overline{\delta}(u_{V,Fl}) = 5.53$ %

A ≡ Author

was increased by approx. 10% in the case of modern, highly perforated packing elements, such as 20–90 mm Hiflow rings made of metal, plastic and ceramic, 17–50 mm Nor-Pac rings, plastic VSP rings (size 2), Ralu rings (sizes $1\frac{1}{2}$ and 2), Tellerette (sizes 1 and 2), Envipac (sizes 1, 2 and 3), Dtnpac (sizes 1 and 2), R-Pac, SR-Pac, Mc-Pac. The same applies to non-perforated Montz packing made of sheet metal and plastic as well as to Impuls packing.

For perforated Mellapak 250 Y packings, Ralu-Pak 250YC with slit perforation, Gempak 200AT and stacked 25 mm metal Białecki rings, the numerical value of the flooding point constant C_{Fl} was found to be:

$$C_{Fl} \approx 0.615 \tag{2-63}$$

The different numerical values, which were found for the constant C_{Fl} (Eqs. (2-61) ÷ (2-63)), indicate that the wall effect cannot be sufficiently expressed by function $f_2(d_h/d_T)$ alone. It is not only the wall distance, but also the shape of the packing wall that needs to be taken into account. This is linked to the resistance coefficient ψ_{Fl} in single-phase flow.

In Fig. 2-16, the $C_{Fl,i}$ values of the experimental data, listed in Tables 2-6, 2-7, 2-8, 2-9 and substituted into Eq. (2-60), are plotted against the resistance coefficient ψ_{Fl} for the respective gas velocity at the flooding point $u_{V,Fl}$, acc. to Eq. (3-14) or (3-26):

$$\psi_{Fl} = K_1 \cdot (\mathrm{Re}_V)_{Fl}^{K_2} \tag{3-14}$$

$$\psi_{Fl} = \psi_0 \cdot (1 - \varphi_P) \tag{3-26}$$

Table 2-9. Data relating to the experimental flooding point values, diagrammed in Fig. 2-17d, in column filled with structured or stacked packing elements. No. of test system acc. to Table 2-2

TP	Packing	Type	Material	Test system	d_S [m]	H [m]	Literature
	TUBE COLUMNS (TS)						
●	**Białecki rings**	25	metal	1	0.025	1.5	[A]
●		52.5		8	0.273	1	[A]
+	**Pall rings**	25	metal	1	0.025	1	[76]
	STACKED PACKINGS						
●	**Białecki rings**	25	metal	1,6	0.150-0.218	1.4-1.7	[A]
●		35		6	0.218	1.5	
○		50		1,6	0.151-0.30	1.0-1.7	
○	**Białecki rings**	50	plastic	1	0.30	1.0	[77], [A]
●	**Hiflow rings**	50	ceramic	1	0.45	1.0	[A]
+	**Pall rings**	50		1,8	0.40-0.487	1.0	[A],[21]
+	**PSL rings**	50	metal	1	0.28	1.4	[A]
	STRUCTURED PACKINGS						
■	**Euroform**		PP	1	0.30	1.4	[40]
■	**Fi-Pak (I-13F)**		metal	6	0.147-0.218	1.4	[A]
▣	**FX (VFF) 260**		PP	1	0.45	2	[A]
□	**Gempak A2 T304**		metal	6,12	0.218	1.4	[A],[A1]
■	**Impuls 50**		ceramic	1	0.385-0.40	1.4	[21]
▲	**Mellapak**	250X	sheet metal	1	1.00	3.5	[26,68,69]
▲		250Y		1,6	0.218-1.00	1.4-3.5	[A],[26,68,69]
▲		500Y		1	1.00	3.5	[68,69]
◆	**Montz**	A1	metal	6,7	0.218	1.4	[A]
◆		A3-500		1,6	0.218-0.45	2.0	[A],[A1]
◇		B1-100	sheet metal	1	0.30	1.4	[A]
◇		B1-200		1,6	0.218-0.30	1.4	[A],[47]
◇		B1-300		1,6	0.218-0.30	1.4	[A],[47]
◆		B2-300		6	0.218	1.4	[A],[A1]
◆		B2-500		6	0.218	1.4	[A],[A1]
◆		C2-200		1	0.30	1.4	[A]
◆		S1	metal	6	0.218	1.4	[A]
✱	**NorPac**	Gr.3	PP	1	0.30	1.4	[A]
✱		Kompakt no.1		1	0.30	1.4	[A]
✱		Kompakt no.2		1	0.30	1.4	[A]
△	**Ralu-Pak**	250 YC	sheet metal	1,6	0.218-0.45	1.4-2.0	[A],[48]
▲	**Sulzer gauze packing**	BX	metal	2,3,6	0.218-0.50	1.4-2.0	[A],[5]
▲	**Honeycomb packing**		PP	1	0.30	1.4	[A]

Number of test points: 196, mean relative deviation $\overline{\delta}(u_{V,Fl}) = 6.06$ %

A ≡ Author

based on Tables 6-1a–6-1c, using the numerical values K_1, K_2 and φ_P determined within the scope of this work. This leads to the following correlation (2-64) for calculating the flooding point constant C_{Fl} for Eq. (2-60):

$$C_{Fl} = \left(C_{Fl,0}\right)_{45^\circ} \cdot \psi_{Fl}^{-1/6} \leq \pm 8\ \% \qquad (2\text{-}64)$$

where $C_{Fl,0} = 0.566$ for a flow angle of gas stream in the packing channels $\alpha = 45^\circ$ is a *dimensionless constant, independent of the type, size and material of the packing.*

Equation (2-64) applies to any types of randomly filled packing elements and structured packings with type Y flow channels (approx. 45°) for resistance coefficients in the range of $\psi_{Fl} = 0.1{-}8$. As shown in Fig. 2-16 and Eq. (2-60), the gas velocity at the

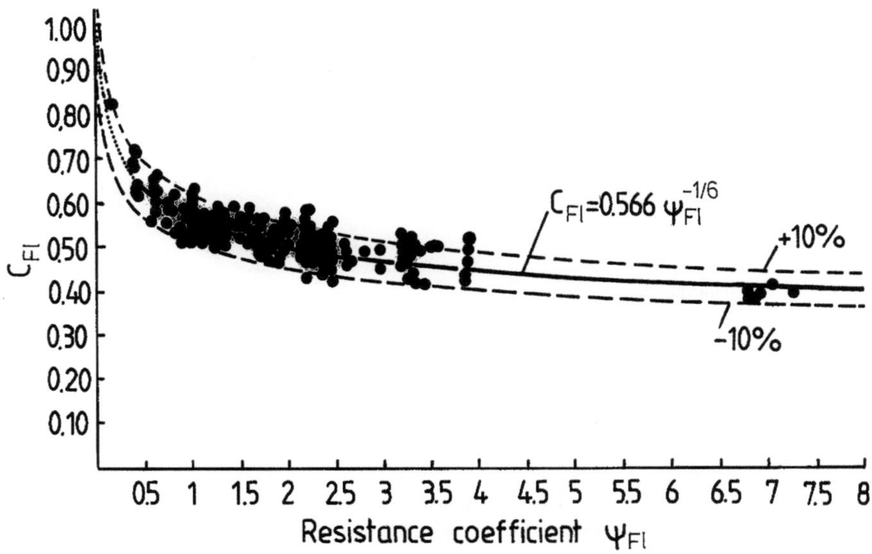

Figure 2-16. Dependency of the constant C_{Fl} on the resistance coefficient at the flooding point ψ_{Fl} plotted on the basis of data in Tables 2-6, 2-7, 2-8 and 2-9

flooding point $u_{V,Fl}$ decreases, as the resistance coefficient ψ_{Fl} increases. The flooding point constant of $C_{Fl} \cong 0.5$ applies to the group of packing elements with a resistance coefficient ψ_{Fl} at the flooding point in the range of 1.8 and 3, see Eq. (2-61). For resistance coefficients in the range of $\psi_{Fl} = 0.9-1.6$, the flooding point constant is assumed to have a mean value of $C_{Fl} \cong 0.55$. The group of perforated packing elements also includes structured packings made of sheet metal, plastic and ceramic (type Y) without wall openings. For practical calculations, the flooding point constant for resistance coefficients ψ_{Fl} in the range of 0.5–0.9 is determined as $C_{Fl} \cong 0.615$. This applies to the group of structured sheet-metal packings (type Y) with slit perforation, e.g. Ralu-Pak 250YC, or perforated packings such as Mellapak 250 Y, Gempak 202 AT, as well as plastic packings 250 Y.

The lowest resistance coefficients ψ_{Fl} are typical of stacked packing elements, honeycomb packings, tube columns as well as for type X packings made of sheet metal and type BX gauze packings with flow channels of 30°. They are in the range of $\psi_{Fl} = 0.1-0.4$.

Metal Raschig rings have been found to have the highest resistance coefficients. A resistance coefficient of approx. $\psi_{Fl} \cong 8.18$ in the turbulent flow range of $Re_V \geq 2100$ results in a flooding point constant of $C_{Fl} \cong 0.4$ for Eq. (2-60).

Influence of the Flow Channel Angle α on the Gas Velocity at the Flooding Point $u_{V,Fl}$

The evaluation of the experimental data shown in Table 2-10 for stacked Hiflow rings, Raschig rings and Pall rings, tube columns with structured Pall rings and Białecki rings,

Table 2-10. Data relating to the experimental flooding point values, diagrammed in Fig. 2-17e, in columns operated at higher pressure. No. of test system acc. to Table 2-2

TP	$d \cdot 10^3$ [m]	Packing	Material	Test system	d_S [m]	H [m]	Literature
▲	10	Berl saddle	ceramic	27	0.086-1.550	0.8	[75]
△	15	Berl saddle	ceramic	27,28,29	0.155	0.8	[75]
+	20	Interpak	metal	27	0.155	0.8	[75]
✳	20	Novalox saddle	ceramic	27	0.155	0.8	[75]
■	15	Pall ring	metal	27	0.155	0.8	[75]
▪ (grey)	35	Pall ring	metal	19	0.50	2.5	[39]
□	15	Pall ring	plastic	27	0.155	0.8	[75]
●	15	Raschig ring	ceramic	27	0.155	0.8	[75]
○	15	Raschig ring	metal	27	0.155	0.8	[75]

Number of test points: 177, mean relative deviation $\overline{\delta}(u_{V,Fl}) = 8.93\ \%$

Hiflow rings, structured Sulzer gauze packings (type BX), as well as type X sheet metal packings and Montz packings X with flow channels of $\alpha = 30°$ resulted in a higher flooding point constant $C_{Fl,0}$, compared to Eq. (2-64), namely:

$$\left(C_{Fl,0}\right)_{30°} = 0.693 \quad \text{for} \quad \alpha = 30° \tag{2-65}$$

A constant numerical value $C_{Fl,0} = 0.566$ for random packings and type Y structured packings with a flow channel angle of $\alpha = 45°$ in relation to the column axis leads to the assumption that gas in a random packing flows through a bundle of parallel flow channels with a diameter of d_h and with a mean angle of approx. $\alpha = 45°$. If the angle of the flow channels α is reduced, the gas velocity at the flooding point increases and reaches its maximum value $C_{Fl,0}$ at $\alpha = 0°$. By substituting the value pairs $(C_{Fl,0})_\alpha$ for $\alpha = 45°$ and $\alpha = 30°$ into correlation (2-64), the following correlation (2-66) can be derived:

$$C_{Fl} = C_{Fl,0} \cdot \cos\alpha \cdot \psi_{Fl}^{-1/6} = 0.80 \cdot \cos\alpha \cdot \psi_{Fl}^{-1/6} \tag{2-66}$$

where:

$C_{Fl,0} = 0.800$ for flow channels with $\alpha = 0°$.
$C_{Fl,0} = 0.693$ for flow channels with $\alpha = 30°$
$C_{Fl,0} = 0.566$ for flow channels with $\alpha = 45°$.

By substituting Eq. (2-65) into Eq. (2-60), it is possible to derive the following Eq. (2-67) for determining the gas velocity at the flooding point $u_{V,Fl}$ in packed columns with any types of column internals.

$$u_{V,Fl} = 0.80 \cdot \cos\alpha \cdot \varepsilon^{6/5} \cdot \psi_{Fl}^{-1/6} \left[\frac{d_T \cdot \Delta\rho \cdot g}{\rho_V}\right]^{1/2} \cdot \left[\frac{d_h}{d_T}\right]^{1/4} \cdot \left(1 - h_{L,Fl}^0\right)^{7/2} \quad \left[ms^{-1}\right] \tag{2-67}$$

The dimensionless flooding point constant $C_{Fl,0} = 0.80$ for Eq. (2-67) applies to various types of packed columns, i.e. for:

(a) randomly filled packed columns
(b) structured packings
(c) stacked packing elements
(d) tube columns stacked with packing elements
(e) empty columns.

The numerical value $C_{Fl,0} = 0.8$ applies to columns in which the angle of the flow channels is assumed to be $\alpha = 0°$. This is the case for tube columns with stacked 25 mm metal Pall rings [76].

It follows from Eq. (2-67) that, as the gas velocity at the flooding point $u_{V,Fl}$ increases, it is possible for increasingly larger droplets to be entrained by the upward gas flow. The gas velocity at the flooding point $u_{V,Fl}$ also increases with the hydraulic diameter d_h and the void fraction ε. In the case of highly perforated packing elements, the gas velocities at the flooding point are higher, due to the lower resistance coefficients ψ_{Fl}. Type X packings with channel angles of $\alpha = 30°$ can always be expected to have higher gas velocities at the flooding point than type Y packings with channel angles of $\alpha = 45°$. In the case of very low liquid loads, e.g. in rectification under vacuum, the liquid hold-up at gas velocities calculated acc. to Eq. (2-67) may be broken down into very small droplets, which leads to the packing being blown empty by the gas.

2.2.4.7
Comparing Experimental Flooding Point Data and SBD Model Acc. to Eq. (2-67)

For the purpose of evaluating the author's own experimental data as well as data taken from literature for columns with $d_S = 0.025-1.4$ m, including data taken by Billet [5] and Bornhütter [66], Sulzer [67–71], Możenski, Kucharski [39] and Krehenwinkel [75] etc., a database was created, which now holds 1,200 experimental items of data for 32 different mixtures from the range of vacuum to normal rectification as well as pressure absorption and/or pressure rectification up to 100 bar, see Table 2-2.

Evaluation of Experimental Flooding Point Data for the Range of Vacuum Rectification and Normal Pressure Range

Figures 2-17a, 2-17b, 2-17c and 2-17d show a comparison between the calculated and experimental gas velocities at the flooding point. More information on the experimental conditions and the test systems can be found in Tables 2-6, 2-7, 2-8 and 2-9.

The gas velocities at the flooding point $(u_{V,Fl})_{calc}$ were determined iteratively, acc. to Eq. (2-67), for the start values $(u_{V,Fl})_{exp}$, $(u_{L,Fl})_{exp}$. Figure 2-17a shows a comparison between the gas velocities $(u_{V,Fl})_{calc}$, calculated using Eq. (2-67), and the experimental values $(u_{V,Fl})_{exp}$ for metal packings. The evaluation was based on approx. 340 experimental values, of which 80–91% were given with a maximum relative error of less than ±8%, using Eq. (2-67). The mean error in the determination of the gas velocity at the flooding point $(u_{V,Fl})_{calc}$ for metal packings was found to be $\bar{\delta}\left(u_{V,Fl}\right) = 4.7\,\%$ during

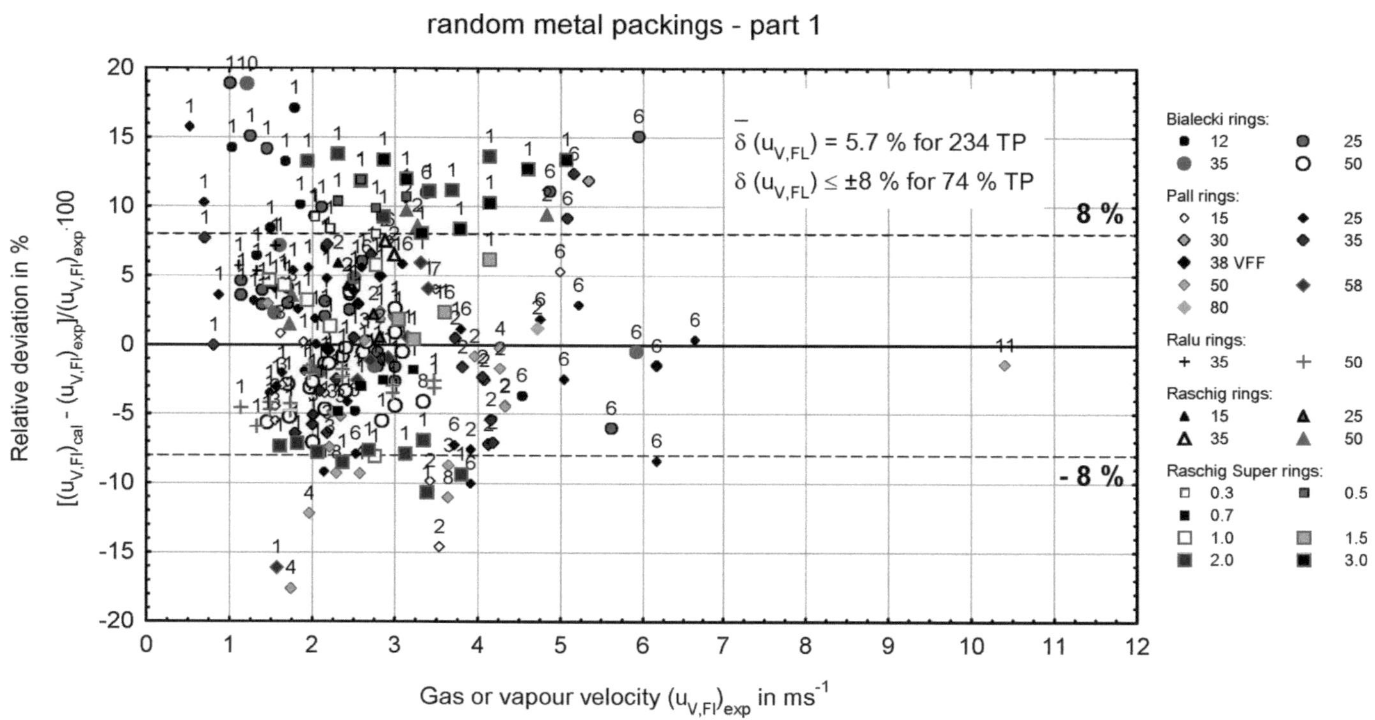

Figure 2-17a. Comparison of experimentally determined gas or vapour velocity at the flooding point $(u_{V,Fl})_{exp}$ and calculated data based on Eq. (2-67), valid for random metal packings of various types, no. of test system as shown in Table 2-2

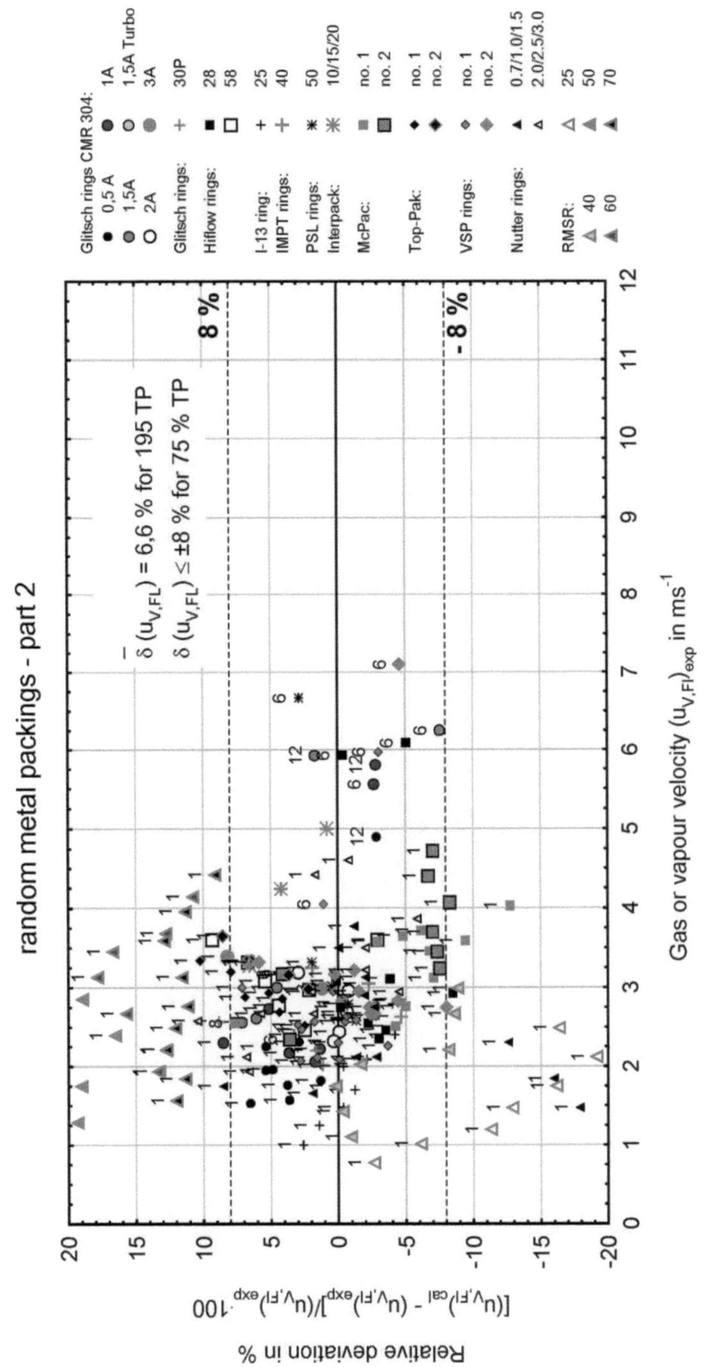

Figure 2-17a. (continued.)

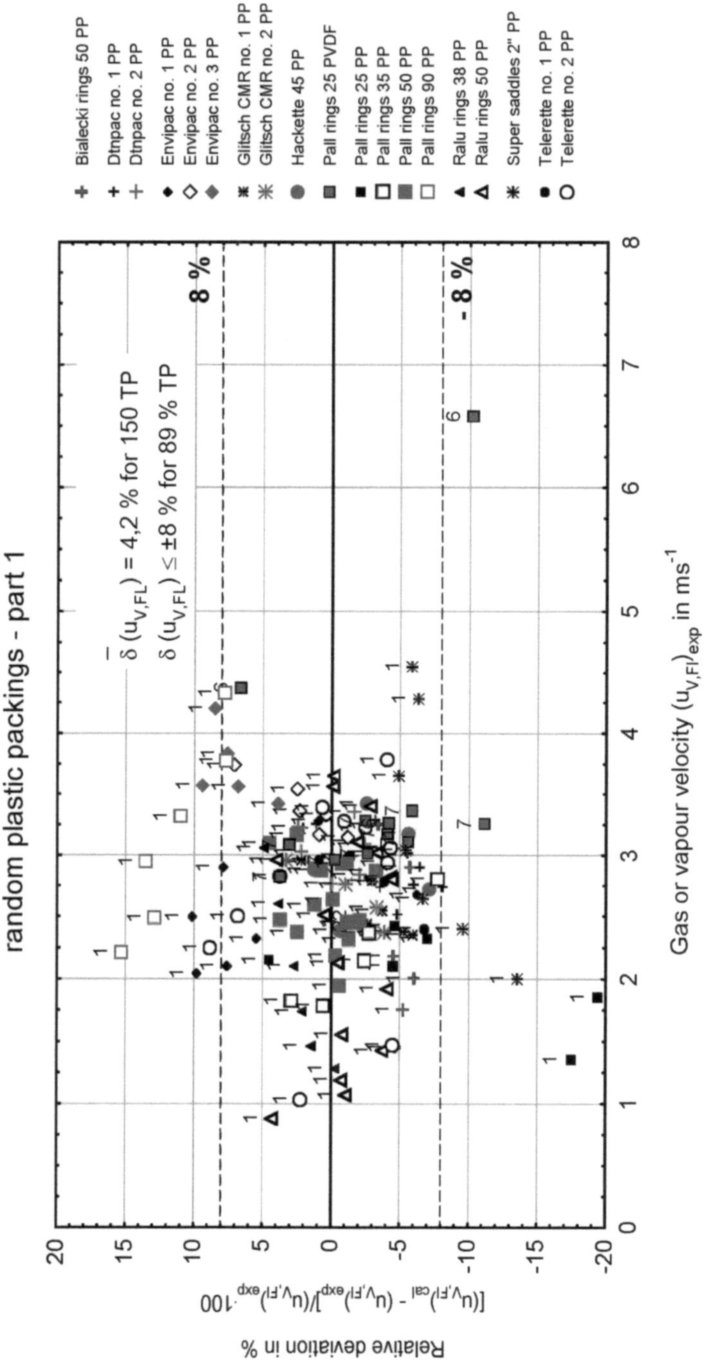

Figure 2-17b. Comparison of experimental determined gas or vapour velocity at the flooding point $(u_{V,Fl})_{exp}$ and calculated data based on Eq. (2-67), valid for random plastic packings of various types, no. of test system as shown in Table 2-2

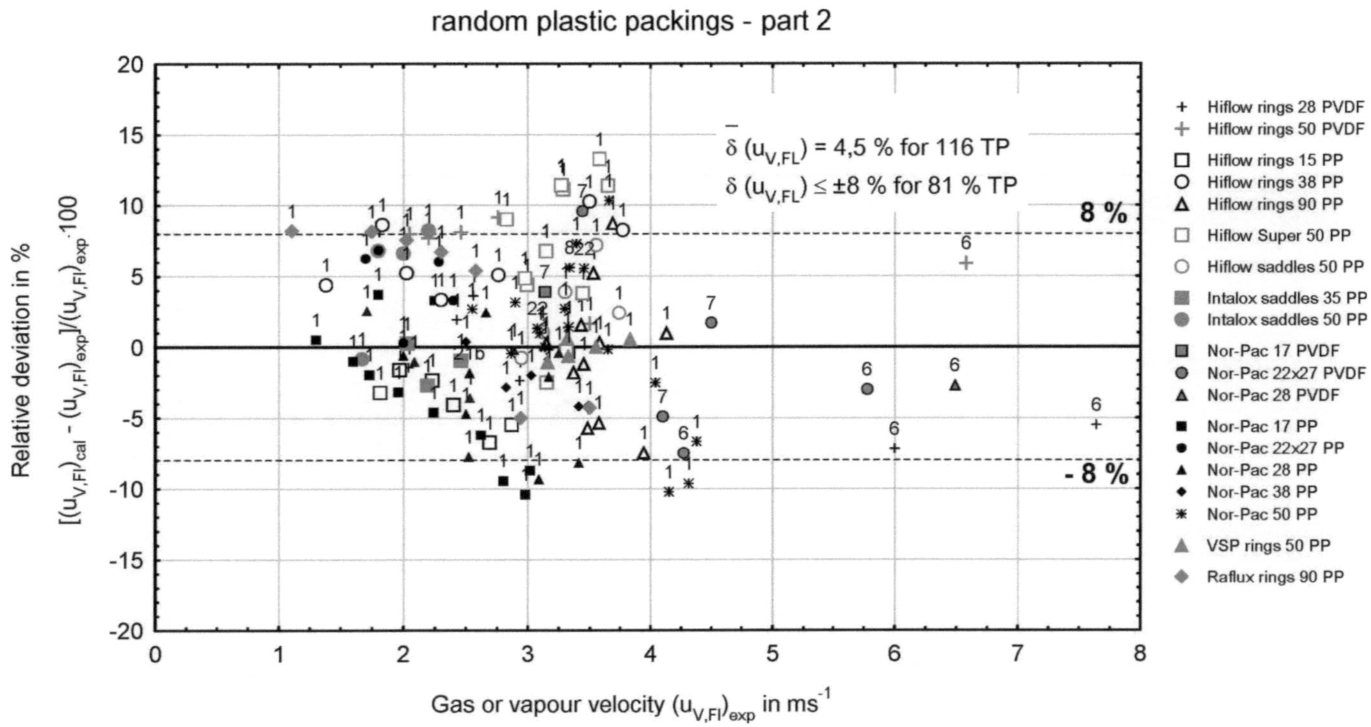

Figure 2-17b. (continued.)

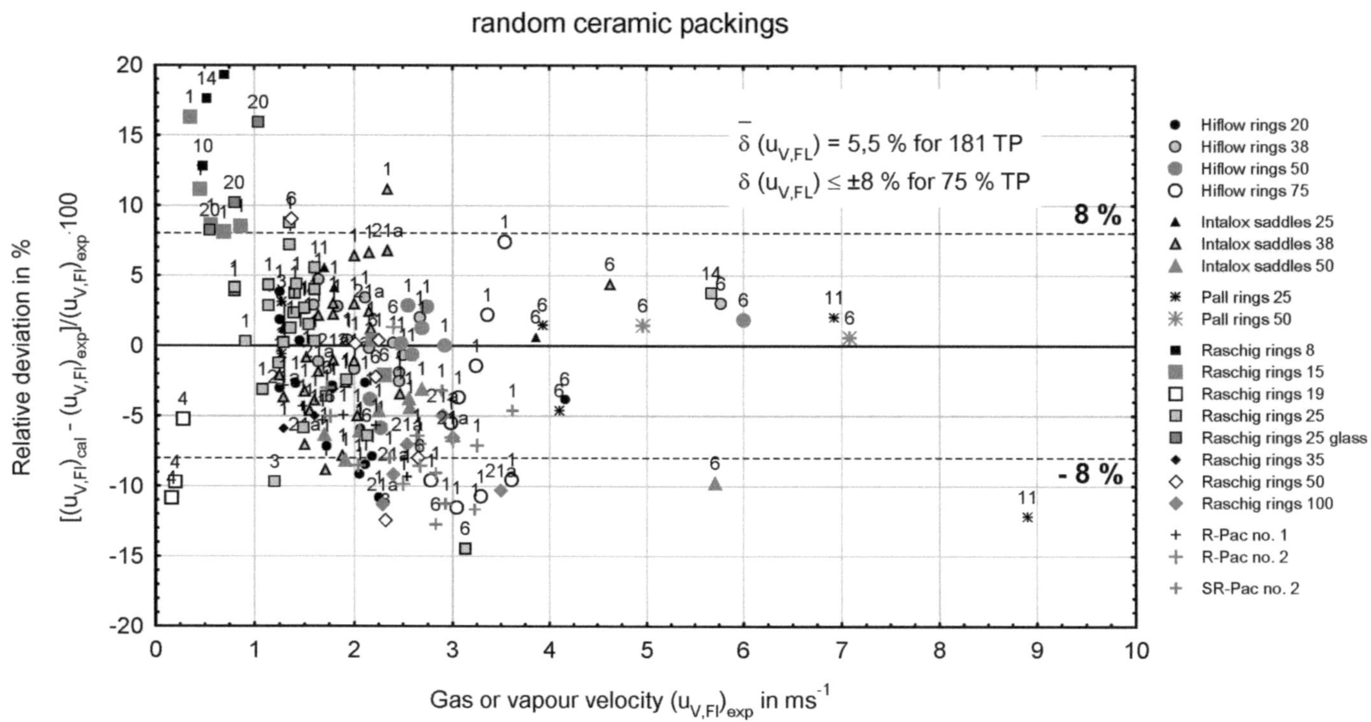

Figure 2-17c. Comparison of experimentally determined gas or vapour velocity at the flooding point $(u_{V,Fl})_{exp}$ and calculated data based on Eq. (2-67), valid for random ceramic packings of various types, no. of test system as shown in Table 2-2

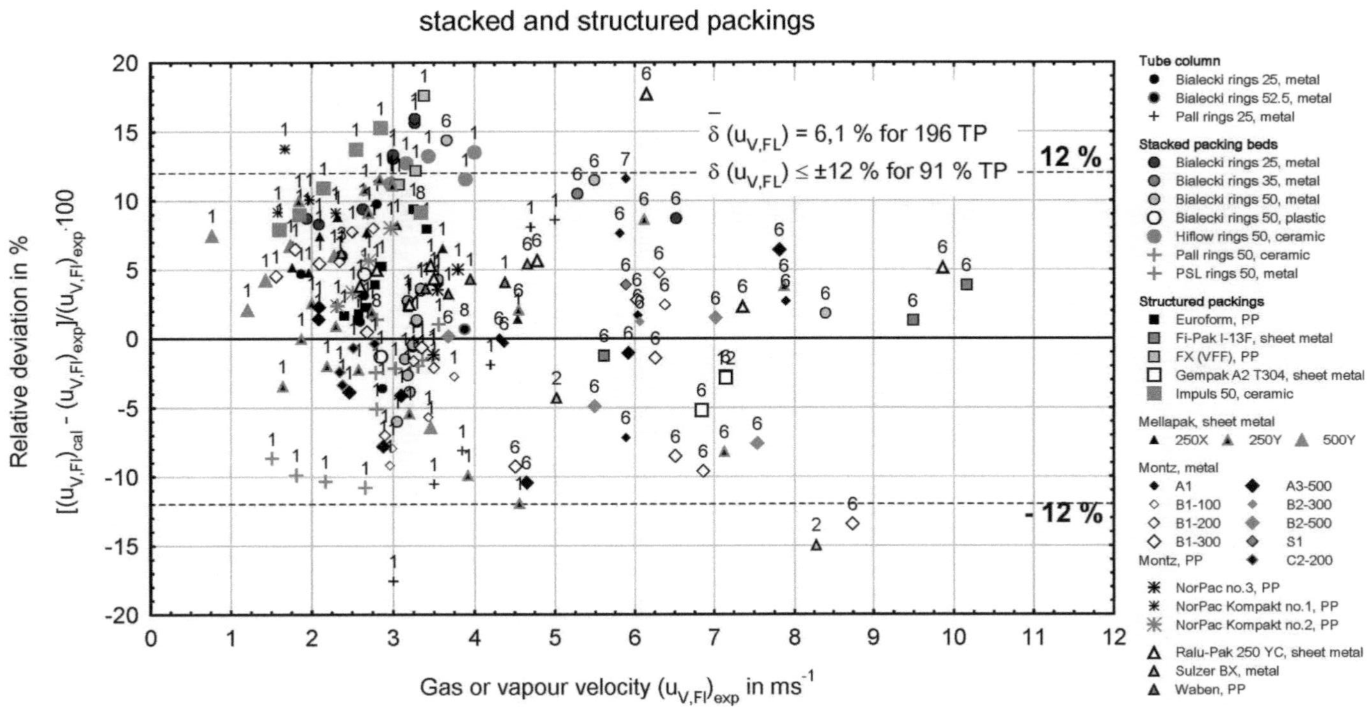

Figure 2-17d. Comparison of experimentally determined gas or vapour velocity at the flooding point $(u_{V,Fl})_{exp}$ and calculated data based on Eq. (2-67), valid for structured packings, tube columns and stacked packings, no. of test system as shown in Table 2-2

the evaluation of the experimental data. This applies to experimental data with $d_S/d \geq 6$ within the validity range of this model, i.e. for gas velocities in the air/water system of $u_{V,Fl} > 0.5\ ms^{-1}$ and $B_L < 5.10^{-3}$, see Fig. 2-17a.

Figures 2-17b, 2-17c and 2-17d show a comparison between the experimentally derived gas velocities at the flooding point $(u_{V,Fl})_{exp}$ and the calculated values $(u_{V,Fl})_{calc}$ for plastic packings (Fig. 2-17b), ceramic packings (Fig. 2-17c) as well as for structured packings and stacked packing elements (Fig. 2-17d). 75–91% of all test points for structured packings are spread over a range of ±12%, see Fig. 2-17d.

The mean relative error $\bar{\delta}\left(u_{V,Fl}\right)$ in the determination of the gas velocity at the flooding point $(u_{V,Fl})_{calc}$, acc. to Eq. (2-67), for the experimental data shown in Figs. 2-17a, 2-17b, 2-17c and 2-17d is given as:

$\bar{\delta}\left(u_{V,Fl}\right) = 4.10\,\%$ for random packings made of plastic – 260 test points
$\bar{\delta}\left(u_{V,Fl}\right) = 5.50\,\%$ for random packings made of ceramic – 181 test points
$\bar{\delta}\left(u_{V,Fl}\right) = 6.10\,\%$ for structured packings, stacked packing elements, tube columns of sheet metal, plastic, ceramic – 196 test points

Equation (2-67) also confirms the test points for diameter ratios of $d_S/d \ll 6$, when the experimentally derived resistance coefficients ψ_{Fl} for smaller columns $d_S/d \ll 6$ are substituted into Eq. (2-67). This is the result of the evaluation of data taken during experiments with the system chlorobenzene/ethylbenzene at 33 and 66.7 bar in a distillation column with $d_S = 0.22$ m ($d_S/d \cong 3.8$) for 50 mm ceramic Intalox saddles, Hiflow rings as well as for 50 mm metal Pall rings and Białecki rings. What is remarkable is that, even in the range of lower $u_{V,Fl}$ values of 0.4–1 ms^{-1}, the new model accurately reflects the experimental vapour velocities $u_{V,Fl}$ for 8 and 15 mm Raschig rings. Deviations greater than ±8% were found mostly in some literature data, for which no detailed information on the packing density N and the geometric data a and ε of the respective packings was available.

Evaluation of Experimental Flooding Point Data for the Pressure Range

Due to the high density of the gas phase, a correction factor for the evaluation of the experimental data has been introduced, as recommended by Mersmann [2], which has proved useful for calculating the droplet velocity u_T, acc. to Eq. (2-75), and the gas velocity at the flooding point $u_{V,Fl}$.

If the density of the continuous phase ρ_V is higher than the density of the ambient air ρ_{air}, the gas velocity at the flooding point $u_{V,Fl}$ is calculated using the correlation:

$$u_{V,Fl} = u_{V,Fl,Eq.(2-67)} \cdot \left(\frac{\rho_V}{\rho_{air}}\right)^{0.18} \quad \left[ms^{-1}\right]. \tag{2-68}$$

Equations (2-67) and (2-68) give the following dimension-dependent final Eq. (2-69) for determining the gas velocity at the flooding point in any type of packed column.

$$u_{V,Fl} = 0.80 \cdot \cos\alpha \cdot \varepsilon^{6/5} \cdot \psi_{Fl}^{-1/6} \left[\frac{d_T \cdot \Delta\rho \cdot g}{\rho_V}\right]^{1/2} \cdot \left[\frac{d_h}{d_T}\right]^{1/4} \cdot (1 - h^0_{L,Fl})^{7/2} \cdot K_{\rho V} \quad \left[ms^{-1}\right] \tag{2-69}$$

where:

$$K_{\rho V} = 1 \quad \text{for} \quad \rho_V \leq \rho_{air\,(1.165\,kgm^{-3})} \text{ and/or} \tag{2-70}$$

$$K_{\rho V} = \left(\frac{\rho_V}{1.165}\right)^{0.18} \quad \text{for} \quad \rho_V > \rho_{air} \tag{2-71}$$

The exponent n = 0.18 in the above equation was determined by evaluating approx. 180 experimental flooding point values in the range of higher pressures up to 100 bar, taken during the course of this work, see Fig. 2-17e.

Figure 2-17e shows a comparison between the experimental flooding point data $u_{V,Fl,exp}$, taken by Krehenwinkel [75], Możenski et al. [39], and the calculation based on Eqs. (2-69) and (2-71) $u_{V,Fl,calc}$. The experimental data is based on various test systems and an operating pressure of up to 100 bar. The experimentally derived gas velocities at the flooding point $u_{V,Fl}$ are in the range of approx. 0.01 to 0.85 ms^{-1}.

A list of individual symbols and systems used can be found in Table 2-10. The comparison shows excellent concurrence of the experiments with the calculation. 80% of the experimental values deviate within ±15% from the graph. After evaluating approx. 180 experimental data points, the mean relative error in the determination of the gas velocity at the flooding point in pressure systems was found to be approx. 8.93%.

An analysis of the experimental results has led to the significant conclusion that it is indeed possible to determine the fluid dynamics of packed columns in the pressure or vacuum range without performing any experimental work, as the results shown by the above correlations are transferable to any pressure range and any type of packing. This is an important conclusion, in particular for practical applications, as experiments in the pressure and negative pressure range are extremely complicated and costly.

The most important result of this work is the ability of the SBD model to describe the flooding in columns with any internals, any test systems in the vacuum, normal pressure and pressure range up to 100 bar accurately enough for practical applications.

In addition, strong proof has been provided, based on a large amount of experimental data using the test system air/water under ambient conditions, that the results are transferable to any test systems and mixtures with extremely divergent properties. Hence, the experimental effort in designing packed columns can be minimised to just a few experiments, using the test system air/water and/or experiments with single-phase flow, in order to determine the resistance coefficient ψ in relation to the Reynolds number of the gas phase.

2.2.4.8
New Dimensionless Correlation for Gas Velocity at Flooding Point Based on SBD Model

Based on Eqs. (2-72) and (2-73), correlation (2-69) is now expressed with an extended Froude number Fr^*_{Fl} at the flooding point. By introducing the dimensionless, extended Froude number of the gas phase

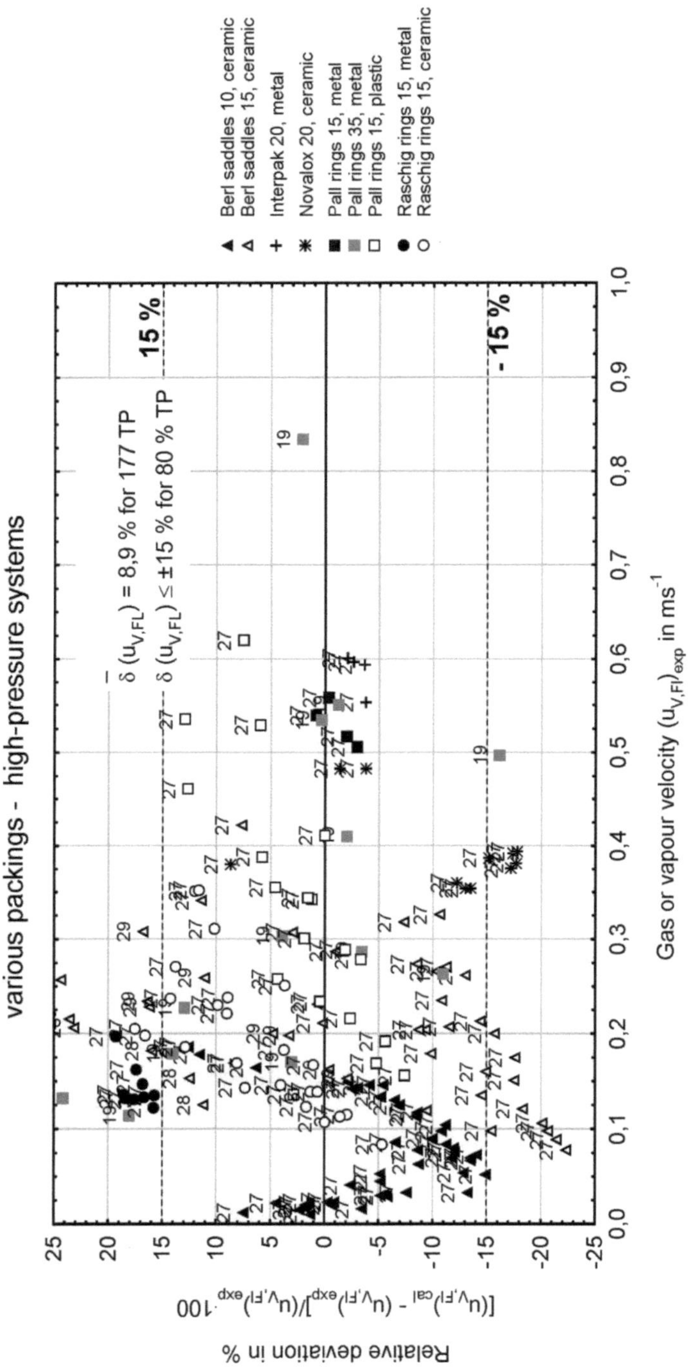

Figure 2-17e. Comparison of experimentally determined gas or vapour velocity at the flooding point $(u_{V,Fl})_{exp}$ and calculated data based on Eq. (2-69), valid for random packings in pressure range up to 100 bar, no. of test system as shown in Table 2-2b

$$Fr^*_{Fl} = \frac{u^2_{V,Fl}}{d_T \cdot g} \cdot \frac{\rho_V}{\Delta\rho} \tag{2-72}$$

it is possible to derive the following flooding point correlation, applicable to any type of column internal, which constitutes the main result of this work:

$$Fr^{*1/2}_{Fl} = 0.8 \cdot \cos\alpha \cdot \psi_{Fl}^{-1/6} \cdot \left[\frac{d_h}{d_T}\right]^{1/4} K_{\rho V} \cdot \varepsilon^{6/5} \cdot \left(1 - h^0_{L,Fl}\right)^{7/2} \tag{2-73}$$

Equation (2-73) is now applicable within the following limits:

$$\frac{d_T^2 \cdot \Delta\rho \cdot g}{\sigma} < 9,\ \eta_V/\eta_L \to 0 \ \text{ as well as } \rho_V/\rho_L << 1 \text{ and } \mathrm{d_h}/d_T > 3 \tag{2-74}$$

and for $Re_L \leq 600$, $Ar = 6.4 \cdot 10^4 - 10^6$, for any types of packing elements and structured packings with resistance coefficients in the range of $\psi_{Fl} \in (0.1-8.5)$.

2.2.4.9
Evaluation of Experimental Data Using Mersmann's Film Model [3]

For the purpose of plotting the flooding curves presented by Mersmann [3], as shown in Figs. 2-5a, 2-5b, 2-5c and 2-5d, the pressure drop of dry random or structured packings $\Delta p_0/H$ was determined using the correlations (3-8) and (3-14), presented in Chap. 3. The empirical constants for determining the resistance coefficient ψ, acc. to Eq. (3-14) or (3-26), in the range of $Re_V \in (500–15{,}000)$ were taken from Tables 6-1a–6-1c.

The experimental results are presented in Figs. 2-5a, 2-5b, 2-5c and 2-5d. They show strong deviations of up to +50% and more, compared to Mersmann's film model [3], in the case of large packing elements, in particular those with lattice structures, as well as structured packings with low irrigation densities. In the case of larger, dimensionless liquid loads $B_L > 2-3 \cdot 10^{-3}$, the experimental data is spread more closely around Mersmann's [3] flooding point curve, and increasingly so for larger-surface packing elements, i.e. those with large geometric surfaces a and small void fractions ε.

In his work, Mersmann [3] only evaluated data for air/liquid systems at ambient conditions for classic packing elements, such as Raschig rings, spheres and Berl saddles. The evaluation of rectification data carried out during the course of this work, also confirms the applicability of the flooding point diagram, acc. to Fig. 2-5, for determining the vapour velocity $u_{V,Fl}$ at the flooding point for small packing elements $d \leq 0.015$ m, e.g. for randomly filled, metal 10 and 20 mm Interpack, 12 mm metal Białecki rings, 17 mm plastic Nor-Pac, 15 mm Pall rings made of metal and plastic as well as 17 mm plastic Hiflow rings, as long as they are within the model's range of validity.

Mersmann's [3] flooding point diagram describes the flooding in packed columns for small packing elements with $d < 25$ mm sufficiently accurately throughout the entire operating range B_L. In the case of lattice and structured packings, the model is only applicable in the range of high liquid loads $B_L > 3-5\cdot10^{-3}$.

2.2.4.10
New Equation for Calculating Individual Droplet Velocity u_T

Equations (2-19), (2-21), (2-56) and (2-67) lead to the following correlation (2-75) for the effective falling velocity of individual droplets in the packing:

$$\begin{aligned} u_T &= \frac{u_{V,Fl}}{\varepsilon^{1.2}\cdot(1-h^0_{L,Fl})^{3.5}}\cdot 0.566 \\ &= 0.80\cdot\cos\alpha\cdot\psi_{Fl}^{-1/6}\cdot\left(\frac{d_h}{d_T}\right)^{1/4}\cdot\left(\frac{d_T\Delta\rho\cdot g}{\rho_V}\right)^{1/2} \quad \left[ms^{-1}\right] \end{aligned} \tag{2-75}$$

The first term in Eq. (2-75) is used to determine the effective droplet velocity u_T, based on the experimental flooding point data, if the values $u_{V,FL}$ and $h^0_{L,Fl}$ as well as the void fraction ε are known. Equation (2-75) therefore constitutes the full form of the correlation based on Eq. (2-24). The second term in Eq. (2-75) is important for various design tasks and constitutes the final equation for the falling velocity of droplets in packed columns.

A similar correlation with d_T, acc. to Eq. (2-26),

$$u_T \approx A_i\cdot\sqrt{\frac{d_T\cdot\Delta\rho\cdot g}{\rho_V}} \tag{2-76}$$

$$u_T \approx A_i\cdot\sqrt{\frac{\sigma_L\cdot\Delta\rho\cdot g}{\rho_V^2}} \tag{2-77}$$

and with $A_i = 1.55$ was found by Mersmann [38] in relation to deformed droplets falling in gases and liquids in columns without internals.

The following correlation (2-78) is applicable to systems with a higher-density gas phase, analogous to Eq. (2-69):

$$u_T = u_{T,\ eqn.(2-75)}\cdot K_{\rho V} = u_{T,\ eqn.(2-75)}\cdot\left(\frac{\rho_V}{\rho_{air}}\right)^{0.18} \tag{2-78}$$

where $K_{\rho V}$ is based on correlation (2-70).

2.2.5
Conclusions Chapter 2.2

1. Based on the evaluation of approx. 1,200 items of experimental flooding point data, the *model of suspended bed of droplets* (SBD model), which is presented in this book, is applicable to any types of randomly filled packing elements, stacked packings, struc-

tured packings and tube columns filled with stacked packing elements, in the range of small, moderate and high phase flow ratios λ_0 in the vacuum and normal pressure range and at high pressures of up to 100 bar.

The latest experiments [66] with random packings have confirmed the droplet formation in packed columns as a result of dripping from the edges and walls of the individual packing elements. This explains why flooding can occur in packed columns at certain phase flow ratios, even in the case of smaller gas velocities, before the packing is filled with films and sprays, as assumed by the film model. Hence, the packed column is flooded earlier than predicted by the film model [3], and the gas velocities at the flooding point, calculated acc. to the film model [3], are much higher for larger lattice packing elements and structured packings than the experimentally derived values, see numerical example 2.1.

The SBD model is based on the assumption that droplet formation occurs at the flooding point and therefore describes the flooding point mechanism accurately, both for higher and smaller gas velocities in pressure absorption and rectification processes, see Fig. 2-17a, 2-17b, 2-17c, 2-17d and 2-17e

The new dimensionless equation for calculating the gas velocity at the flooding point (2-73), based on the *SBD model*, is as follows:

$$Fr_{Fl}^{*1/2} = 0.8 \cdot \cos\alpha \cdot \psi_{Fl}^{-1/6} \cdot \left[\frac{d_h}{d_T}\right]^{1/4} K_{\rho V} \cdot \varepsilon^{6/5} \cdot \left(1 - h_{L,Fl}^0\right)^{7/2},$$

with the dimensionless extended Froude number, acc. to Eq. (2-72)

$$Fr_{Fl}^* = \frac{u_{V,Fl}^2}{d_T \cdot g} \cdot \frac{\rho_V}{\Delta\rho},$$

which in practice can also take a dimension-dependent form acc. to Eq. (2-69)

$$u_{V,Fl} = 0.80 \cdot \cos\alpha \cdot \varepsilon^{6/5} \cdot \psi_{Fl}^{-1/6} \left[\frac{d_T \cdot \Delta\rho \cdot g}{\rho_V}\right]^{1/2} \cdot \left[\frac{d_h}{d_T}\right]^{1/4} \cdot (1 - h_{L,Fl}^0)^{7/2} \cdot K_{\rho V} \ \left[ms^{-1}\right]$$

where: $K_{\rho V} = 1$ for $\rho_V \leq \rho_{air\,(1.165\ kgm^{-3})}$ and/or

$$K_{\rho V} = \left(\frac{\rho_V}{1.165}\right)^{0.18} \quad \text{for} \quad \frac{\rho_V}{\rho_{air}} > 1$$

acc. to Eqs. (2-70) and (2-71).

Equation (2-69) applies to any types of random and structured packings made of any material, provided the resistance coefficient ψ_{Fl} at the flooding point for single-phase flow is known. The validity of Eq. (2-69) for internals in the range of ψ_{Fl} = 0.1–8.2 has been verified. The calculation of the gas velocity $u_{V,FL}$ at the flooding point, acc. to Eq. (2-67) and/or the extended Eq. (2-69), is performed iteratively. The simplification of Eq. (2-67) leads to Eq. (2-60).

2. Based on the dimensionless Eq. (2-73) and/or (2-69), it is possible to determine the gas and/or vapour velocity with a mean relative error of ~ 6% for systems used in vacuum and normal pressure rectification as well as in absorption processes, if the physical properties ρ_L, σ_V, η_L, η_V, ε_L, the geometric packing data a and ≤ and the liquid hold-up at the flooding point $h^0_{L,Fl}$ are known. In the case of pressure systems, the gas velocity at the flooding point can be determined with an accuracy of ±15%, see Fig. 2-17e.
3. If the phase flow ratio at the flooding point λ_0 is known, it is possible to determine the liquid hold-up $h^0_{L,Fl}$, acc. to Eq. (2-46)

$$h^0_{L,Fl} = \frac{\sqrt{\lambda_0^2(m+2)^2 + 4\lambda_0(m+1)(1-\lambda_0)} - (m+2)\lambda_0}{2\cdot(m+1)(1-\lambda_0)} \quad \left[m^3 m^{-3}\right] \tag{2-46}$$

The parameter m in this equation is calculated using Eq. (2-43) for turbulent liquid flow $Re_L \geq 2$

$$m = -0.82 + \frac{\lambda_0}{\lambda_0 + 0.5} \tag{2-43}$$

and Eq. (2-44) for laminar liquid flow $Re_L < 2$.

$$m = -0.90 + \frac{\lambda_0}{\lambda_0 + 0.5} \tag{2-44}$$

In the case of larger packing elements, at moderately negative and normal pressure, the liquid flow is practically turbulent, which means that Eqs. (2-46) and (2-42) are mostly applicable. Hence, it is possible to determine the flooding point without having to ascertain any experimental hold-up data at the flooding point.
4. The evaluation of the experimental flooding point data acc. to Eq. (2-73) and/or (2-69) was based on roughly 1,200 items of experimental data for approx. 200 random and structured packings and tube columns. The columns, internals and operating parameters were varied within the following ranges:

$$\begin{array}{lclcll}
0.1 & \leq & \psi_{Fl} & \leq & 8.5 & [-] \\
0.025 & \leq & d_S & \leq & 1.4 & m \\
0.59 & \leq & \varepsilon & \leq & 0.988 & m^3m^{-3} \\
0.008 & \leq & d & \leq & 0.100 & m \\
54 & \leq & a & \leq & 750 & m^2m^{-3} \\
0.45 & \leq & H & \leq & 6.0 & m \\
0.013 & \leq & p_T & \leq & 100 & bar \\
0 & \leq & \lambda_0\cdot 10^3 & \leq & 1000 & [-] \\
0.01 & \leq & u_V & \leq & 18 & ms^{-1} \\
0.1 & \leq & F_{V,Fl} & \leq & 5.5 & \sqrt{Pa} \\
0 & < & u_L & \leq & 56\cdot 10^{-3} & ms^{-1} \\
0 & < & Re_L & \leq & 600 & [-]
\end{array} \tag{2-79}$$

Table 2-2 contains a list of 32 mixtures with widely different physical properties, which were used for evaluating the experimental flooding point data.

5. Based on the *SBD model*, the gas and/or vapour velocity at the flooding point can be determined in the following way: Firstly, the phase flow ratio at the flooding point λ_0 is calculated as a start value, acc. to Eq. (2-1), for $V_{V,Fl} = V_V$. Subsequently, the liquid hold-up at the flooding point $h^0_{L,Fl}$ is calculated in the first iteration, acc. to Eq. (2-46). Now Eq. (2-69) is used to determine the gas and/or vapour velocity at the flooding point $u_{V,Fl}$, based on the physical properties of the system ρ_V, ρ_L, σ_L, η_V, η_L, the geometric data a and ε of the respective packing elements and the mean resistance coefficient, acc. to Table 6-1a–6-1c.

   Finally, it must be ascertained whether or not the assumption $Re_L \geq 2$ is applicable. If Re_L is below 2, the new calculation of the hold-up at the flooding point is performed, based on Eq. (2-42), followed by the calculation of the gas velocity at the flooding point. Only a few iterations are necessary to determine a sufficiently accurate value for $u_{V,Fl}$ and thus for the variables λ_0 and $h^0_{L,Fl}$.

   Further to the correlations presented above, there are a number of calculation examples at the end of this chapter, which illustrate how the model can be used to determine the vapour velocity at the flooding point for 50 mm randomly filled Pall rings for the system ethyl benzene/styrene under vacuum, and the gas velocity at the flooding point under pressure for 25 mm randomly filled Białecki rings for the air/water system, for structured gauze packing BX as well as for 15 mm Pall rings made of plastic.

6. The *film gas thrust shear force model*, developed by Mersmann (1965) [3], for determining the gas velocity at the flooding point is most suitable for applications using classic and/or large-surface packing elements with a $\geq$ 300 m^2m^{-3} throughout the entire operating range. This is reflected by the shading in the flooding point diagram developed by Mersmann [3], see Fig. 2-18. The *film gas thrust shear force model*, on

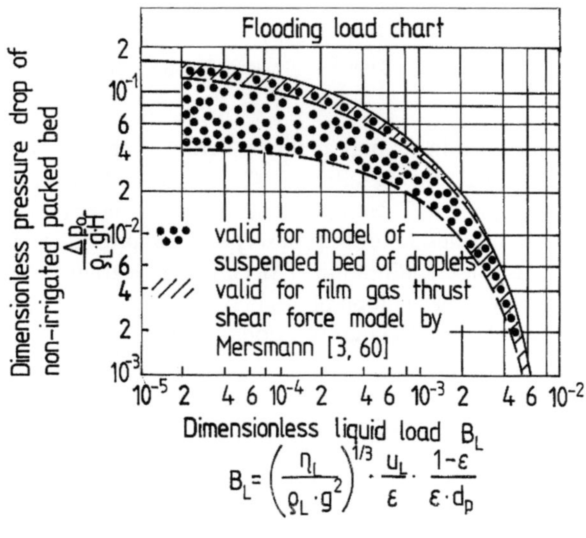

Figure 2-18. Flooding line depending on the dimensionless liquid load with ranges of validity of the SBD model, based on the results of this work, and the film gas thrust shear force model, developed by Mersmann [3, 60, 91]

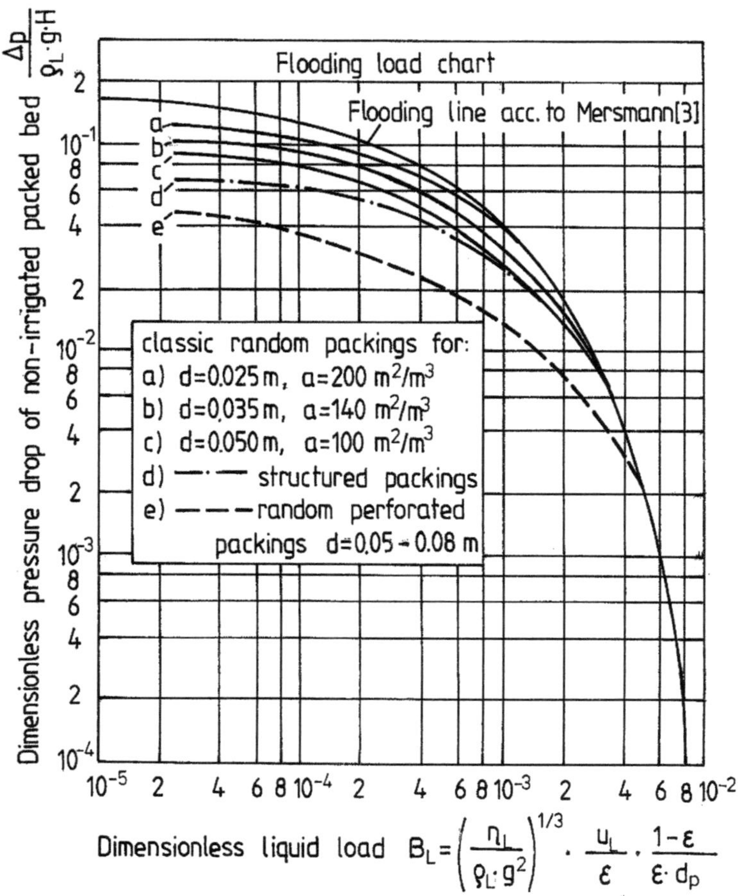

Figure 2-19. Flooding line for various packings. Compilation of results from Figs. 2-5a–d

the other hand, only applies to perforated packing elements, if the dimensionless liquid loads are high, $B_L \geq 5 \cdot 10^{-3}$, as shown by Figs. 2-5a, 2-5b, 2-5c and 2-5d as well as Fig. 2-19. These operating conditions mostly occur in the case of extremely high liquid loads, in pressure absorption and pressure rectification.

A further modification of Mersmann's diagram, based on the model presented in the first German edition of this book [90, 91], can be found in Bornhütter's work (1991) [66, 87] and by Grabbert, Bonitz (1998) [92].

Farther capacity diagrams for different packings are shown in Annex to this chapter, Figs. 2-22, 2-23, 2-24, 2-25, 2-26, 2-27, 2-28, 2-29, 2-30, 2-31, 2-32, 2-33, 2-34, 2-35, 2-36, 2-37 and 2-38.

2.3 Determining Column Diameter

If the gas capacity factor $F_{V,Fl}$ at the flooding point is determined in relation to a given packing element, it is possible to ascertain the operating point marked by the gas capacity factor F_V. The following applies:

$F_{V,U} < F_V < F_{V,Fl}$. Depending on the separation process, the gas capacity factor F_V is set at approx. 10–80% of the value at the flooding point. This leads to the following correlation for the column diameter d_S:

$$d_S = \sqrt{\frac{4 \cdot A_S}{\pi}} = \sqrt{\frac{4}{\pi}} \cdot \sqrt{\frac{V}{\sqrt{\rho_V} \cdot F_V}} \qquad [m] \qquad (2\text{-}80)$$

where V is the gas capacity in kgs^{-1}.

2.4 Lower Loading Line

Vacuum distillation columns are mostly operated at very low liquid loads u_L, in particular in the rectifying section. For this reason, it is important to ascertain the lower loading line $u_{L,U}$, Fig. 2-20, as well as the flooding point. Kirschbaum [19] and Schmidt [7, 32] have looked at methods to determine the lower loading line.

The lower loading line is largely dependent on the physical properties of the mixture to be separated as well as on the material and geometric dimensions of the packing

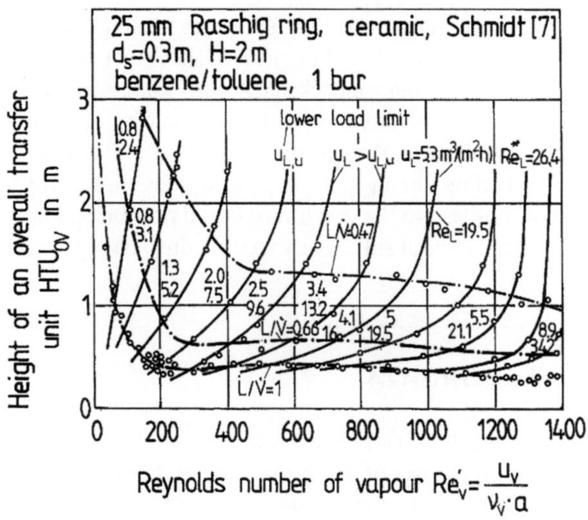

Figure 2-20. Height of an overall transfer unit for various L/V ratios as a function of the Reynolds number of vapour Re_V and liquid $Re*_L$ acc. to Schmidt [32]. As shown in this diagram, the lower loading line is below the flooding line

elements, whereas interfacial instabilities do not appear to have an effect on the lower loading line.

Schmidt [32] has presented various equivalent correlations for determining the lower loading line $u_{L,U}$. The simplest equation, which is also the clearest and easiest one to apply, is as follows:

$$u_{L,U} = 7.7 \cdot 10^{-6} \cdot \frac{C_L^{2/9}}{(1 - T_L)^{1/2}} \cdot \left[\frac{g}{a}\right]^{1/2} \quad \left[ms^{-1}\right], \tag{2-81}$$

The relative error of this equation is ±20%, which has been verified for vacuum conditions up to 20 mbar and for liquid/vapour ratios of $L/V = 0.5-1.25$.

The liquid number is given as:

$$C_L = \frac{\rho_L \cdot \sigma_L^3}{\eta_L^4 \cdot g} \tag{2-82}$$

and the shear stress number is:

$$T_L \cong 0.9 \cdot \left[\frac{u_V}{u_{V,Fl}}\right]^{2.8} = 0.9 \cdot \left[\frac{F_V}{F_{V,Fl}}\right]^{2.8} \tag{2-83}$$

Equation (2-81) was derived by Schmidt [7, 32], based on his own experimental results and those found in literature, using the systems ethanol/water, dichloroethane/toluene and benzene/toluene under normal pressure and vacuum of up to 20 mbar. It is applicable to 25–50 mm ceramic Raschig rings and metal Pall rings.

Schmidt [7, 32] noticed the dewetting phenomenon, which causes the separation efficiency below the loading line $u_{L,U}$ to drop. The phenomenon is amplified by a decrease in the specific liquid load u_L. Equation (2-81) shows that, in the case of mixtures with a decreasing liquid number C_L, the lower loading line $u_{L,U}$ shifts towards the lower values of the specific liquid load u_L, which has a positive effect on the elasticity of the column. The operating range of packed columns using such systems is therefore larger, Fig. 2-21.

In the case of mixtures with low viscosity η_L and high surface tension σ_L, such as aqueous mixtures, the specific liquid load at the flooding point $u_{L,Fl}$ can fall below the lower loading line $u_{L,U}$, i.e. $u_{L,Fl} < u_{L,U}$, as shown in Fig. 2-21 [7, 32]. These operating conditions lead to the downward liquid flow being impounded and entrained by the gas (vapour), without sufficiently wetting the packing surface.

2.4.1
Conclusions Section 2.4

Based on Schmidt's [32] correlation (2-81) for calculating the lower loading line $u_{L,U}$ and the knowledge of the loading line of the type of packing used, it is easy to determine by calculation whether the specific liquid load u_L, selected for the separation task of the column, is above the lower loading line $u_{L,U}$, $u_{L,U} < u_L < u_{L,Fl}$.

Figure 2-21. Lower loading line $u_{L,U}$ and flooding line for 25 mm Pall ring column acc. to Schmidt [32]. This diagram highlights the overlapping of both limit lines and shows the operating range of the packed column for a given system

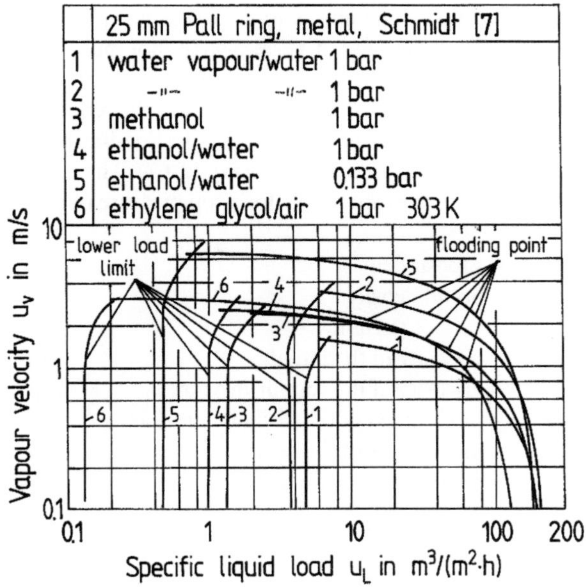

List of Numerical Examples – Chapter 2 "Flooding Point"

2.1 Determining the vapour capacity factor $F_{V,Fl}$ at the flooding point, acc. to Eq. (2-67) and the column diameter $d_S$for metal, randomly filled 50 mm Pall rings for the separation of the mixture ethyl benzene/styrene under vacuum.

2.2 Determining the vapour capacity factor $F_{V,Fl}$, acc. to Eq. (2-67), for metal, randomly filled 25 mm Białecki rings for the air/water test system at ambient conditions.

2.3 Determining the vapour capacity factor $F_{V,Fl}$, acc. to Eq. (2-67), for Sulzer gauze packing BX for the separation of the mixture ethyl benzene/styrene at a top pressure of 66.7 mbar.

2.4 Determining the vapour capacity factor $F_{V,Fl}$ at the flooding point, acc. to eqns (2-69) and (2-71), and the column diameter d_S for randomly filled 15 mm plastic Pall rings for the separation of the mixture methanol/nitrogen (N_2) at 30 bar.

2.5 Determining the vapour capacity factor $F_{V,Fl}$ at the flooding point, acc. to Eq. (2-69), and the relative column load $F_V/F_{V,Fl}$ in an existing column of a demethaniser being upgraded to a structured Mellapak 350Y sheet metal packing at an operating pressure of 32 bar.

Numerical Example 2.1 – Sections 2.2.4 and 2.3

A column filled with metal 50 mm Pall rings is used to separate 1,000 kg of the mixture ethyl benzene/styrol with $x_F = 0.5895$ mol mol^{-1} per hour, at a top pressure of

$p_T = 66.7$ mbar. The aim is to achieve a top product with a mole fraction of higher-volatility ethyl benzene of $x_D = 0.9618$ mol mol^{-1} and a bottom product with a mole fraction of $x_W = 0.0018$ mol mol$^{-1}x_W$. What column diameter is required, if the column is operated at 46.3% of the flood load and the reflux ratio is $r = 6.28$?

The vapour capacity factor $F_{V,Fl}$ at the flooding point is determined using the method described in Sect. 2.2.4 as well as Eq. (2-67). The following variables are given, with the physical properties applicable for a top pressure of $p_T = 66.7$ mbar.

– vapour density	ρ_V
– liquid density	$\rho_L = 835.2\ \text{kgm}^{-3}$
– surface tension	$\sigma_L = 25.1\ \text{mNm}^{-1}$
– viscosity of the liquid	$\eta_L = 0.437 \cdot 10^{-3}\ \text{kgm}^{-1}\ \text{s}^{-1}$
– viscosity of the vapour mixture	$\eta_V = 7.14 \cdot 10^{-6}\ \text{kgm}^{-1}\ \text{s}^{-1}$
– molar mass of the distillate product	$M_D = 106.1\ \text{kg kmol}^{-1}$
– molar mass of the feed	$M_F = 105.34\ \text{kg kmol}^{-1}$
– relative vapour capacity	$F_V/F_{V,Fl} = 0.463$

Solution

A. Calculating the flows and flow rates at the column top according to the task definition

1. *Feed rate:*

$$\begin{aligned} F &= 1000\ kgh^{-1} \\ \dot{F} &= F/M_F \\ &= 1000/105.34 = 9.493\ kmolh^{-1} \end{aligned}$$

2. *Distillate rate:*

$$\begin{aligned} \dot{D} &= \dot{F} \cdot \frac{x_F - x_W}{x_D - x_W} \\ &= 9.493 \cdot \frac{05895 - 0.0018}{0.9618 - 0.0018} = 5.812\ kmolh^{-1} \end{aligned}$$

3. *Vapour rate at column top:*

$$\begin{aligned} \dot{V} &= (r+1) \cdot \dot{D} \\ &= (6.28 + 1) \cdot 5.812 = 42.3\ kmolh^{-1} \end{aligned}$$

with a molar mass of $M_D = 106.1$ kg kmol^{-1}:

$$\begin{aligned} V &= \dot{V} \cdot M_D \\ &= 42.3 \cdot 106.1 = 4488.9\ kgh^{-1} \end{aligned}$$

4. *Liquid rate L at column top:*

$$\begin{aligned} L &= r \cdot \dot{D} \cdot M_D \\ &= 6.28 \cdot 5.812 \cdot 106.1 = 3872.2\ kgh^{-1} \end{aligned}$$

B. Determining the vapour capacity factor $F_{V,Fl}$

1. The phase flow ratio at the flooding point λ_0 is given as:

$$\lambda_0 = \frac{\dot{V}_L}{\dot{V}_V} = \frac{L \cdot \rho_V}{\rho_L \cdot V} = \\ = \frac{3872.2 \cdot 0.257}{835.2 \cdot 4488.9} = 2.654 \cdot 10^{-4}$$

2. The liquid hold-up $h^0_{L,Fl}$ at the flooding point is determined, acc. to Eq. (2-47), using a start value of $\lambda_0 = 2.654 \cdot 10^{-4}$. As a result, $h^0_{L,Fl}$ is given as:

$$h^0_{L,FL} = \frac{\sqrt{1.44 \cdot \lambda_0^2 + 0.8 \cdot \lambda_0 \cdot (1 - \lambda_0)} - 1.2 \cdot \lambda_0}{0.4 \cdot (1 - \lambda_0)}$$

$$= \frac{\sqrt{1.44 \cdot \left(2.654 \cdot 10^{-4}\right)^2 + 0.8 \cdot 2.654 \cdot 10^{-4} \cdot \left(1 - 2.654 \cdot 10^{-4}\right)} - 1.2 \cdot 2.654 \cdot 10^{-4}}{0.4 \cdot (1 - 2.654 \cdot 10^{-4})}$$

$$= 3.565 \cdot 10^{-2} m^3 m^{-3}$$

3. The vapour velocity at the flooding point for 50 mm metal Pall rings is determined acc. to Eq. (2-67). The following values apply for the geometric data a and ε as well as the constants K_1, K_2 for $Re_V > 2100$ in relation to Eq. (3-14), acc. to Table 6-1a: $N_0 = 6100\ Nm^{-3} \Rightarrow a_0 = 110\ m^2\ m^{-3}$, $\varepsilon_0 = 0.952\ m^3\ m^{-3}$, $K_1 = 3.23$, $K_2 = -0.0343$ for Eq. (3-15) to calculate the resistance coefficient.
   The application of Eq. (2-67) also requires the following parameters:

   – the hydraulic diameter

$$d_h = 4 \cdot \varepsilon/a = 4 \cdot 0.952 \big/ 110 = 0.0346\ \text{m}$$

   – the droplet diameter

$$d_T = \sqrt{\sigma/\Delta\rho \cdot g} = \\ = \sqrt{0.0251/\left((835.2 - 0.257) \cdot 9.81\right)} = 1.75 \cdot 10^{-3} m$$

4. Firstly, Eq. (2-67) for $\alpha = 45^\circ$ is used to estimate the gas velocity at the flooding point $u_{V,Fl}$ for $\psi_{Fl} \cong 2.42$ with $Re_V \geq 2100$ and $h_{L,Fl} = 0.0365\ m^3\ m^{-3}$.

$$u_{V,Fl} = 0.566 \cdot \psi_{Fl}^{-1/6} \left[\frac{d_h}{d_T}\right]^{1/4} \cdot \varepsilon^{6/5} \cdot \left[\frac{d_T \Delta\rho \cdot g}{\rho_V}\right]^{1/2} \cdot \left[1 - h^0_{L,Fl}\right]^{3.5}$$

$$= 6.43\ ms^{-1}$$

As a result, the vapour velocity u_V with $F_V/F_{V,Fl} = 0.463$ is given as:

$$\begin{aligned} u_V &= 0.463 \cdot u_{V,Fl} \\ &= 0.463 \cdot 6.43 = 2.98\ ms^{-1} \end{aligned}$$

the column cross section A_S is:

$$A_S = \frac{V_w}{3600 \cdot \rho_V \cdot u_V} = \frac{4488{,}9}{3600 \cdot 0{,}257 \cdot 2{,}98} = 1{,}628\ m^2$$

and, finally, the column diameter d_S is given as:

$$d_S = \sqrt{\frac{4 \cdot A_S}{\pi}} = 1.44m;\ \text{based on the assumption that } d_S = 1.45\text{m}$$

5. Based on the specific liquid load u_L

$$\begin{aligned} u_L &= \frac{L}{3600 \cdot \rho_L \cdot A_S} = \frac{3872.2}{3600 \cdot 835.2 \cdot 1.65} \\ &= 7.8 \cdot 10^{-4}\ ms^{-1} \end{aligned}$$

the Reynolds number Re_L is calculated as:

$$\begin{aligned} Re_L &= \frac{u_L}{\nu_L \cdot a} = \frac{u_L \cdot \rho_L}{\eta_L \cdot a} = \\ &= \frac{7.8 \cdot 10^{-4} \cdot 835.2}{0.437 \cdot 10^{-3} \cdot 110} = 13.55 > 2 \end{aligned}$$

As a result, the use of Eq. (2-47) for calculating the liquid hold-up $h^0_{L,Fl}$ at the flooding point is justified.

6. Based on a specific liquid load of $u_L = 7.8\ 10^{-4}\ ms^{-1}$, the vapour velocity at the flooding point $u_{V,Fl}$ is calculated as $u_{V,Fl} = 6.69\ ms^{-1}$, acc. to Eq. (2-67), using iterative calculation. The phase flow ratio at the flooding point λ_0 is given as $1.1 \cdot 10^{-4}$ for $\psi_{Fl} = 2.34$ and for $h^0_{L,Fl} = 2.38\ 10^{-2}\ m^3\ m^{-3}$.
The calculated value for $u_{V,Fl}$ is only 4.1% higher than the velocity of 6.43 ms^{-1} obtained by approximate calculation (point 4).
The vapour capacity factor at the flooding point $F_{V,Fl}$ is now given as:

$$F_{V,Fl} = u_{V,Fl} \cdot \sqrt{\rho_V} = 3.39\ ms^{-1}\sqrt{\text{kgm}^{-3}}\ \left[\sqrt{\text{Pa}}\right]$$

The following numerical value was found by Billet [5], Chap. 2, for the vapour capacity factor $F_{V,Fl}$ with $\dot{L}/\dot{V} = 1$ for the same system and a marginally different phase flow ratio, compared to the one specified in the task definition:

$$F_{V,Fl} = 3.3\ ms^{-1}\sqrt{kgm^{-3}}\ \left[\sqrt{Pa}\right]$$

Relative error $\delta(F_{V,Fl})$:

$$\delta(F_{V,Fl}) = \frac{(F_{V,Fl})_{calc} - (F_{V,Fl})_{exp}}{(F_{V,Fl})_{exp}} \cdot 100$$

$$= \frac{3.39 - 3.3}{3.3} \cdot 100 = 2.73\%$$

Further calculations were performed, based on other methods, to determine the flood load factor $F_{V,Fl}$, acc. to this task definition. The results were as follows:

1. Eckert's method [9]:	$F_{V,Fl} = 5.0$	$Pa^{0.5}$	$\delta(F_{V,Fl}) = 51.5\%$
2. Mersmann's method [3]:	$F_{V,Fl} = 4.98$	$Pa^{0.5}$	$\delta(F_{V,Fl}) = 50.9\%$
3. Billet's method [5]:	$F_{V,Fl} = 3.55$	$Pa^{0.5}$	$\delta(F_{V,Fl}) = 7.6\%$
4. Equation (2-67):	$F_{V,Fl} = 3.39$	$Pa^{0.5}$	$\delta(F_{V,Fl}) = 2.73\%$
5. Exp. value:	$F_{V,Fl} = 3.3$	$Pa^{0.5}$	acc. to Billet [5] for $(\dot{L}/\dot{V}) = 1$.

Numerical Example 2.2 – Chapter 2

The aim is to determine the gas velocity at the flooding point $u_{V,Fl}$ in a column randomly filled with 25 mm metal Białecki rings with a diameter of $d_S = 0.15$ m. The column is operated at a gas velocity of $u_V = 1\ ms^{-1}$ and a specific liquid load of $u_L = 0.0111\ ms^{-1}$, using the test system air/water at 1 bar and 293 K. The technical data of the Białecki rings is as follows:

$$N = 55000\ m^{-3}$$
$$\Rightarrow a = 238\ m^2\ m^{-3},\ \varepsilon = 0.94\ m^3 m^{-3}.$$

The physical properties are valid for 1 bar and 293 K:

$\eta_V = 18.2 \cdot 10^{-6}\ m^2\ s^{-1}$	$\rho_L = 998.2\ kgm^{-3}$
$\eta_L = 10^{-3}$ Pas	$\rho_L = 1.17\ kgm^{-3}$
$\sigma_L = 72.4 \cdot 10^{-3} Nm^{-1}$	$g = 9.80665\ ms^{-2}$

Solution

1. The following parameters are required for calculating the gas velocity at the flooding point $u_{V,Fl}$, acc. to Eq. (2-67):

   1.1 K_3, K_4 coefficients for determining the resistance coefficient for the dry packing at the flooding point for $Re_V > 2100$.
   Based on Eq. (3-16) and Table 6-1a, the numerical values for the model constants K_3, K_4, [20] for $Re_V \geq 2100$ are as follows:

   $K_3 = 4.13$, $K_4 = -0.0522$.

   The start value is assumed to be a mean value of $\psi_{Fl,m} \cong 2.60$, acc. to Table 6-1a.

1.2 Hydraulic diameter d_h:

$$d_h = 4 \cdot \frac{\varepsilon}{a} = 4 \cdot \frac{0.942}{238} = 1.583 \cdot 10^{-2} m$$

1.3 Droplet diameter:

$$d_T = \sqrt{\frac{\sigma_L}{(\rho_L - \rho_V) \cdot g}} = \sqrt{\frac{0.0724}{(988.2 - 1.17) \cdot 9.80665}}$$

$$= 2.72 \cdot 10^{-3} m$$

2. The Reynolds number Re_L is given as:

$$\mathrm{Re}_L = \frac{u_L}{\nu_L \cdot a} = \frac{u_L \cdot \rho_L}{\eta_L \cdot a} =$$

$$= \frac{11.1 \cdot 10^{-3} \cdot 998.2}{10^{-3} \cdot 238} = 46.55 > 2$$

The liquid hold-up $h^0_{L,Fl}$ is determined using correlation (2-47). The start value for the phase flow ratio at the flooding point λ_0 is assumed to be:

$$\lambda_0 = \frac{u_L}{u_V} = 11.1 \cdot 10^{-3} \quad where \quad u_V = 1\ ms^{-1}$$

This leads to a value of $h^0_{L,Fl} = 0.2056\ m^3\ m^{-3}$, acc. to Eq. (2-47). Based on the first iteration, the gas velocity at the flooding point $(u_{V,Fl})_1$ is given as the numerical value of $(u_{V,Fl})_1 = 1.7132\ ms^{-1}$. Subsequently, a new phase flow ratio of $\lambda_0 = 11.1 \cdot 10^{-3} / 1.713 = 6.46 \cdot 10^{-3}$ is created, for which the $(u_{V,Fl})_2$ value is now calculated iteratively. If the difference between the value $(u_{V,Fl})_{i+1}$, based on iteration *i+1*, and the value $(u_{V,Fl})_I$, based on iteration *I*, is less than 0.01 ms^{-1}, the iteration is finished.

The iterative calculation leads to the following numerical value for $u_{V,Fl}$:

$$u_{V,Fl} = 1.776\,\mathrm{ms}^{-1},$$

which is practically identical to the value $u_{V,Fl} = 1.75\ ms^{-1}$ for $u_L = 11.1 \cdot 10^{-3}\ ms^{-1}$, shown in Figs. 2-2a and 2-2b, which was experimentally derived. The relative error is therefore:

$$\delta\left(u_{V,Fl}\right) = \frac{1.75 - 1.776}{1.75} \cdot 100\% = -1.49\ \%$$

The column is operated at approx.

$$\frac{F_V}{F_{V,Fl}} = \frac{u_V}{u_{V,Fl}} = \frac{1}{1.776} = 0.563$$

i.e. at a flood load of 56.3%. The phase flow ratio at the flooding point

$$\lambda_0 = \frac{u_L}{u_{V,Fl}} = \frac{11.1 \cdot 10^{-3}}{1.776} = 6.25 \cdot 10^{-3}$$

now gives the following liquid hold-up at the flooding point:

$$h^0_{L,Fl} = 15.3 \cdot 10^{-2}\, \mathrm{m^3 m^{-3}}.$$

Based on Table 2-3, the experimentally derived $h^0_{L,Fl}$ value is 16.2%.
This gives a relative error of:

$$\delta\left(h^0_{L,Fl}\right) = \frac{15.3 - 16.2}{16.2} \cdot 100\% = -5.55\%$$

for $Re_{V,Fl} = 2553.7$, hence:

$$\psi_{Fl} = 4.13 \cdot 2553.7^{-0.0522} = 2.745$$

Example 2.3 – Chapter 2

The aim is to determine the gas velocity at the flooding point for the Sulzer gauze packing BX, using the system ethyl benzene/styrene at 66.7 mbar. The column diameter is d_S = 0.5 m, the reflux rate at the column top is L = 1491 kg/h, as shown in Fig. 6-33 [5]; $\dot{L}/\dot{V} = 1$. The angle of the flow channels for BX packings is $\alpha = 30°$. The gas velocity at the flooding point, which was experimentally derived by Billet [5], is $u_{V,Fl} = 7.50\ \mathrm{ms^{-1}}$.

Solution

Based on a mean value of $\psi_m \in (2100\text{–}10{,}000)$ $\psi_m = 0.374$ and the geometric data a and ε, shown in Table 6-1c, the numerical values for Eq. (2-67) are as follows:

$$d_h = 4 \cdot \frac{\varepsilon}{a} = 4 \cdot \frac{0.95}{500} = 0.0076\ \mathrm{m}$$

$$d_T = 1.75 \cdot 10^{-3}\ \mathrm{m}$$

$$u_L = \frac{1491}{0.52 \cdot \frac{\pi}{4} \cdot 835.2 \cdot 3600} = 2.52 \cdot 10^{-3}\ \mathrm{ms^{-1}}$$

$$\lambda_0 = \frac{L}{V} \cdot \frac{\rho_V}{\rho_L} = 1 \cdot \frac{0.257}{835.2} = 3.08 \cdot 10^{-4}$$

For $\lambda_0 = 3.08 \cdot 10^{-4}$, m is calculated using Eq. (2-43):

$$m = -0.82 + \frac{\lambda_0}{\lambda_0 + 0.5} = -0.82 + \frac{3.08 \cdot 10^{-4}}{3.08 \cdot 10^{-4} + 0.5} \cong -0.82$$

and the liquid hold-up at the flooding point, acc. to Eq. (2-47), is given as:

$$h^0_{L,Fl} = 3.834 \cdot 10^{-2}\ m^3 m^{-3}$$

After the first iteration, the gas velocity at the flooding point, acc. to Eq. (2-66), is calculated as:

$$u_{V,Fl} = 0.8 \cdot \cos 30° \cdot 0.374^{-1/6} \cdot \left[\frac{7.6 \cdot 10^{-3} m}{1.75 \cdot 10^{-3} m}\right]^{0.25} \cdot 0.95^{1.2} \cdot$$

$$\cdot \left[\frac{1.75 \cdot 10^{-3} \cdot 835 \cdot 9.81}{0.257}\right]^{1/2} \cdot (1 - 0.03834)^{3.5} =$$

$$= 7.222\ ms^{-1}$$

In the second iteration, the numerical values for λ_0 and Re_V are calculated as:

$$\lambda_0 = \left(\frac{u_L}{u_V}\right)_{Fl} = \frac{2.52 \cdot 10^{-3}}{7.677} = 3.28 \cdot 10^{-3} \quad \Rightarrow \quad h^0_{L,Fl} = 4 \cdot 10^{-2}\ m^3 m^{-3}$$

and

$$Re_V = \frac{u_V \cdot d_P}{(1-\varepsilon) \cdot \nu_V} = \frac{6 \cdot u_V}{a \cdot \nu_V} = \frac{6 \cdot 7.220 \cdot 0.257}{500 \cdot 7.14 \cdot 10^{-6}} = 3116.8$$

Based on Eq. (3-14) and Table 6-1c, the resistance coefficient of the dry packing ψ is calculated as:

$$\psi = 1.21 \cdot 3116.8^{-0.14} = 0.3823$$

The gas velocity at the flooding point $u_{V,Fl}$, acc. to Eq. (2-67) is given as:

$$u_{V,Fl} = 0.8 \cdot \cos 30° \cdot 0.3823^{-1/6} \cdot \left[\frac{7.61}{1.75}\right]^{0.25}$$

$$\cdot 0.95^{1.2} \cdot \left[\frac{1.75 \cdot 10^{-3} \cdot 835 \cdot 9.81}{0.257}\right]^{1/2} \cdot (1 - 0.039)^{3.5}$$

$$= 7.18\ ms^{-1}$$

At 0.58%, this value only differs marginally from the numerical value given by the first iteration, which means that the value for the gas velocity at the flooding point can be assumed to be:

$$u_{V,Fl} \cong 7.18 \text{ ms}^{-1}$$

The relative error in the determination of the gas velocity at the flooding point $u_{V,Fl,exp}$ is therefore:

$$\delta\left(u_{V,Fl}\right) = \frac{7.18 - 7.50}{7.50} \cdot 100\% = -6.18\ \%$$

Numerical Example 2.4 – Chapter 2

A column with a diameter of $d_S = 0.155$ m is filled with 15 mm plastic Pall rings for the separation of the mixture methanol/nitrogen at 30 bar and 251 K. What is the column load, assuming a liquid flow of 464 kgh^{-1} and a gas flow of 472 kgh^{-1}? The physical properties and technical data of the packing are as follows:

Physical properties:

gas density:	$\rho_L = 41.06 \text{ kgm}^{-3}$
liquid density:	$\rho_L = 831.0 \text{ kgm}^{-3}$
surface tension:	$\sigma_L = 24.17 \cdot 10^{-3} \text{Nm}^{-1}$
viscosity:	
– liquid	$\eta_L = 122 \cdot 10^{-5}$ Pas
– gas phase	$\eta_V = 16.2 \cdot 10^{-6}$ Pas

Technical packing data:

(acc. to Krehenwinkel, Chapter 2 [75])	15 mm Pall ring – PP
column diameter:	$d_S = 0.155$ m
packing density:	$N = 247600 \text{ m}^{-3}$
specific weight:	$G = 130 \text{ kgm}^{-3}$
geometric surface of packing elements:	$a = 375 \text{ m}^2 \text{ m}^{-3}$
void fraction:	$\varepsilon = 0.846 \text{ m}^3 \text{ m}^{-3}$

Loads:

gas phase:	$V = 472 \text{ kgh}^{-1}$
liquid phase:	$L = 464 \text{ kgh}^{-1}$
specific column loads:	$u_V = 0.169 \text{ ms}^{-1}$
	$u_L = 29.6 \text{ m}^3 \text{ m}^{-2} \text{ h}^{-1}$
	$= 8.22 \cdot 10^{-3} \text{ ms}^{-1}$

Solution

The gas velocity at the flooding point $u_{V,Fl}$ is determined for higher gas densities as $\rho_V = 1.2\ \mathrm{kg/m^3}$, based on Eq. (2-69).

The phase flow ratio at the flooding point $\lambda_{0,1}$

$$\lambda_{0,1} = \frac{u_L}{u_V} = \frac{8.22 \cdot 10^{-3}}{0.169} = 4.864 \cdot 10^{-2}$$

based on Eq. (2-46), and for m= −0.7313, acc. to Eq. (2-43), is given as:

$$h^0_{L,Fl} = 33.19 \cdot 10^{-2}\ \mathrm{m^3\,m^{-3}}$$

The application of Eq. (2-69) requires the following variables:

– the hydraulic diameter of the packing d_h:

$$d_h = 4 \cdot \frac{\varepsilon}{a} = 4 \cdot \frac{0.846}{375} = 9.024 \cdot 10^{-3}\mathrm{m}$$

– the droplet diameter d_T, based on the Eq. (2-26):

$$d_T = \sqrt{\frac{\sigma}{(\rho_L - \rho_V)}} = \sqrt{\frac{0.0242}{(831 - 41.06) \cdot 9.81}} = 1.77 \cdot 10^{-3}\ \mathrm{m}$$

– the numerical value of the Reynolds number of the gas phase Re_V, which is calculated using Eq. (3-10):

$$Re_V = \frac{u_V \cdot d_p}{(1-\varepsilon) \cdot \nu_V} \cdot K$$

with a constant particle diameter d_p

$$d_p = 6 \cdot \frac{1-\varepsilon}{a} = 2.464 \cdot 10^{-3} m$$

and the wall factor K

$$K = \left(1 + \frac{4}{d_S \cdot a}\right)^{-1} = 0.936$$

as well as the kinetic viscosity of the gas phase

$$\nu_V = \frac{\eta_V}{\rho_V} = \frac{16.2 \cdot 10^{-6}}{41.06} = 3.945 \cdot 10^{-7}\quad \mathrm{m^2 s^{-1}}$$

and for

$$u_V = 0.169\,\text{ms}^{-1}$$

which gives:

$$\text{Re}_V = \frac{0.169 \cdot 2.464 \cdot 10^{-3}}{(1 - 0.846) \cdot 3.945 \cdot 10^{-7}} \cdot 0.936 = 6415.6 \geq 2100$$

Based on Table 6-1a, the numerical values K_3 and K_4, used for determining the resistance coefficient ψ in the turbulent flow range $\text{Re}_V > 2100$, are given as:

$$K_3 = 3.23 \quad K_4 = -0.0343$$

Hence:

$$\psi = 3.23 \cdot 6415.6^{-0.0343} = 2.391$$

Based on Table 6-1a, the packing form factor φ_P, used for determining the resistance coefficient ψ acc. to Eq. (3-27) leads to the same value:

$$\psi = \frac{522.4}{\text{Re}_V} + 2.306 = \frac{522.4}{6415.6} + 2.306 = 2.387 \approx 2.39$$

Based on Eq. (2-71) for $\rho_V >> \rho_{V,air}$, the following applies:

$$K_{\rho V} = \left(\frac{\rho_V}{1.165}\right)^{0.18} = 1.90$$

Equation (2-69) now leads to the following result:

$$u_{V,Fl} = 1.90 \cdot 0.566 \cdot 2.391^{-1/6} \left[\frac{9.024 \cdot 10^{-3}}{1.77 \cdot 10^{-3}}\right]^{1/4} \cdot$$

$$\cdot 0.846^{1.2} \cdot \left[\frac{1.77 \cdot 10^{-3} \cdot (831.0 - 41.06) \cdot 9.81}{41.06}\right]^{1/2} \cdot [1 - 0.3319]^{7/2} =$$

$$= 0.161\ ms^{-1}$$

Note

The column floods, as the experimental gas velocity u_V in the column $u_V = 0.169\ \text{ms}^{-1}$, acc. to [75], is approximately equal to the gas velocity at the flooding point $u_{V,Fl} = 0.161\ \text{ms}^{-1}$, based on Eq. (2-69), with a relative deviation of: $\delta(F_{V,Fl}) = 4.72\%$.

Numerical Example 2.5 – Chapter 2

The trays of a demethaniser, operated at approx. 32 bar, are upgraded to Mellapak 350Y. The aim is to determine the maximum loading capacity of the upgraded packed column in the rectifying section with a diameter of $d_S = 1.2$ m and in the stripping section with $d_S = 2.282$ m [72]. The operating data is taken from the article by Ghelfi, Kreis, Alvarez and Hunkeler [72].

Solution

1.

Table 2-11. Physical properties and operating data of the upgraded column

Column diameter $\mathbf{d_S}$ [m]	$\mathbf{d_S = 1.2}$	$\mathbf{d_S = 2.282}$
Top pressure **p** [bar]	30.8	28.7
Temperature $\mathbf{t_V = t_L}$ [°C]	19.8	−2.7
Gas density $\boldsymbol{\rho_V}$ [kgm^{-3}]	51.46	63.10
Liquid density $\boldsymbol{\rho_L}$ [kgm^{-3}]	394.6	392.2
Surface tension $\boldsymbol{\sigma_L}$ [mNm^{-1}]	2.5	1.8
Gas viscosity $\boldsymbol{\eta_V}$ [Pas]	$8.9 \cdot 10^{-6}$	$9.0 \cdot 10^{-6}$
Liquid viscosity $\boldsymbol{\eta_L}$ [Pas]	$0.083 \cdot 10^{-3}$	$0.078 \cdot 10^{-3}$
Mass stream of gas **V** [kgh^{-1}]	12438	42930.7
Mass stream of liquid **L** [kgh^{-1}]	16.060	93731.9
Specific liquid load $\mathbf{u_L}$ [ms^{-1}]	$10 \cdot 10^{-3}$	$16.21 \cdot 10^{-3}$
Gas velocity $\mathbf{u_V}$ [ms^{-1}]	0.0534	0.0462

2. The technical data of the Mellapak 350Y, acc. to Table 6-1c, is as follows:

   $a = 350\ m^2\ m^{-3}$
   $\varepsilon = 0.965\ m^3\ m^{-3}$
   $K_1 = 5.756,\ K_2 = -0.321,\ K_3 = 1.3662,\ K_4 = -0.133$
   $C_{Fl,0} = 0.566$

   The predicted maximum loading capacity of the Mellapak 350Y is calculated iteratively, acc. to Eq. (2-69), in the first iteration for:

$$\lambda_{0,I} \leq \left(\frac{u_L}{u_{V,Fl}}\right) = \frac{L \cdot \rho_V}{\rho_L \cdot V}$$

   The following applies to the rectifying section with $d_S = 1.2$ m:

$$\lambda_{0,I} = \frac{16.060 \cdot 51.46}{12438 \cdot 394.6} = 0.1684$$

and with $d_S = 2.282$ m:

$$\lambda_{0,I} = \frac{93731.9 \cdot 63.10}{42930.7 \cdot 392.2} = 0.3513$$

Now the liquid hold-up $h^0_{L,Fl}$ at the flooding point is determined using Eq. (2-46) with the parameter m, based on Eq. (2-43):

$$m = -0.82 + \frac{\lambda_{0,I}}{\lambda_{0,I} + 0.5} = -0.82 + \frac{0.1684}{0.1684 + 0.5} = -0.568 \qquad (d_S = 1.2\ m)$$

and

$$m = -0.82 + \frac{0.3513}{0.3513 + 0.5} = -0.4073 \qquad (d_S = 2.282\ m)$$

Equation (2-46) now leads to the following result, valid for $d_S = 1.2$ m and $\lambda_0 = 0.1684$:

$$h^0_{L,Fl} = \frac{\sqrt{0.1684^2 \cdot (-0.568 + 2)^2 + (4 \cdot 0.1684 \cdot (-0.568 + 1) \cdot (1 - 0.1684))}}{2 \cdot (-0.568 + 1) \cdot (1 - 0.1684)} -$$

$$-\frac{(-0.568 + 2) \cdot 0.1684}{2 \cdot (0.568 + 1) \cdot (1 - 0.1684)} = 0.427\ m^3 m^{-3}$$

and valid for $d_S = 2.282$ m and $\lambda_{0,I} = -0.4073$:

$$h^0_{L,Fl} = 0.47\ \mathrm{m^3\ m^{-3}}$$

3. The gas velocity at the flooding point $u_{V,Fl}$ is calculated, using Eq. (2-69) for:

$$d_h = 4 \cdot \frac{\varepsilon}{a} = 4 \cdot \frac{0.965}{350} = 0.01103\ m$$

$$d_T = \sqrt{\frac{\sigma_L}{\Delta\rho \cdot g}} = \sqrt{\frac{0.0025}{(394.6 - 51.46) \cdot 9.81}} = 8.62 \cdot 10^{-4}\ m \qquad (d_S = 1.2\ m)$$

and/or

$$d_T = 7.47 \cdot 10^{-4}\ m \quad (d_S = 2.282\ m)$$

The resistance coefficient ψ, for the dry column $d_S = 1.2$ m and $u_V = 4 \cdot V/(\pi \cdot d_S{}^2 \cdot \rho_V) = 0.0594\ \mathrm{ms^{-1}}$, is given as:

$$Re_V = \frac{u_V \cdot d_p}{(1 - \varepsilon) \cdot \nu_V} = \frac{6 \cdot u_V}{a \cdot \nu_V} = \frac{6 \cdot 0.0594 \cdot 51.46}{350 \cdot 8.9 \cdot 10^{-6}} = 5887.2$$

$$\Rightarrow \psi = K_3 \cdot Re_V^{K_4} = 0.4306$$

and based on $d_S = 1.2$ m and $u_V = 0.0462\ ms^{-1}$:

$$Re_V = 5556.6 \qquad \text{and} \qquad \psi = 0.343$$

Now the gas velocity at the flooding point $u_{V,Fl}$ can be calculated, using Eq. (2-69), for a column diameter of $d_S = 1.2$ m:

$$u_{V,Fl} = 0.566 \cdot 0.4306^{-1/6} \cdot 0.97^{1.2} \cdot \left(\frac{11.03 \cdot 10-3}{0.862 \cdot 10-3}\right)^{1/4} \cdot$$

$$\cdot \left(\frac{8.62 \cdot 10-4 \cdot (394.6-5.46) \cdot 9.81}{51.46}\right)^{1/2} \cdot$$

$$\cdot \left(\frac{51.46}{1.17}\right)^{0.18} \cdot (1-0.43)^{3.5} = 0.0809\ \text{ms}^{-1}$$

The flood gas load factor is therefore given as:

$$F_{V,Fl,I} = u_{V,Fl,I} \cdot \sqrt{\rho_V} = 0.5805\ \sqrt{\text{Pa}}$$

Following the creation of new phase flow ratios at the flooding point $\lambda_{0,II}$ and $\lambda_{0,III}$, further iterations lead to the final result, for a column diameter of $d_S = 1.2$ m:

$$u_{V,Fl} = 0.096\ ms^{-1} \qquad \text{and} \qquad F_{V,Fl} = 0.69\ \sqrt{Pa}$$

and on $d_S = 2.282$ m:

$$u_{V,Fl} = 0.056\ ms^{-1} \qquad \text{and} \qquad F_{V,Fl} = 0.44\ \sqrt{Pa}$$

The column load in the rectifying section is as follows, for a column diameter of $d_S =$ 1.2 m:

$$\frac{F_V}{F_{V,Fl}} = \frac{u_V}{u_{V,Fl}} = \frac{0.0594}{0.096} \cdot 100 = 61.9\%$$

and in the stripping section, for a column diameter of $d_S = 2.282$ m:

$$\frac{F_V}{F_{V,Fl}} = \frac{u_V}{u_{V,Fl}} = \frac{0.0462}{0.056} \cdot 100 = 82.5\%$$

According to information found in literature, the maximum column load [72] is given as:

$$\frac{F_V}{F_{V,Fl}} = 58\% \qquad \text{for} \quad d_S = 1.2\ m$$

and

$$\frac{F_V}{F_{V,Fl}} = 72\% \qquad \text{for} \quad d_S = 2.282\ m$$

This leads to the following flood gas load factors $F_{V,Fl}$:

$$F_{V,Fl} = 0.735\ \sqrt{Pa} \qquad \text{for} \quad d_S = 1.2\ m$$

and

$$F_{V,Fl} = 0.509\ \sqrt{Pa} \qquad \text{for} \quad d_S = 2.282\ m$$

The relative deviations $\delta(F_{V,Fl})$ are therefore:

$$\delta(F_{V,Fl}) = -6.12\% \qquad \text{for} \quad d_S = 1.2\ m$$

and

$$\delta(F_{V,Fl}) = -13.5\% \qquad \text{for} \quad d_S = 2.282\ m$$

Note

The load values of the Mellapak pressure column, which were calculated using the SBD model, are safe and only marginally different from the values found in literature, given the extreme physical properties of the separated system and the operating conditions under high pressure.

Annex Chapter 2

Flood Load Diagrams for Various Random and Structured Packings

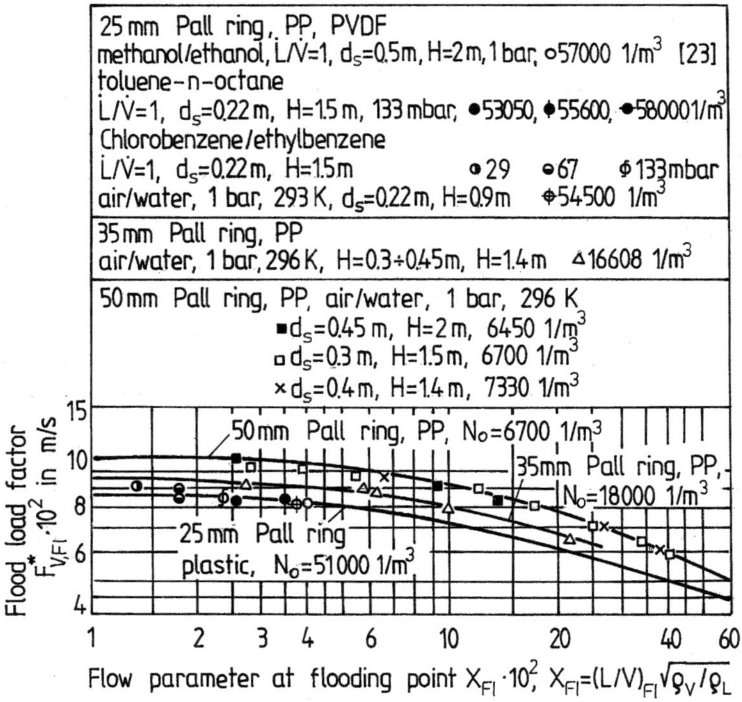

Figure 2-22. Capacity diagram $F^*_{V,Fl} = f(X_{Fl})$ for random 25–50 mm Pall rings made of plastic

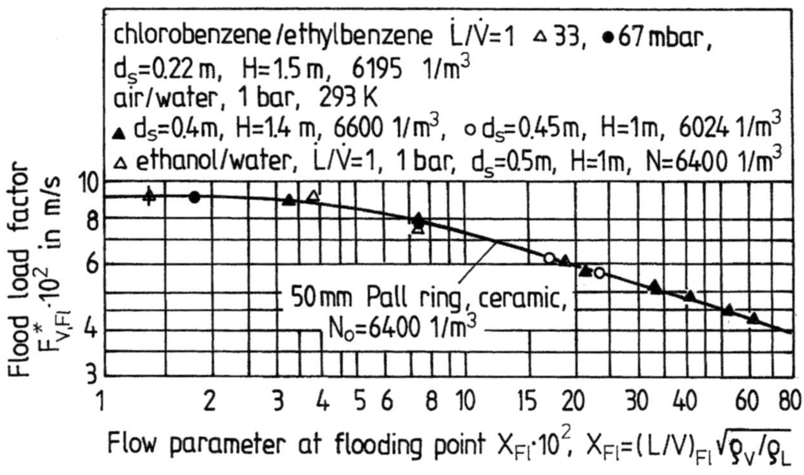

Figure 2-23. Capacity diagram $F^*_{V,Fl} = f(X_{Fl})$ for random 50 mm Pall rings made of ceramic

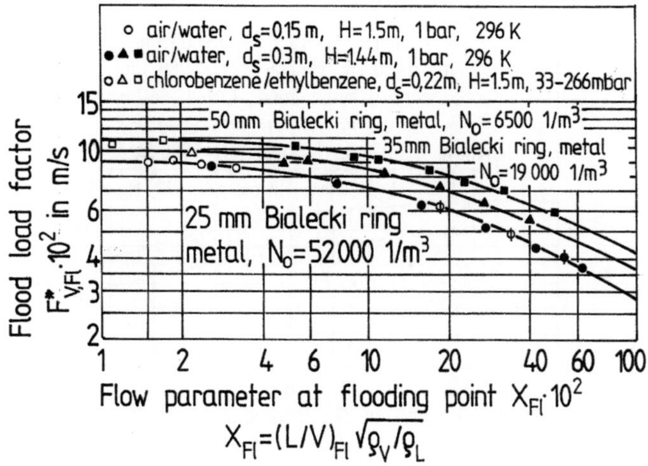

Figure 2-24. Capacity diagram $F^*_{V,Fl} = f(X_{Fl})$ for random 25—50 mm metal Pall rings

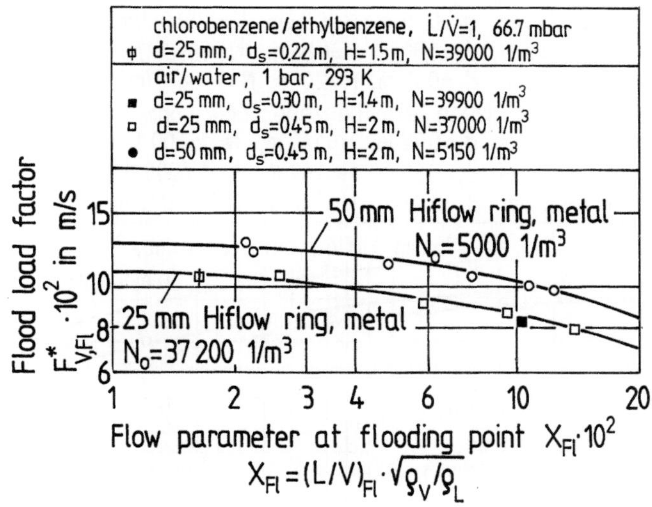

Figure 2-25. Capacity diagram $F^*_{V,Fl} = f(X_{Fl})$ for random 25—50 mm metal Hiflow rings

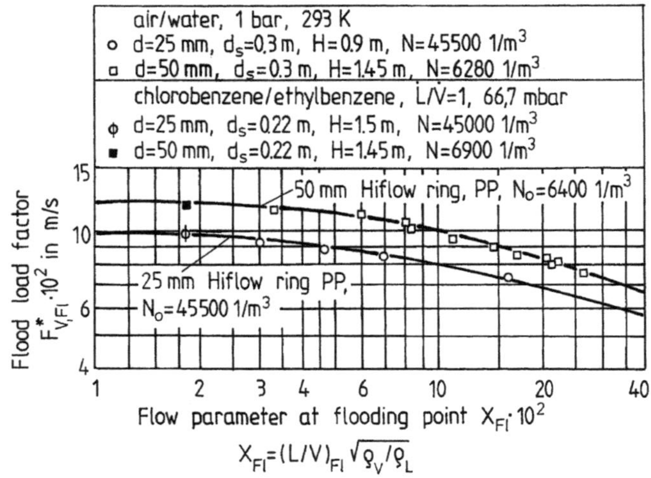

Figure 2-26. Capacity diagram $F^*_{V,Fl} = f(X_{Fl})$ for random 25–50 mm Hiflow rings made of plastic

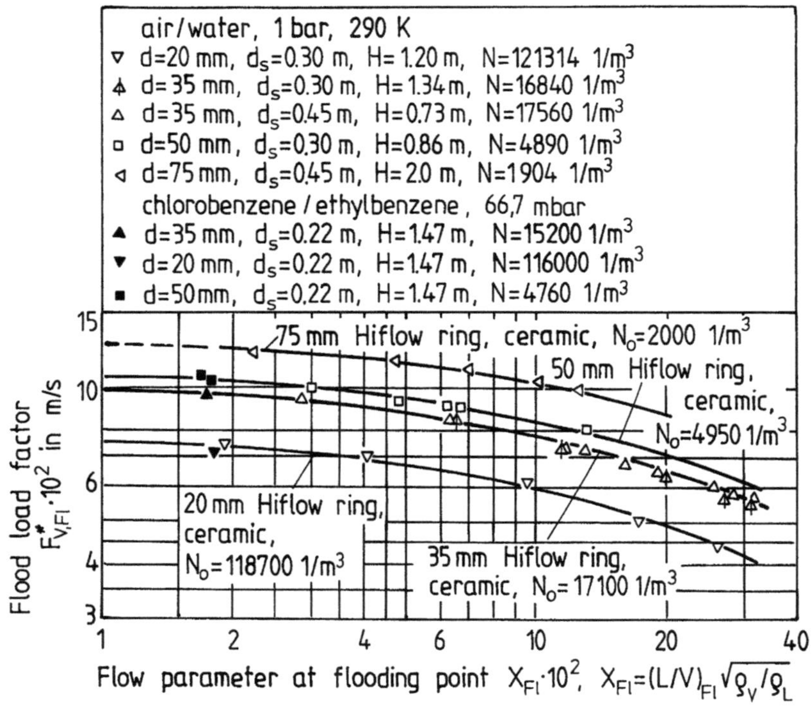

Figure 2-27. Capacity diagram $F^*_{V,Fl} = f(X_{Fl})$ for random 25–75 mm Hiflow rings made of ceramic

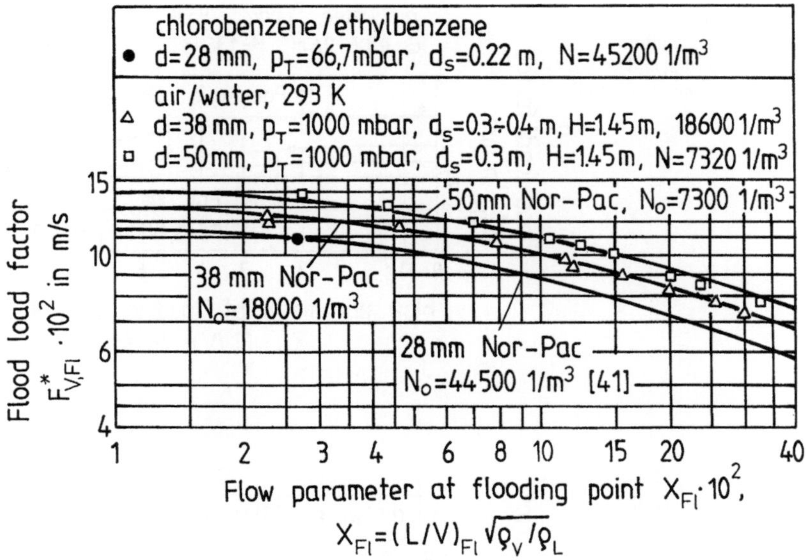

Figure 2-28. Capacity diagram $F^*_{V,Fl} = f(X_{Fl})$ for random 28–50 mm Nor-Pac packing made of plastic

Figure 2-29. Capacity diagram $F^*_{V,Fl} = f(X_{Fl})$ for random metal VSP rings, sizes 1 and 2

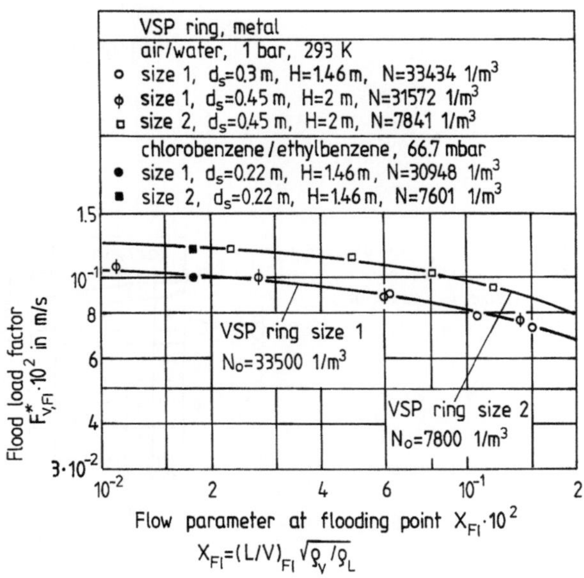

Figure 2-30. Capacity diagram $F^*_{V,Fl} = f(X_{Fl})$ for random Mc-Pac, sizes 1 and 2, and 50 mm Pall rings made of metal

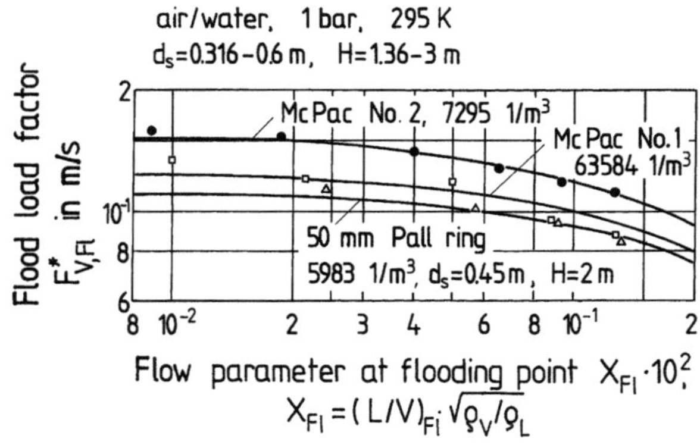

Figure 2-31. Capacity diagram $F^*_{V,Fl} = f(X_{Fl})$ for random Dtnpac, sizes 1 and 2, made of plastic

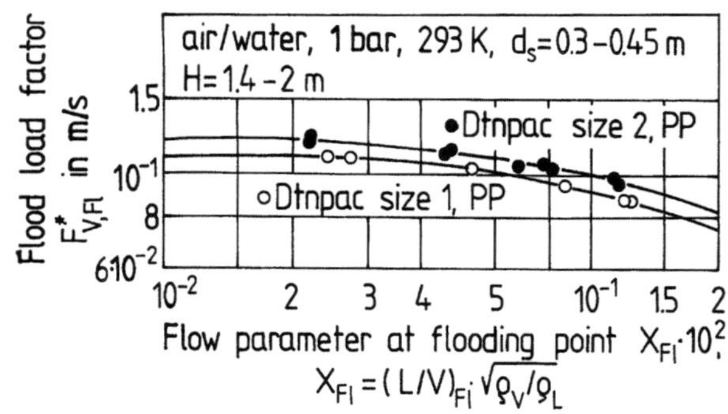

Figure 2-32. Capacity diagram $F^*_{V,Fl} = f(X_{Fl})$ for random Envipac, sizes 1–3, made of plastic

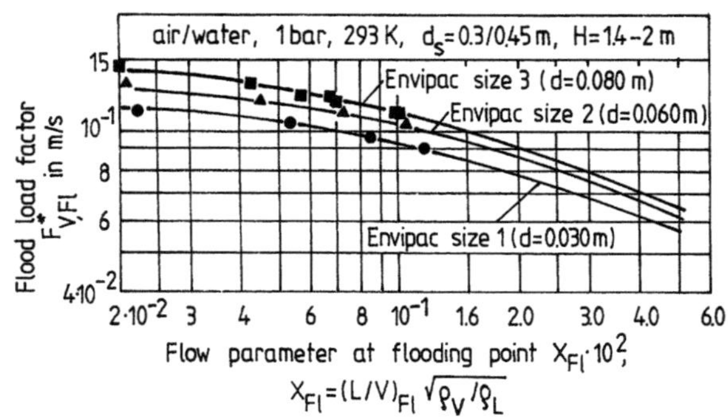

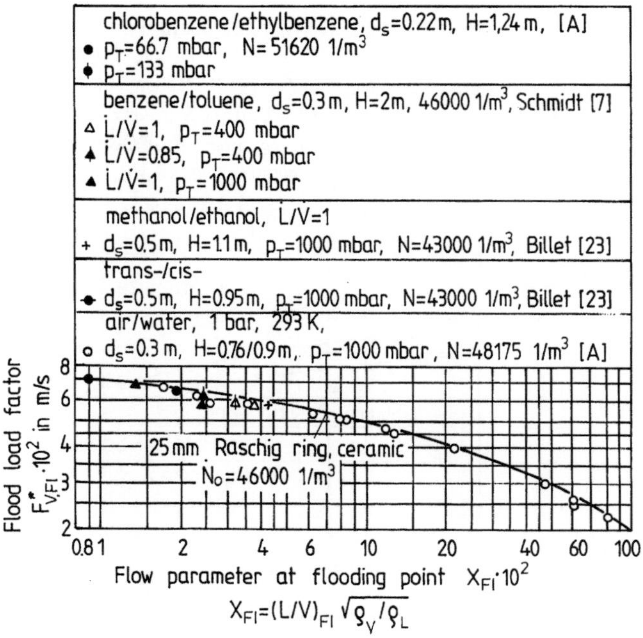

Figure 2-33. Capacity diagram $F^*_{V,Fl} = f(X_{Fl})$ for random 25 mm Raschig rings made of ceramic

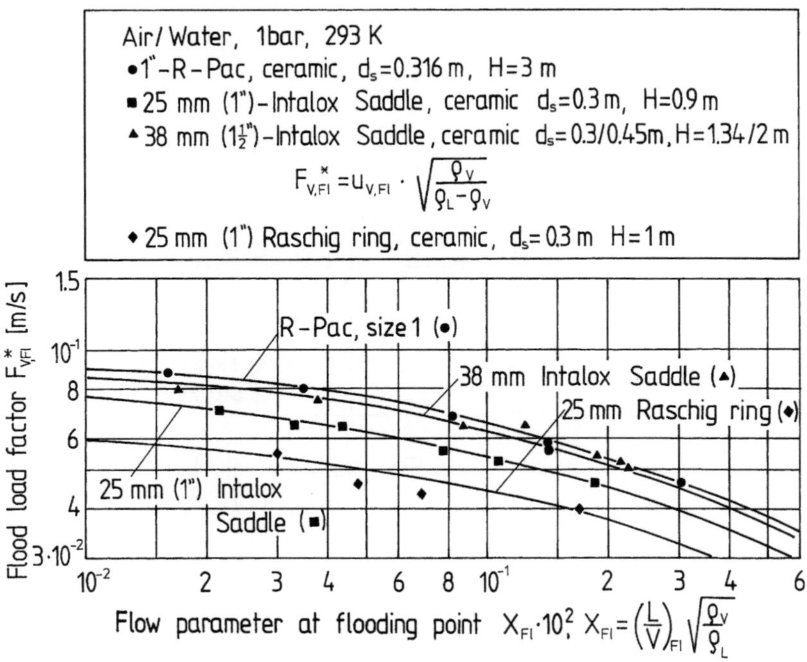

Figure 2-34. Capacity diagram $F^*_{V,Fl} = f(X_{Fl})$ for random R-Pac size 1, 25–38 mm Intalox saddles and 25 mm Raschig rings, made of ceramic

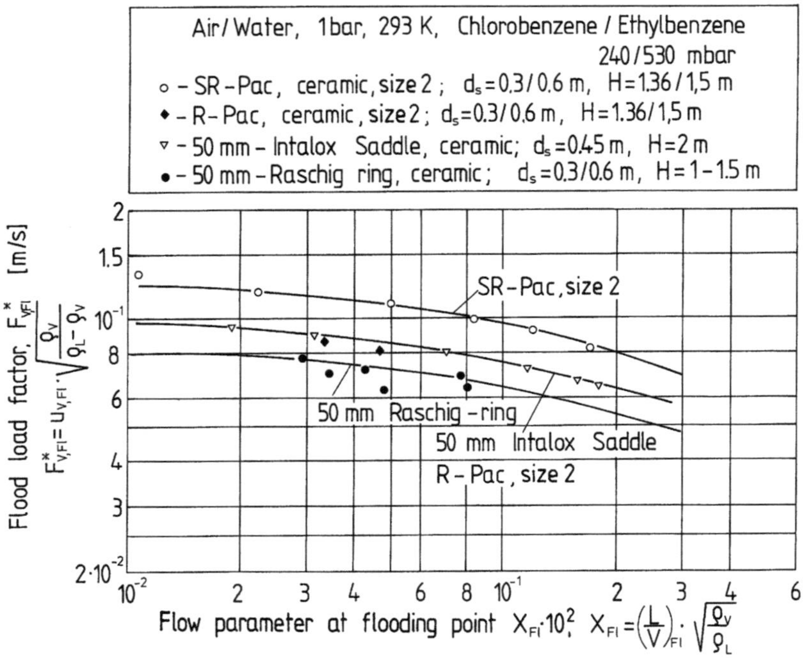

Figure 2-35. Capacity diagram $F^*_{V,Fl} = f(X_{Fl})$ for random R-Pac size 2, SR-Pac, 50 mm Intalox saddles and 50 mm Raschig rings, made of ceramic

Figure 2-36. Capacity diagram $F^*_{V,Fl} = f(X_{Fl})$ for stacked 25 and 50 mm metal Białecki rings

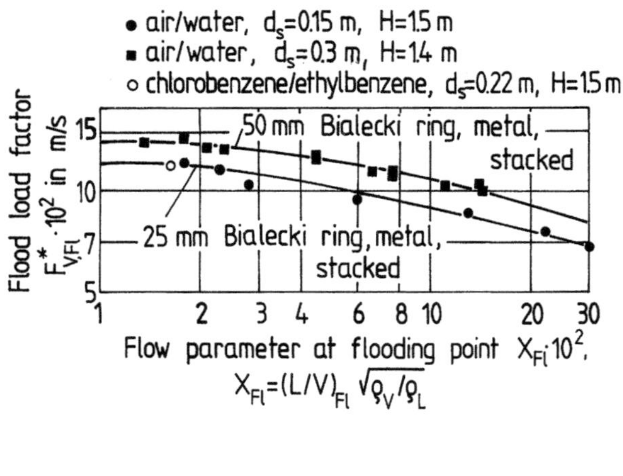

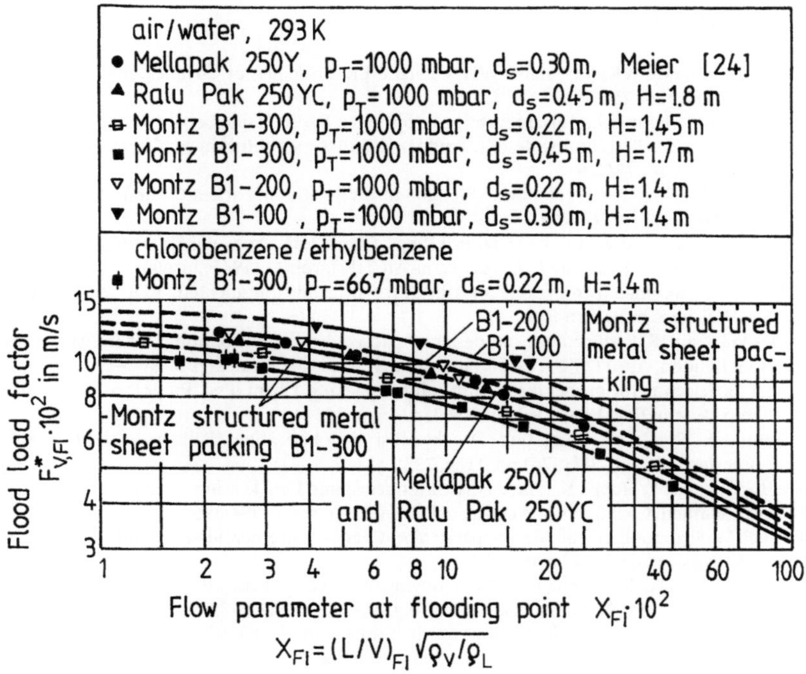

Figure 2-37. Capacity diagram $F^*_{V,Fl} = f(X_{Fl})$ for various structured packingss

Figure 2-38. Capacity diagram $F^*_{V,Fl} = f(X_{Fl})$ for GEA-H2 plastic structured packings

Structured packing GEA-H2, PP						
TP	Type	a [m²/m³]	system	p_T [mbar]	d_s [m]	H [m]
□	NET 38-150	80	air, 293 K	1000	0.60	1.80 / 3.15
△	FKP 327	125				3.04
○	FKP 319	150				1.2 / 3.05

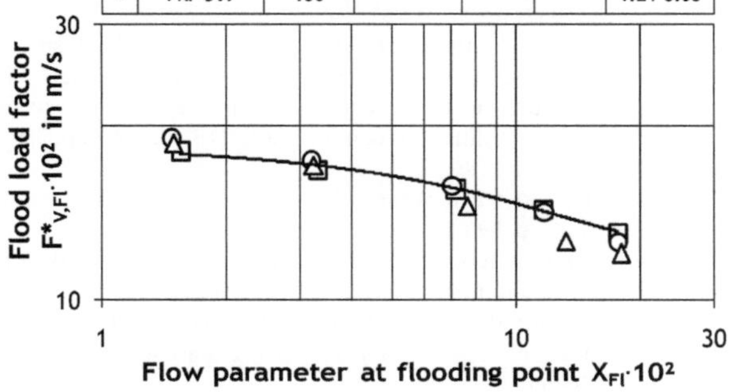

References Chapter 2

1. Walker WH, Levis WK, Mc Adams WH, Gilliland ER. Principles of Chemical Engineering 3rd ed., Mc Graw-Hill, New York (1937)
2. Mersmann A. Thermische Verfahrenstechnik. Springer, Berlin/Heidelberg (1980)
3. Mersmann A. Zur Berechnung des Flutpunktes in Füllkörperschüttungen. Chem.-Ing.-Techn., vol. 37 (1965) no. 3, p. 218/226
4. Sherwood TK, Shipley GH, Holloway FA. Flooding Velocities in Packed Column. Ing. Eng. Chem., vol. 30 (1938) no. 7, p. 765/769
5. Billet R. Recent Investigations of Metal Pall Rings. Chem. Eng. Prog., vol. 63 (1967) no. 9, p. 53/65
6. Lobo WE, Friend L, Hashmall F, Zenz F. Trans AICHE-J, vol. 41 (1945), p. 693
7. Schmidt R. Zweiphasenstrom und Stoffaustausch in Schüttungsdichten. VDI-Verlag, Düsseldorf (1972), VDI-Forschungsheft 550
8. Bolles WL, Fair JR. Performance and design of packed distillation columns. 3rd Int. Symp. on Destillation, London, April (1979). EFCE Publications Series No. 3, vol. 2, p. 3.3./35–89
9. Eckert JS. Selecting the proper distillation column packing. Chem. Eng. Progr., vol. 66 (1970) no. 3, p. 39 resp.. Chem. Eng. Progr., vol. 59 (1963) no. 5, p. 76
10. Schumacher R. Gasbelastung und Druckverlust von berieselten Füllkörperschüttungen. "vt"-Verfahrenstechnik, vol. 10 (1976) no. 11, p. 727/732
11. Weiß S, Schmidt E, Hoppe K. Obere Belastungsgrenze und Druckverlust bei der Destillation in Füllkörperkolonnen. Chem.Techn., Leipzig, vol. 27 (1975) no. 7, p. 394/396
12. Reichelt W. Strömung in Füllkörperapparaten bei Gegenstrom einer flüssigen und einer gasförmigen Phase. Verlag Chemie, Weinheim (1974)
13. Lapidus L, Elgin JC. Mechanics of vertical-moving fluidized systems. AIChE-J., vol. 1 (1957), p. 63/68
14. Richardson JF, Zaki WN. Trans. Inst. Chem. Eng., vol. 32 (1954), p. 35/53
15. Hu S, Kinter RC. The fall of single drops through water. AICHE J. (1955) no. 1, p. 42
16. Eckert JS, Walter LF. What affects packed bed distillation. Hydrocarbon Processing, vol. 43 (1964) no. 2, p. 107/114
17. Maćkowiak J, Billet R. New method of packed column design for liquid-liquid. Processes with Random and Stacked Packings. Ger. Chem. Eng., vol. 9 (1986) no. 1, p. 48/64 resp., Chem.-Ing.-Tech., vol. 57 (1985) no. 1, p. 56/57
18. Billet R, Maćkowiak J, Pajk M. Hydraulic and mass transfer in filled tube columns. Chem. Eng. Processing, vol. 2 (1985), no. 1
19. Kirschbaum E. Destillier- und Rektifiziertechnik. Springer, Berlin/Heidelberg/New York (1969)
20. Maćkowiak J. Einfluss der Form eines Füllkörpers auf die Hydraulik und den Stoffübergang bei der Absorption. Dissertation TU-Wrocław (Polen) (1975)
21. Billet R, Maćkowiak J, Ługowski Z, Filip S. Development and performance of impulse packing for gas/liquid-systems. Fette, Seifen, Anstrichmittel, vol. 85 (1983) no. 10, p. 383/391
22. Wiggert U. Druckverlust und Flüssigkeitsinhalt in berieselten Schüttungen. Report – Max Plack Institut (1959) no. 92, p. 243
23. Billet R. Stand, Entwicklung und Aussichten der Destillation und Rektifikation im Vergleich zu anderen Trennmethoden. Chemie-Technik, vol. 2 (1974), p. 355/361
24. Meier W, Hunkeler R, Stöcker WD. Performance of a new regular tower packing 'Mellapak'. 3rd Int. Symp. on Distillation, London, April (1979). vol. 2, p. 3.3/1–17 resp.. Chem.-Ing.-Tech., vol. 51 (1979) no. 2, p. 119/122
25. Eckert JS, Foote EH, Walter LF. What affects packing performance. Chem. Eng. Process, vol. 62 (1966) no. 1, p. 59/67
26. Informationsmaterial der Firma Sulzer/Winterthur. Trennkolonnen für Destillation und Absorption Packungen, Kolonnen, Anlagen. d./22.13.06.20-V.85-20
27. Informationsmaterial der Firma VFF, Ransbach/Baumbach., Westerwald (1986)
28. Informationsmaterial der Firma Raschig, Raschig GmbH. Ludwigshafen/Rhein no. 6 (1985) and VPR-5903/86
29. Bańczyk L, Grobelny A, Jarzynowski M. Porównawcze badania ceramiczynch siodełek. Intalox i pierścieni Raschiga (orig. Polish). Inż. Chem. i Procesowa vol. 4 (1983) no. 1, p. 3/14. as well as Inż. Chem. i Procesowa, vol. 4 (1983) no. 2, p. 267/279
30. Yilmaz T. Flüssigkeitsseitiger Stoffübergang in berieselten Füllkörperschüttungen. Chem.-Ing. -Techn., vol. 45 (1973) no. 5, p. 253/259

31. Strigle RR, Porter JKE. Metal Intalox – A new distillation packing. 3rd Int. Symp. on Distillation, London (1979)
32. Schmidt R. The lower capacity limits of packed column. I. Chem. E. Symp. Series No. 56, EFCE Publications. Series No. 3, vol. 2, p. 3.1/1–3.1/13
33. Mersmann A, Beyer von Morgenstern L, Deixler A. Deformation, Stabilität und Geschwindigkeit fluider Partikel. Chem.-Ing.-Tech., vol. 55 (1983) no. 11, p. 865/867
34. Mersmann A, Deixler A. Packungskolonnen. Chem.-Ing.-Tech., vol. 56 (1986) no. 1, p. 19/31
35. Gieseler M. Durchströmungsverhalten von Packungen und Schüttungen aus Koksen unterschiedlicher Größe und Gestalt in Gegenwart einer Flüssigkeit hoher Viskosität. Dissertation TU Clausthal (1972)
36. Blaß E, Kurtz R. Der Einfluss grenzflächenenergetischer Größen auf den Zweiphasen-Gegenstrom durch Raschigring-Füllkörpersäulen; Teil 1: Flüssigkeitsinhalt. "vt"-verfahrenstechnik, vol. 10 (1976) no. 11, p. 721/724
37. Blaß E, Kurtz R. Der Einfluss grenzflächenenergetischer Größen auf den Zweiphasen-Gegenstrom durch Raschigring-Füllkörpersäulen; Teil 2: Druckverlust und Flutpunkt. "vt"-verfahrenstechnik, vol. 11 (1977) no. 1, p. 44/48
38. Mersmann A. Zum Flutpunkt in Flüssig-Flüssig-Gegenstromkolonnen. Chem.-Ing.-Tech., vol. 52 (1980) no. 12, p. 933/942
39. Możeński C, Kucharski E. Hydraulika wypełnień pod zwiększonym ciśnieniem (orig. Polish). Inż. Chem. i Procesowa, vol. 3 (1986) no. 3, p. 373/384
40. Billet R, Maćkowiak J. Neuartige Füllkörper aus Kunststoffen für thermische Stofftrennverfahren. Chemie-Technik, vol. 9 (1980) no. 5, p. 219/226
41. Billet R, Maćkowiak J. Wirksamkeit von Kunststoff-Füllkörper bei der Absorption,. Desorption und Vakuumrektifikation. "vt"-verfahrenstechnik, vol. 15 (1982) no. 2, p. 67/74
42. Billet R, Maćkowiak J. Neues Verfahren zur Auslegung von Füllkörperkolonnen für die Rektifikation. „vt"-verfahrenstechnik, vol. 17 (1983) no. 4, p. 203/211
43. Billet R, Maćkowiak J. How to Use the Absorption Data for Design and Scale-up of Packed Columns. Fette, Seifen, Anstrichmittel, vol. 86 (1984) no. 9, p. 349/358
44. Billet R, Maćkowiak J. Hiflow-Ring ein Hochleistungsfüllkörper für Gas-Flüssig-Systeme. Teil 1: Ausführung in Kunststoff. Chemie-Technik, vol. 13 (1984) no. 12, p. 37/46
45. Billet R, Maćkowiak J. Hiflow-Ring ein Hochleistungsfüllkörper für Gas-Flüssig-Systeme. Teil 2: Ausführung in Metall. Chemie-Technik, vol. 14 (1985) no. 4, p. 91/99
46. Billet R, Maćkowiak J. Hiflow-Ring ein Hochleistungsfüllkörper für Gas-Flüssig-Systeme. Teil 3: Ausführung in Keramik. Chemie-Technik, vol. 14 (1985) no. 5, p. 195/206
47. Billet R, Maćkowiak J. Application of Modern Packings in Thermal Separation Processes. Chem.Eng. Technol. vol. 11 (1988), p. 213/227
48. Billet R, Maćkowiak J. Hochwirksame metallische Packung für Gas- und Dampf-Flüssig-Systeme. Chem.-Ing.-Tech., vol. 57 (1985) no. 11, p. 976/978
49. Vogt M. Modifizierte Darstellung der Flutgrenze in Füllkörperschüttungen. Chem.-Ing.-Tech., vol. 57 (1985) no. 4, p. 332/333
50. Kafarow WW. Osnovy massopjeredatschi, (orig. Russian). Verlag für chemische Literatur, Moskau (1972), p. 391
51. Kleinhückenkotten H. Untersuchung zur Auslegung von Füllkörperkolonnen mit geschütteter Füllung. „vt"-verfahrenstechnik, vol. 2 (1975) no. 6, p. 275/279
52. Won-Hi Hong, Brauer H. Stoffaustausch zwischen Gas und Flüssigkeit in Blasensäulen. VDI-Forschungsheft no. 624 (1984), VDI-Verlag Düsseldorf
53. Stichlmair J. Berechnung des Druckverlustes bei der Durchströmung ruhender Feststoffschüttungen.. Vortrag auf der Internen Sitzung des Fachausschusses "Thermische Zerlegung von Gas- und Flüssigkeitsgemischen" der GVD-VDI-Gesellschaft Verfahrenstechnik und Chemieanlagen in Bochum (1983)
54. Informationsmaterial der Firma Ceilcote. Tellerette-Manual, Ceilcote Company, 140 Sheldon Road/Berea, Ohio 44017
55. Stichlmair J. Grundlagen der Dimensionierung des Gas/Flüssigkeits-Kontaktapparates, BODENKOLONNE. Verlag Chemie (Reprotext), Weinheim, New York (1978)
56. Billet R, Maćkowiak J. 'Hydraulisches Verhalten und Stoffübertragung einer neuartigen Kolonnenpackung für Gas-Flüssig-Systeme'. Chemie-Technik, vol. 11 (1982), p. 1107–1114
57. Levich VG. Physicochemical hydrodynamics. Prentice-Hall, Ing. Englewood Cliffs, New York (1962)

58. Covelli B, Mülli R. Mitschleppen von Flüssigkeitstropfen über einer Sprudelschicht. Chem.-Ing.-Tech., vol. 51 (1981) no. 11, p. 877/879
59. Soo SL. Fluid dynamics of multiphase systems. BlaisdeU Publishing Company (1967), Waltham, M
60. Mersmann A. persönliche Mitteilung (1987)
61. Zenz FA. Petroleum Refiner, vol. 36 (1957) no. 8, p. 147/155
62. Strom JR., Kinter RC. Wall effect for the fall of single drops. AIChE-J., vol. 4 (1958), p. 153/156
63. Reinhart A. Das Verhalten fallender Tropfen. Dissertation no. 3412, Eidgenössische TH in Zürich (1964)
64. Kovalenko VS. For calculation of falling velocity of droplets (orig. Russian). Meor. Osnovy khim. tekhnol., Moscow, vol. 12 (1978) no. 3, p. 464/466
65. Clift R, Grace JR, Weber ME. Bubbles, Drops and Particles. Acad. Press, New York, San Francisco, London (1978)
66. Bornhütter H. Stoffaustausch von Füllkörperschüttungen unter Berücksichtigung der Flüssigkeitsströmungsform. Dissertation, TU München (1991)
67. Plüs RC, Ender Ch. A new high-efficiency packing „Mellaring-VSP". Presentation at Achema 1991 – Frankfurt a.M. on 9.6.1991. Conference volume published by Dechema
68. Süess A, Spiegel L. Hold-up of Mellapak structured packings. Sulzer Bros AG (Schweiz) Ltd., Separation Column, Winterthur
69. Spiegel L, Meier L. Structured packings – Capacity and pressure drop at very high liquids loads. Chemical plants + Processing, vol. 1 (1995) and. Chemical Eng. Technology – Wiley VCM-Verlag (1995)
70. Gierczycki A. Hydraulika aparatów wypełnionych z wybranym wypełnieniem (orig. Polish). Inż. Chem. i. Procesowa vol. 4 (1995) no. 4, p. 507–518
71. Viricny V, Stanek V. An experimental Set-up to Messure Flow Transierts in Counter-Current Packed Bed Column,. Chem. Biochem. Eng. vol. 10 (1996), no. 2, p. 55–61
72. Ghelfi L, Kreis H, Alvarez JA, Hunkeler H. Structured Packing in Pressure Columns. Poster-Int. Conference and Exibition on Destillation and Absorption. Birmingham, England, 7-9 September (1992) (c) 1994 by Sulzer Chemtech Ltd., Wintertur
73. Piche S, Larachi F, Grandjean B. Floading Capacity in Packed Towers,. Database, Correlations and Analysis,. Ind. Eng. Chem. Res. 2 vol. 40 (2001), p. 476–487
74. Kuźniewska-Lach I. Bestimmung der Flutpunktgeschwindigkeit in Füllkörperkolonnen. Inż. Chem. i Procesowa (orig. Polish) vol. 17 (1996) no.2, p. 267–277
75. Krehenwinkel H. Experimentelle Untersuchungen der Fluiddynamik und der Stoffübertragung in Füllkörperkolonnen bei Drücken bis 100 bar. Dissertation, TU Berlin, (Dezember 1986)
76. Maćkowiak J, Suder S. The new tube column with Pall-Rings for countercurrent and cucurrent processes. Vortrag Vesprem (Ungarn) 1979 and. Chem.-Ing.-Techn., vol. 50 (1978) no. 7, p. 550–551
77. Maćkowiak J, Suder S. Hydraulika i wymiana masy w kolumnie wypełnionej układanymi pierścieniami Białeckiego (orig. Polish). Inż. Chem. i. Procesowa vol. 8 (1977) no. 3, p. 651–664
78. Maćkowiak J. Hydraulische Untersuchungen von geordneten 50 mm keramischen Raschigringen. Internal study –TU-Wrocław-1975
79. Maćkowiak J. Mc-Pac – ein neuer metallischer Füllkörper für Gas-Flüssigkeitssysteme. Chem.-Ing.-Techn. vol. 73, no. 1+2 (2001) p. 74–79
80. Maćkowiak J. „Mc-Pac – Nowe metalowe wypełnienie dla układów gaz-ciecz". Inż. Chemiczna i Procesowa, vol. 21 (2000) p. 679–689
81. Bylica J, Jaroszyński M. Badania porównawcze hydrauliki wypełnień usypowych i konstrukcyjnych (orig. Polish). Inż. Chem. i Procesowa, (1995) no. 3, p. 421–439
82. Maćkowiak J, Ługowski Z. Geringe Apparatevolumina und Betriebskosten mit neuen keramischen Füllkörpern – R-Pac und SR-Pac. Verfahrenstechnik, vol. 29 (1995), no. 6, p. 19–22
83. Maćkowiak J, Szust J. Hydraulika i wymiana masy w kolumnach wypełnionych ceramicznymi pierścieniami R-Pac i SR-Pac w układach gaz/ciecz (orig. Polish). Inż. Chem. i Procesowa, vol. 18 (1997), no. 4, p. 675–691
84. Bańczyk L, Woźniak A, Szymkowiak E. Hydraulika kolumny wypełnionej ceramicznymi pierścieniami Białeckiego (orig. Polish). Inż. Chem. i Procesowa, vol. 7 (1977), no. 1, p. 261–274
85. Bańczyk L, Woźniak A, Jarzynowski M, Grobelny A. Hydraulika kolumny wypełnionej ceramicznymi pierścieniami IChN oraz Raschiga (orig. Polish). Inż. Chem. i Procesowa, vol. 9 (1979), no. 1, p. 15–28

86. Produktinformationen – Firma Raschig-Ludwigshafen-vt (1996). Raschig- Super Ring no.1 u.2, Metall und Kunststoff
87. Bornhüter K, Mersmann A. Druckverlust und Flutpunkt in Füllkörperschüttungen. CIT-Chem.-Ing.-Tech. vol. 64 (1992) no.3, p. 305/305
88. Kister MZ. Destillation design. McGraw-Hill, Inc., New York (1992)
89. Billet R. Packed towers. VCH-Weinheim, New York (1995)
90. Maćkowiak J. Determinition of flooding gasvelocity and liquid hold-up at flooding in packed columns for gas/liquid-systems. Chem. Eng. Tech. 13 (1991) p. 184–196
91. Maćkowiak J, Mersmann A. Zur maximalen Belastbarkeit von Kolonnen mit modernen Füllkörpern und Packungen für Gas/Flüssigkeitssysteme. Chem.-Ing.-Tech. vol. 63 (1991) no.5, p. 503–506
92. Grabbert G, Bonitz R. Fluiddynamik bei der Zweiphasenströmung durch Füllkörper- und Packungskolonnen. TU Bergakademie Freiberg, Freiberger Forschungshefte, A 842 (1998)
93. Linek V, Moucha T, Rejl F. Hydraulic and mass transfer characteristics of packings for absorption and distillation columns. Rauschert-metall-sattel-rings. Trans IChemE, Vol 79, Part A, October 2001
94. Schultes M. Raschig Super-Ring. A new Fourth Generation Packing Offers New Advantages. Trans IChemE, Vol 81, Part A, January 2003
95. Metal Random Packing – Superior random packings combined with innovative mass transfer technology. Technical Information by Sulzer Chemtech Ltd., 22.64.06.40 – IV.06 – 50, Switzerland
96. Pall-ring, Technical Information by Raschig, 2008
97. Technical Information by rvt PE GmbH

CHAPTER 3

Pressure Drop of Dry Packed Columns

3.1 Introduction

In the counter-current two-phase flow of gas and liquid, a pressure drop occurs in the packed column. It is important to know the pressure drop in the gas phase in single- and two-phase counter-current flow through random or structured packings, in order to assess the operating mode of packed columns. The total pressure drop in the packing $\Delta p/H$ per 1 m packing height is often derived from the product of the pressure drop $\Delta p_0/H$ of the dry packing and the quotient $\Delta p/\Delta p_0$, which takes into account the influence of the specific liquid load u_L, [1–5]:

$$\frac{\Delta p}{H} = \frac{\Delta p_0}{H} \cdot \left[\frac{\Delta p}{\Delta p_0}\right] \tag{3-1}$$

This chapter first takes a closer look at the calculation of the pressure drop of dry packings $\Delta p_0/H$ for various types of packings, followed by the calculation of the pressure drop of irrigated packings $\Delta p/H$ in Chap. 4. The application of the SBD model, presented in Chap. 2, requires the knowledge of the resistance coefficient ψ, for the purpose of accurately determining the gas velocity at the flooding point $u_{V,Fl}$, which is why the accurate calculation of this parameter is extremely important for design of packed columns.

3.2 Law of Resistance for Single-Phase Flow in Packed Columns

Deriving the Equation

The flow of gas through a random or structured packing can be regarded as a flow through a bundle of identical channels with a height H and diameter d_h, see Fig. 3-1. Therefore, the laws applicable to gas flow through an empty tube must also apply in this case.

J. Maćkowiak, *Fluid Dynamics of Packed Columns*,
Chemische Technik/Verfahrenstechnik, DOI 10.1007/b98397_3,

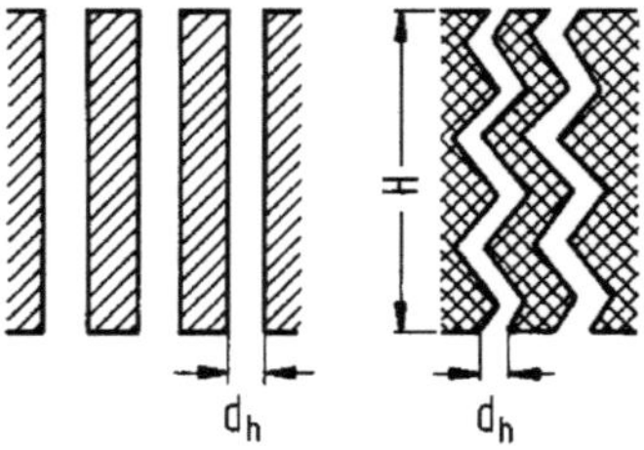

Figure 3-1. Hydraulic model of random packed bed or structured packing as bundle of channels of the same length and diameter d_h in single-phase flow

This can be described in mathematical terms, using the fundamental equation by Darcy and Weißbach [1–3]:

$$\frac{\Delta p_0}{H} = \lambda \cdot \frac{\bar{u}_V^2 \cdot \rho_V}{2 \cdot d_h} [Pa/m]. \tag{3-2}$$

In this work, the dependency of the resistance coefficient λ on the Reynolds number Re_V of the gas phase, which characterises the flow, is presented on the basis of the channel model, which describes the packing as a bundle of parallel tubes with equal diameters, see Fig. 3-1 and [1–12, 19–29].

Other models, developed by Reichelt [1, 2], Brauer [3, 4], Ergun [5], Kast [11] and Billet, Schultes [37], are not significantly different from the channel model and will not be discussed further in this book.

The size of the individual, hypothetical flow channels d_h in the packing is indirectly determined in the channel model, by calculating the particle diameter d_p of all the different types of packing elements, using the packing-specific variables a and ε, based on the following correlation [1–5, 10, 11]:

$$\begin{aligned} d_P &\equiv 6 \cdot \tfrac{V_1}{A_1} = 6 \cdot \tfrac{V_1 \cdot N}{A_1 \cdot N} = 6 \cdot \tfrac{V_F}{A_F} \\ &= 6 \cdot \tfrac{1-\varepsilon}{a} \quad [\mathrm{m}] \end{aligned} \tag{3-3}$$

In the case of spheres, d_p is equal to the geometric diameter d. The relative void fraction of a packing ε is given by:

$$\varepsilon \equiv \frac{V_S - V_F}{V_S} = 1 - \frac{V_F}{V_S} = 1 - \frac{V_1 \cdot N}{V_S} \quad \left[\mathrm{m}^3 \mathrm{m}^{-3}\right]. \tag{3-4}$$

V_1 is the volume and A_1 is the surface of an individual packing element, whereas V_F is the material volume and A_F is the surface of all N packing elements in the column volume. N $[\mathrm{m}^{-3}]$ is defined as the packing density, which is given by manufacturers as one of the most important packing-related variables. If the packing density of the packed bed is known and found to be different from the manufacturer's data, then the other geometric packing data ε and a can be easily recalculated.

The numerical Example 3.1 at the end of this chapter illustrates the calculation of the geometric packing data a and ε for different packing densities N, if the standard packing data a_0, ε_0 and N_0 are provided by the manufacturer.

The work of Brauer [3, 4] and Reichelt [1, 2] as well as [6, 19–24] show that it make sense to take into account the influence of the column wall on the hydraulic diameter d_h of packed columns with random and stacked packing elements.

The correlation between the hydraulic diameter d_h and the particle diameter d_P, taking into account the peripheral zone and/or the wall area A_F, is given as [3, 4]:

$$d_h \equiv 4 \cdot \frac{V_S - V_F}{A_F - A_W} = \frac{2}{3} \cdot \frac{\varepsilon}{1-\varepsilon} \cdot d_P \cdot K \quad [\mathrm{m}]. \tag{3-5}$$

The wall factor K can also be easily calculated, based on the geometric packing data and the size of the column d_S [3], using Eq. (3-6):

$$\frac{1}{K} = 1 + \frac{2}{3} \cdot \frac{1}{1-\varepsilon} \cdot \frac{d_P}{d_S} \quad [-] \tag{3-6}$$

Equation (3-6) shows that values of $K < 1$ are expected for columns with small diameters d_S. $K \approx 1$ only applies to large column diameters d_S.

These considerations are applicable to random as well as stacked packing elements.

In the case of structured packings, the influence of the column wall on the pressure drop is not taken into account for larger columns, i.e. $K = 1$. This is described in more detail in the following parts of this chapter.

By replacing the effective gas velocity $\bar{u}_V$ with the quotient u_V/ε and then substituting Eqs. (3-5) and (3-6) into Eq. (3-2), the single-phase flow pressure drop Δp_0 per 1 m packing height is given by Eq. (3-7); [1–4]:

$$\frac{\Delta p_0}{H} = \frac{3}{4} \cdot \lambda \cdot \frac{1-\varepsilon}{\varepsilon^3} \cdot \frac{u_V^2 \rho_V}{d_P \cdot K} \quad \left[Pam^{-1}\right] \tag{3-7}$$

Substituting in Eq. (3-7) the resistance coefficient ψ for *0.75·λ* and the gas capacity factor F_V for the product $u_V \cdot \sqrt{\rho_V}$ [7] gives Eq. (3-8):

$$for\ \psi \equiv 0.75 \cdot \lambda \quad \Rightarrow \quad \frac{\Delta p_0}{H} = \psi \cdot \frac{1-\varepsilon}{\varepsilon^3} \cdot \frac{F_V{}^2}{d_P \cdot K} \quad \left[\mathrm{Pam}^{-1}\right], \tag{3-8}$$

which is used to determine the pressure drop of the dry packing. This equation can only be applied in the whole flow range, if the resistance coefficient ψ is known.

The resistance coefficient ψ is dependent on the Reynolds number Re_V of the gas phase, i.e. $\psi = f(Re_V)$. Based on the assumed model, the Reynolds number Re_V is defined as:

$$\mathrm{Re}_V = \frac{3}{2} \cdot \frac{\bar{u}_V \cdot d_h}{\nu_V}. \tag{3-9}$$

Analogously, by replacing the effective gas velocity $\bar{u}_V$ with u_V/ε and then substituting Eq. (3-5) into Eq. (3-9), it is possible to calculate the modified Reynolds number of the vapour and/or gas, taking into account the wall influence [3, 4, 6].

$$\mathrm{Re}_V = \frac{u_V \cdot d_p}{(1-\varepsilon) \cdot \nu_V} \cdot K \tag{3-10}$$

The resistance coefficient ψ for a given packed bed, whose packing-specific geometric data a and ε is known, can be determined for any given experimental data $\Delta p_0/H$ and factor F_V using the transposed Eq. (3-8):

$$\psi = \frac{d_P \cdot K \cdot \varepsilon^3}{1-\varepsilon} \cdot \frac{\left[\frac{\Delta p_0}{H}\right]}{F_V^2}. \tag{3-11}$$

Based on experimental pressure drop data $\Delta p_0/H$ of the dry packing in a column with a diameter d_S, it is possible to determine the resistance coefficient ψ for different vapour capacity factors F_V and the packing density N.

This correlation $\psi = f(\mathrm{Re}_V)$ is known as the *law of resistance* [1–5]. Empirical laws of resistance can be found in literature for various types of packed columns, such as the law of resistance [3, 4] for spheres:

$$\psi = \frac{160}{\mathrm{Re}_V} + \frac{3.1}{\mathrm{Re}_V^{0.1}} \tag{3-12}$$

and the well-known model developed by Ergun [5] and quoted by Brauer [3]:

$$\psi = \frac{150}{\mathrm{Re}_V} + 1.75, \tag{3-13}$$

which was experimentally developed for ceramic Raschig rings, Intalox saddles and Berl saddles.

The augend in Eq. (3-13) determines the resistance coefficient of the laminar gas flow, whereas the addend takes into account the turbulent gas flow through the packing.

Further models for calculating the resistance coefficient ψ can be found in the work of Reichelt [1, 2], Brauer [3, 4], Schmidt [9], Teutsch [10], Kast [11], Bemer and Kalis [12] as well as Krehenwinkel [18] and Billet, Schultes [37]. They were developed mainly for ceramic Raschig rings, Intalox saddles and Berl saddles/spheres, based on experiments with air as the flow medium. Valuable data and models for determining the resistance coefficient for numerous classic, but also new lattice and structured packings for high-pressure absorption can be found in the work of Billet, Schultes [37], Krehenwinkel [18] and Bornhütter [17] for large lattice packings.

Despite the existence of a number of models, the experimental data from these authors clearly differs, which highlights the uncertainties that can be found in literature when it comes to determining the resistance coefficient ψ for identical types of packings. It must be assumed that they did not take into account all the important factors influencing the gas flow in the packing.

Hence, an accurate calculation of the pressure drop, on the basis of these models, cannot be guaranteed.

This work assumes the empirical correlation [6, 19, 20, 22] for determining the resistance coefficient ψ:

$$\psi = K_1 \cdot \mathrm{Re}_V^{K2} \tag{3-14}$$

The constant K_1 and exponent K_2 take on different values for the transition range $\mathrm{Re}_V < 2100$ and the range of the fully developed turbulent flow $\mathrm{Re}_V \geq 2100$. However, the law of resistance, as represented here, does not differentiate between gas flow in a random packing, filled with ring or saddle packing elements, and a structured packing. If there is incomplete flow through the free volume of the packing, this has a direct effect on the resistance coefficient ψ and does not require a complicated correction of the particle diameter d_P, as pointed out in the earlier work of other authors, e.g. [1–4].

The following example, based on experimental data for Pall rings, shows which parameters have a substantial influence on the resistance coefficient during gas flow through random and stacked packing elements.

3.2.1 Determining the Resistance Coefficient ψ for Pall Rings

Metal Pall rings belong to the group of classic packing elements, which are still the most widely used packings today. There is a large amount of experimental and accurate data available, taken by Billet [13] and [7, 8, 14, 16, 17], which illustrates the influence of various parameters on dry pressure drop.

Figure 3-2 shows the influence of the column diameter d_S and the physical properties of the mixtures investigated on the pressure drop of the dry packing $\Delta p_0/H$ for 50 mm Pall rings, at a constant packing density of approx. 6400 m^{-3}. The pressure drop $\Delta p_0/H$ decreases with the column diameter d_S and is lower in columns with small diameters than in larger-diameter columns.

Figure 3-3 shows the significant influence of the packing density N on the pressure drop $\Delta p_0/H$. As the packing density N increases, the pressure drop also increases, due to the larger specific packing surface a and the smaller void fraction ε of the packing, see Eq. (3-8).

The resistance coefficient ψ, as shown in Figs. 3-2 and 3-3, was determined using experimental pressure drop data for different columns equipped with metal Pall rings, with column diameters of $d_S = 0.150-0.8$ m, and with various systems. As can be seen in Figs. 3-4a and 3-4b, the resistance coefficient is independent of the packing density N and the column diameter d_S for diameter ratios of $d_S/d \geq 6$ throughout the entire flow

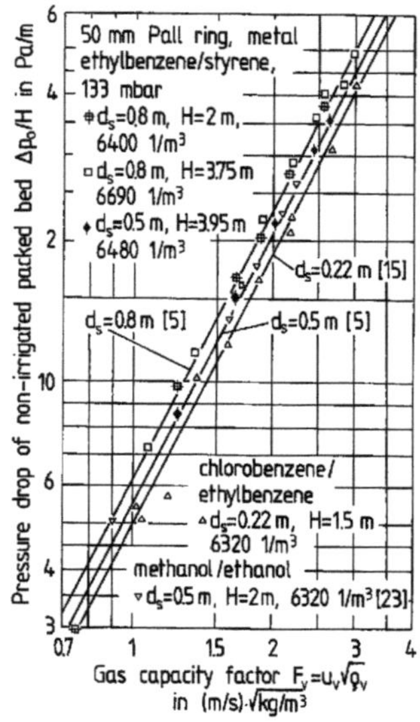

Figure 3-2. Influence of column diameter d_S on single-phase flow pressure drop $\Delta p_0/H$, valid for 50 mm metal Pall rings [8, 13]

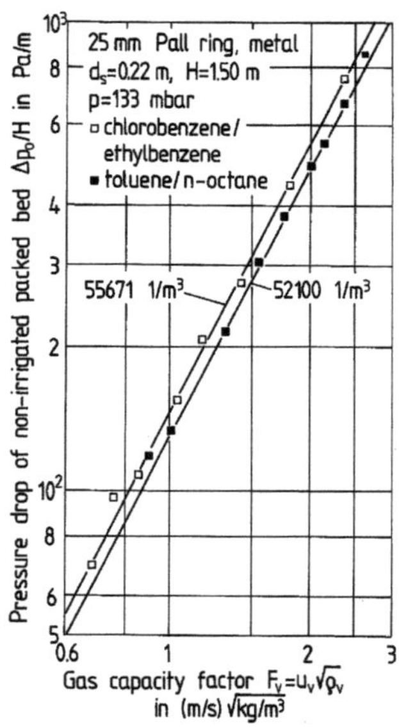

Figure 3-3. Influence of packing density on single-phase flow pressure drop, valid for 25 mm metal Pall rings [A]

range. The experimental values in relation to turbulent gas flow, in the range of $Re_V \geq 2100$, are given by Eq. (3-15a)

$$\psi = 3.23 \cdot Re_V^{-0.034} \quad for \quad Re_V \geq 2100 \tag{3-15a}$$

with a relative error of ±8%, which is sufficiently accurate for practical applications.

This takes into account all parameters that have a relevant influence on the resistance coefficient ψ.

Figure 3-4a. Resistance coefficient ψ as function of modified Reynolds number of vapour Re_V for 15–80 mm metal Pall rings, valid for various test systems, prepared on the basis of literature data [8, 13]

15 ÷ 80 mm Pall ring, metal					
TP	$10^3 \cdot d$ [m]	system	p_T [mbar]	d_S [m]	lit.
○	15	ethylbenzene / styrene	133	0.8	[13]
△	25			0.8	
▽	35			0.5	
◆	35			0.8	
⊕	50			0.8	
⊟	50			0.5	
◇	80			0.8	
▲	25	air	1000	0.15	[A]
▲	25			0.22	
▼	35			0.22	
□	50			0.3	
▼	35	methanol / ethanol	1000	0.5	[13]
▽	35	chlorobenzene / ethylbenzene	66.7	0.22	[A]

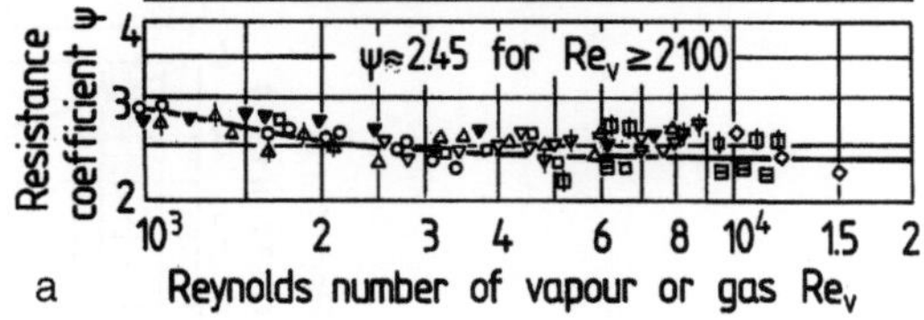

a

In the transition range, for $400 \leq Re_V < 2100$, the modified Reynolds number Re_V has a much stronger influence on the resistance coefficient ψ. In this range, the experimental data, plotted in Figs. 3-4a and 3-4b, are given by the empirical Eq. (3-15b):

$$\psi = 10 \cdot Re_V^{-0.18} \quad \textit{for} \quad Re_V < 2100. \tag{3-15b}$$

The influence of physical properties, operating pressure and constructive parameters on the resistance coefficient ψ is expressed sufficiently accurately by the law of resistance, according to Eq. (3-15a,b). Experiments based on air flow in packings are therefore sufficient to allow the law of resistance $\psi = f(Re_V)$ to be transferred to any type of system under vacuum, in the normal and high pressure range. Equation (3-15a) is applicable to the range of Reynolds numbers of $2100 \leq Re_V \leq 16000$ and diameter ratios of $d_S/d \geq 6$.

Figures 3-5 and 3-6 show the between the resistance coefficient ψ and the modified Reynolds number of the gas phase for Pall rings made of plastic and ceramic. For theses materials, the same correlations (3-15a,b) for the determination of the resistance coefficient are applicable.

According to Figs. 3-4a and 3-4b, the fully developed turbulent flow starts at $Re_V \cong$ 2100 to 10000. Based on the rule of thumb, this corresponds to a vapour capacity factor F_V of:

$$F_V \cong 0.8 \ ms^{-1}\sqrt{kgm^{-3}} \quad \text{for 50 mm Pall rings}$$

$$F_V \cong 1.0 \ ms^{-1}\sqrt{kgm^{-3}} \quad \text{for 35 mm Pall rings}$$

Figure 3-4b. Resistance coefficient ψ as function of modified Reynolds number of vapour Re_V for ceramic Pall rings, valid for various test systems [A]

15 ÷ 35 mm Pall ring, ceramic

TP	$10^3 \cdot d$ [m]	system	p_T [mbar]	d_S [m]	H [m]	N [1/m³]	lit.
▿	15	chlorobenzene/ethylbenzene	67	0.22	1.5	236000	[A]
✢	15		133	0.22	1.5	236000	
♦	25		33/67/133	0.22	1.5	53200	
o	25		133	0.22	1.5	55670	
ɸ	25		266	0.15	1.7	57180	
▵	35		67	0.22	1.45	19050	
•	25	toluene/n-octane	133	0.22	1.5	52000	[A]
▼	15	air, 293 K	1000	0.3	0.87	250000	[A]
⊖	25		1000	0.3	0.82	53500	
▲	35		1000	0.3	1.2	19650	
x	15	methanol/nitrogen	7150 - 8450	0.155	0.8	210000	[9]

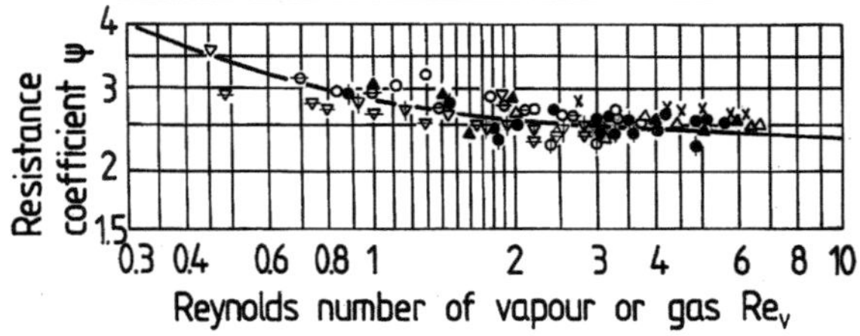

$$F_V \cong 1.2\ ms^{-1}\sqrt{kgm^{-3}} \quad \text{for 25 mm Pall rings}$$

In the case of larger packing elements, used for rectification, with diameters of 35, 50 and 80 mm, it can often be assumed that the column operates in the turbulent range and the resistance coefficient ψ is only marginally dependent on the Reynolds number Re_V of the vapour phase.

Figure 3-5. Resistance coefficient ψ as function of modified Reynolds number of vapour or gas Re_V for 25–50 mm plastic Pall rings, valid for various test systems [A]

25 ÷ 50 mm Pall ring, plastic

TP	$10^3 \cdot d$ [m]	system	p_T [mbar]	d_s [m]	H [m]	N [1/m³]
o	25	air, 293 K	1000	0.22	0.9	54500
•	25			0.3		55180
ɸ	25	toluene /n-octane	133	0.22	1.5	53050
⊕	25					55583
♦	25					58066
●	25	chlorobenzene/ethylbenzene	66.7	0.22	1.5	54300
▵	35	air, 291 K	1000	0.3	1.2	16284
▲	35			0.45		17140
▫	50			0.3	1.35	6425
▪	50			0.45	0.85	

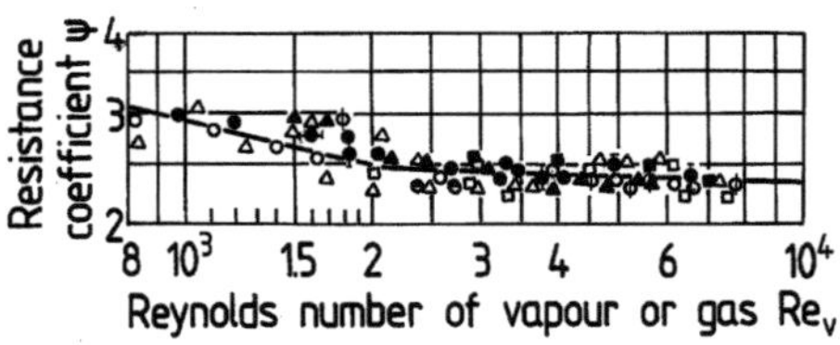

Figure 3-6. Resistance coefficient ψ as function of modified Reynolds number of vapour or gas Re_V, valid for: (**a**) 25 mm ceramic Pall rings [A]; (**b**) 50 mm ceramic Pall rings [A]

25 mm Pall ring, ceramic						
TP	$10^3 \cdot d$ [m]	system	p_T [mbar]	d_S [m]	H [m]	N [1/m³]
●	25	chlorobenzene/ ethylbenzene	66.7	0.22	1.47	36621

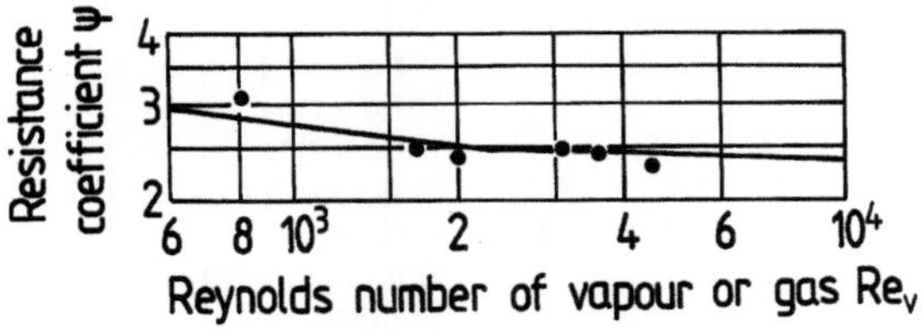

50 mm Pall ring, ceramic					
TP	$10^3 \cdot d$ [m]	system	d_S [m]	H [m]	N [1/m³]
▲	50	air 293 K, 1 bar	0.3	1	6450
△			0.4	1.1	6562
x			0.4	1	6042
⩚			0.3	1.1	6338
■			0.45	0.85	6026
□	50	water vapour	0.5	1	-

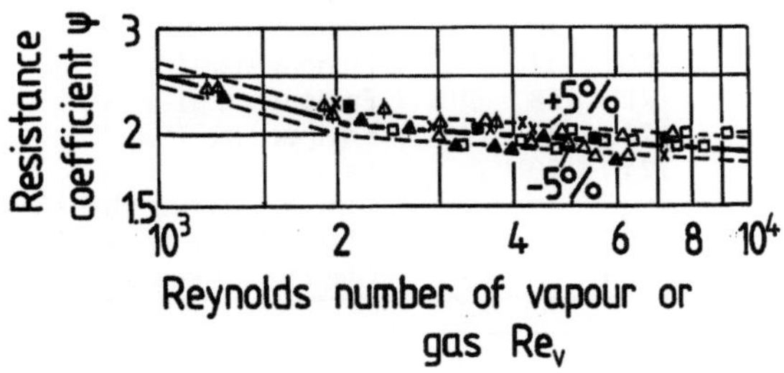

3.2.2
Determining the Resistance Coefficient for Other Random Packings

As shown in Fig. 3-7, the experimentally derived resistance coefficients ψ, applicable to different rectification systems, have been plotted against the Reynolds number of the vapour phase Re_V for a number of random packings. Figure 3-7 shows that the resistance coefficients ψ *decrease, as the Reynolds number* Re_V increases, and range between 0.65 and 9.

There are qualitatively similar curve progressions, comparable with the ones shown in Figs. 3-4, 3-5 and 3-6, where the resistance coefficient ψ for turbulent gas flow at $Re_V \geq 2100$ is only marginally dependant on the Reynolds number Re_V.

line	$10^3 \cdot d$ [m]	packing type	material
1	58	Hiflow ring	metal
1	28	Hiflow ring	metal
3	28	Hiflow ring	pp
4	50 ÷ 90	Hiflow ring	pp
5	20 ÷ 50	Hiflow ring	ceramic (1988)
6	75	Hiflow ring	ceramic
7	50	Hiflow ring	pp
8	50	Hiflow ring	pp
		Białecki ring	metal
9	25 ÷ 53	Pall ring	metal
		Glitsch CMR ring size 2A	pp
10	50	Białecki ring	metal, stacked
11	35	Białecki ring	metal, stacked
12	28 ÷ 50	NSW ring	plastic
13	15 ÷ 50	Raschig ring	ceramic
14	25 ÷ 50	Intalox saddle	ceramic
15a	38	Intalox saddle	pp
15b	50	Intalox saddle	pp
16	45	Hackette	pp
	58	Pall ring, VFF	metal
17	45	Top-Pak size 1	metal
	80	Top-Pak size 2	metal
18	45	Dtnpac size 1	pp
19	70	Dtnpac size 2	pp
20a	32	Envipac size 1	pp
20b	58	Envipac size 1	pp
20c	80	Envipac size 1	pp
21	50	Ralu ring	pp
22	-	Glitsch CMR ring size 2A	metal
		Glitsch CMR ring size 3A	metal
23	-	Glitsch CMR ring size 1A	pp

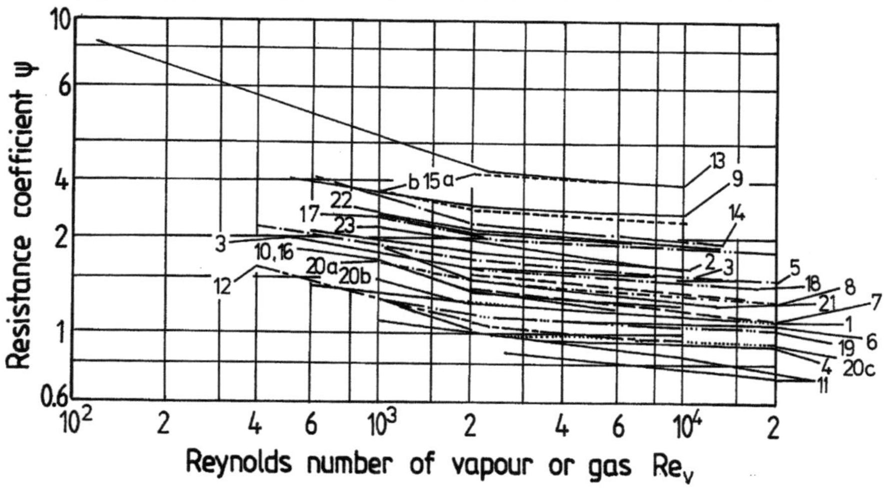

Figure 3-7. Resistance coefficient ψ as function of modified Reynolds number of vapour or gas Re_V, valid for various packing shapes, prepared on the basis of figures in the annex to Chap. 3

Hence, the principles valid for Pall rings, in relation to the influence of packing density N, system, operating pressure and column diameter d_S, are also applicable here. The spread of the experimental values around the plotted curve is given as $\delta(\psi) \leq 5-10\%$.

Figure 3-7 also shows the dependency between the resistance coefficient ψ and the size of the packing elements. This correlation can be expected, particularly if the individual packing elements are of the same type but differ in terms of the open wall area of the individual elements. In the case of Pall rings, the proportion of open wall area is approx. 28%, regardless of the packing diameter. For metal Hiflow rings, the proportion is 44.8%, for $d = 0.028$ m, and 65.7%, for a ring diameter of $d = 0.058$ m. Hence, the resistance coefficient ψ for 58 mm Hiflow rings is smaller than the value for the same type of packing with a diameter of just 28 mm, see Sect. 3.3.

For metal Białecki rings [6], the following correlations for determining the resistance coefficient ψ were found:

$$\psi = 10.17 \cdot Re_V^{-0.17} \quad \text{for} \quad Re_V < 2100 \tag{3-16a}$$

$$\psi = 4.13 \cdot Re_V^{-0.0522} \quad \text{for} \quad Re_V \geq 2100 \tag{3-16b}$$

The resistance coefficients for Białecki rings are marginally higher than those applicable to Pall rings. This is due to the fact that the open area in the packing is approx. 23% for Białecki rings, which is lower compared to Pall rings (approx. 28%).

Tables 3-1, 3-2 and 6-1a-c list the numerical values for the constants K_1, K_3 and the exponents K_2, K_4 of Eq. (3-14) for determining the resistance coefficients for a number of random and structured packings. The tables can be found at the end of the respective chapters. The value pairs K_3, K_4 for Eq. (3-14) are applicable in the turbulent flow range, for $2100 \leq Re_V \leq 20000$, whereas the values K_1, K_2 are valid for the transition range, for the Reynolds numbers $400 < Re_V < 2100$.

3.2.3 Determining the Resistance Coefficient ψ for Structured Packings

Influence of the Column Diameter

Experiments using columns with diameters of $d_S = 0.22$ m, 0.3 m and 0.45 m, and using air flow in non-perforated sheet-metal packing produced by Montz, type B1-300Y, as well as Ralu-Pak 250 YC with slit perforation, have shown that the column diameter d_S has a significant influence on the pressure drop $\Delta p_0/H$ of the dry packing [14, 15, 18, 25]. The results can be found in Fig. 3-8a,b. Figure 3-8c shows the pressure drops of the dry, perforated Mellapak 250Y as a function of the gas capacity factor F_V, relating to columns with diameters of $d_S = 0.15$ m [18], $d_S = 0.22-0.3$ m [A] and $d_S = 1.0$ m [25]. According to Fig. 3-8a, the pressure drop $\Delta p_0/H$ in the column equipped with Montz packing B1-300 is reduced by 25%, if the column diameter is increased from 0.22 m to 0.45 m, whilst the gas capacity F_V remains the same. The difference in pressure drop $\Delta p_0/H$ shown in Fig. 3-8c is even more noticeable. The pressure drop in the column

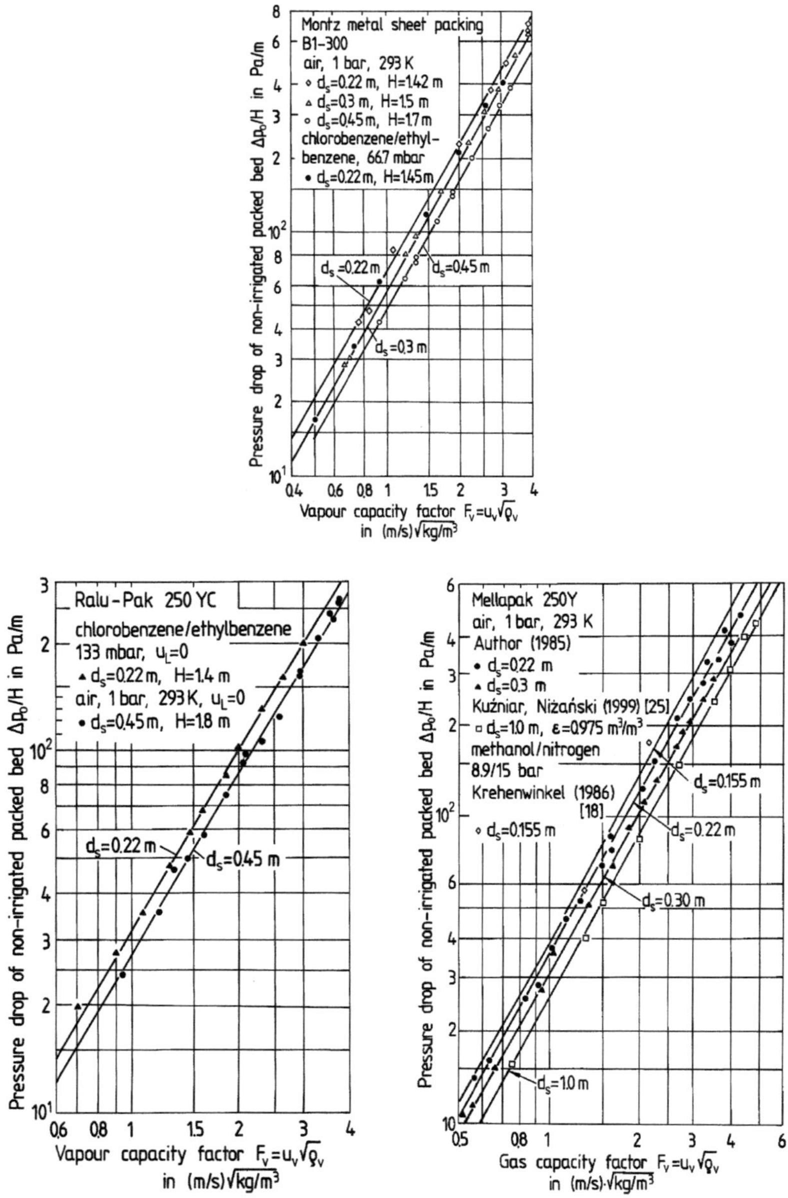

Figure 3-8. Influence of column diameter d_S on single-phase flow pressure drop $\Delta p_0/H$, valid for: (**a**) sheet-metal packing produced by Montz type BI-300 [15], (**b**) sheet-metal packing type Ralu-Pak 250 YC, produced by Raschig [15], (**c**) sheet-metal packing type Mellapak 250 Y, produced by Sulzer [18, A, 25], (**d**) schematic representation of the installation of structured packings in columns with small and large diameter

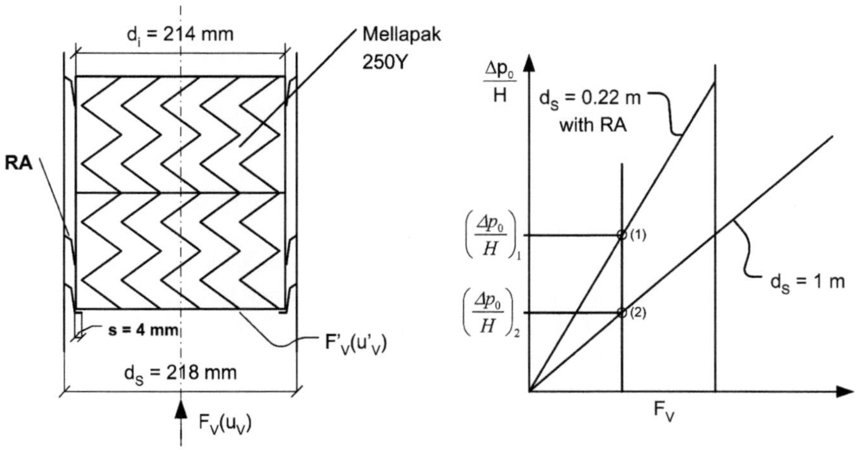

$d_S = 0.218$ m with RA | **$d_S = 1$ m without RA**

$$\left(\frac{\Delta p_0}{H}\right)_1 = \psi_1 \left(\frac{1-\varepsilon}{\varepsilon^3}\right) \frac{F_V'^2}{d_P} \qquad F_V' \neq F_V$$

$$\left(\frac{\Delta p_0}{H}\right)_2 = \psi_2 \left(\frac{1-\varepsilon}{\varepsilon^3}\right) \frac{F_V^2}{d_P} \qquad F_V' = F_V$$

$$m = u_V' \cdot \rho_V \cdot (d_i - 2s)^2 \frac{\pi}{4} = u_V \cdot \rho_V \cdot d_S^2 \frac{\pi}{4} \qquad \bigg/ \left(\frac{\pi}{4} \cdot \sqrt{\rho_V}\right)$$

$$F_V' \cdot (d_i - 2s)^2 = F_V \cdot d_S^2$$

$$F_V' = F_V \frac{d_S^2}{(d_i - 2s)^2} = F_V \frac{1}{\left(\frac{d_i - 2s}{d_S}\right)^2} = F_V \frac{1}{\varphi_1}$$

$$\varphi_1 = \left(\frac{d_i - 2s}{d_S}\right)^2 = \left(\frac{214 - 2 \cdot 4}{218}\right)^2 = 0.893$$

$$\left(\frac{\Delta p_0}{H}\right)_1 = \psi_1 \left(\frac{1-\varepsilon}{\varepsilon^3 \cdot d_P}\right) \frac{F_V^2}{\varphi_1^2}$$

$$\left(\frac{\Delta p_0}{H}\right)_2 = \psi_2 \left(\frac{1-\varepsilon}{\varepsilon^3 \cdot d_p}\right) \frac{F_V^2}{\varphi_2}$$

$$\frac{\Delta p_{0,1}}{\Delta p_{0,2}} = \frac{\psi_1}{\varphi_1^2} \cdot \frac{\varphi_2^2}{\psi_2} = \frac{\psi_1}{\psi_2} \cdot \frac{1}{\varphi_1^2} = const. \quad for \quad \varphi_2 \approx 1$$

$$\Rightarrow \qquad \psi_2 \sim \frac{\psi_1}{\varphi_1^2}$$

Figure 3-8. (continued)

equipped with Mellapak 250Y is reduced by 35%, if the column diameter is increased from $d_S = 0.155$ m to $d_S = 1$ m.

In contrast to randomly filled packing elements, the pressure drop for single gas flow $\Delta p_0/H$ is generally reduced here as the column diameter d_S is increased, whilst the gas capacity in the packing system remains the same. It follows from this that the resistance coefficient ψ is dependent on the column diameter d_S in packed columns up to a certain diameter d_S. According to the manufacturer Sulzer [16], it is only after the column exceeds a diameter of $d_S \cong 0.5-1.0$ m that the resistance coefficient ψ no longer changes. The experimental results, presented in Fig. 3-8a–c, confirm these findings.

Figure 3-8c also indicates that the single-phase flow pressure drop $\Delta p_0/H$ in large columns remains the same, even if the gas capacity factors F_V are significantly increased. This leads to a change in the fluid dynamic characteristics of each packed column. Based on Eq. (2-69), the gas velocities at the flooding point $u_{V,Fl}$ of these columns are higher than those in smaller-diameter columns.

A larger plant with $d \approx 1$ m would therefore be more suitable for the experimental determination of the law of resistance for structured packings, based on the assumption that $\psi \neq f(d_S)$. The wall factor K, according to Eq. (3-8) and applied in Eq. (3-14), for considering the influence of the wall on the resistance coefficient, does not yield the correct results in the case of structured packings.

The higher pressure drop $\Delta p_0/H$, which occurs in smaller columns with the same gas capacity factor F_V, is caused by the seal rings at the individual packing elements. They are installed between the column wall and the packing elements, in order to prevent the liquid in two-phase flow operation from flowing towards the column wall by channelling it towards the centre of the column, see Fig. 3-8d. The possibility of predicting the law of resistance in larger-diameter columns is therefore of great practical value when designing columns with structured packing, if the experimental results on the pressure drop for single-phase flow in smaller plants are available.

This task can be approximately solved, based on the example of Mellapak 250Y, by correcting the resistance coefficient ψ, in accordance with the model presented in Fig. 3-8d. This is based on the assumption that, for the same gas flow $\dot{V}$, the seal rings have practically no influence on the pressure drop in large columns with a diameter of $d_{S,2}$, whereas in smaller columns with a diameter of $d_{S,1}$, the single-phase flow pressure drop $\Delta p_0/H$ is the same, even if the gas capacity factors $F_{V,1}$ are lower than those in larger columns. Hence, the following correlation applies approximately to Mellapak 250Y, according to Eq. (3-17):

$$\psi_A \cong \psi_1 \cdot \frac{\varphi_2^2}{\varphi_1^2} \tag{3-17}$$

where

$$\varphi_1 = \left(\frac{d_{i,1} - 2s}{d_{S,1}}\right)^2 \text{ and } \varphi_2 = \left(\frac{d_{i,2} - 2s}{d_{S,2}}\right)^2. \tag{3-18}$$

φ_1 is the ratio of the effective area of the individual packing elements, minus the area of the seal ring, to the cross-sectional area of the smaller test column. An analogous correlation applies for φ_2, in relation to the larger column with $d_{S,2} > d_{S,1}$.

The substitution of the constructive data for Mellapak 250Y and columns for $d_{S,1} = 0.22$ m, as shown in Fig. 3-8c, into Eq. (3-18) gives the conversion formula (3-19), applicable to Mellapak 250Y, for calculating the predicted resistance coefficient for large columns ψ_A:

$$\psi_A \underset{d_S \to 1}{=} \psi_1 \cdot \frac{\varphi_1^2}{\varphi_2^2} \underset{d_S \to 1}{=} \frac{\psi_1}{1.26 \cdot \varphi_2^2} = 0.794 \cdot \frac{\psi_1}{\varphi_2^2} \tag{3-19}$$

The numerical Examples 3.2 and 4.2 illustrate how the approximation correlations (3-17) to (3-19) for converting the resistance coefficient and determining the pressure drop of Mellapak 250 Y can be applied to larger columns. In order to apply Eqs. (3-17), (3-18) and (3-19) to other packings, the geometric data of the respective packing, seal ring and column must be known.

Hence, due to the installation of seal ring in the test columns, the fluid dynamic characteristics of structured packings can only be transferred to other column diameters from $d_S \geq 1$ m.

Influence of Perforation and Flow Channel Angle in Packing Elements on the Single Flow Pressure Drop

The pressure drop of the dry sheet-metal Montz packing is shown in Fig. 3-8e for type B1-300Y, and in Fig. 3-8f for type B1-500Y. The figures indicate that the influence of the perforation in the individual packing elements increases with the specific packing surface area a. This results in an enhanced separation efficiency of the packing. Whilst the differences in the pressure drop for the B1-300Y version are small, they are more noticeable for the B1-500Y version. At the same time, experiments were carried out to determine the separation efficiency of non-perforated and perforated packings, types B1-300Y and B1-500Y, which showed that the packing with the higher separation efficiency also had the higher pressure drop. Due to the small differences in the pressure drop of the B1-300Y packings, the differences in the separation efficiency were also smaller. Investigations carried out on the B1-500Y packing show that the separation efficiency of the perforated B1-500Y packing in the main operating range was found to be higher by a factor ≈ 2, compared to the non-perforated packing. The maximum loading capacity of the packing with the lower pressure drop was higher, as would be expected, according to Eq. (2-69). Using the above equation, it is also possible to predict the increase of the maximum loading capacity of type X packings with flow channel angle α of 30°. In addition to the pressure drop of the type BX gauze packing, Fig. 3-8f also shows the relevant data for the type X sheet-metal packing with perforation. The difference in the pressure drop of the dry packing between type B1-500Y, type X of the same packing and the BX-500 packing is significant. This manifests itself in the loading capacity and the separation efficiency of the individual packings.

Law of Resistance

The correlation between the resistance coefficient ψ and the Reynolds number Re_V, with the wall factor $K = 1$, is shown in Fig. 3-9 for non-perforated, and Fig. 3-10 for perforated

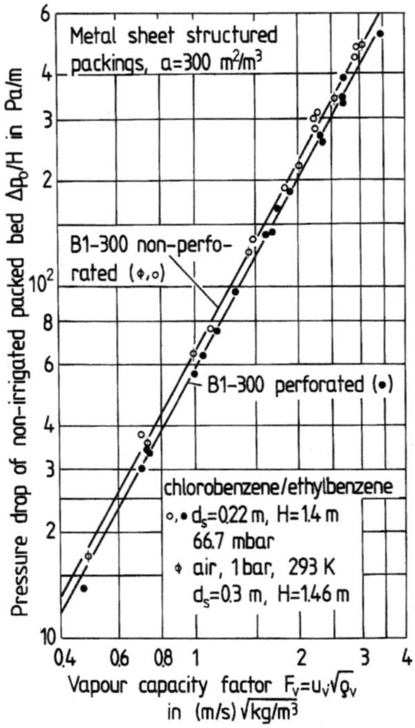

Figure 3-8e. Influence of perforation on single-phase flow pressure drop of structured packings, type B1-300, produced by Montz

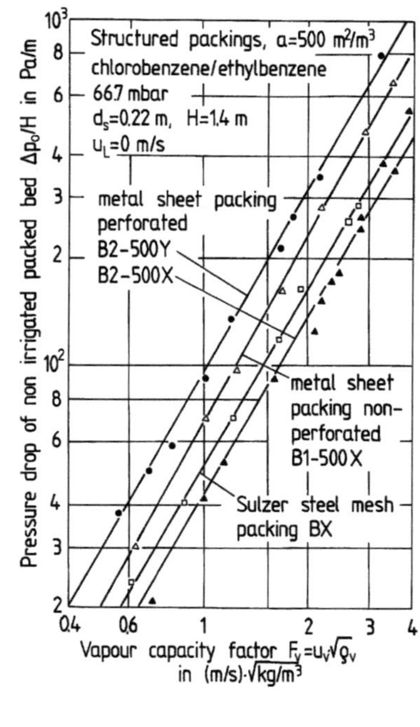

Figure 3-8f. Influence of perforation on single-phase flow pressure drop of structured packings, type B1-500, produced by Montz with various flow channel angles and materials (sheet metal, steel mesh)

Figure 3-9. Resistance coefficient ψ as function of Reynolds number of vapour or gas Re_V, valid for non-perforated Montz packing

Non-perforated structured packings						
TP	packing	type	system	p_T [mbar]	d_S [m]	H [m]
◆	Montz packing, pp	C1-200	air, 293 K	1000	0.3	1.42
▲	Montz packing, metal sheet	B1-300	chlorobenzene/ ethylbenzene	33	0.22	1.4
▲				66.7		
▲				133		
△			air, 293 K	1000	0.22	1.4
△					0.30	1.48
△					0.45	1.48
▽		B1-200	chlorobenzene/ ethylbenzene	66.7	0.22	1.4
▼			air, 293 K	1000	0.3	1.6

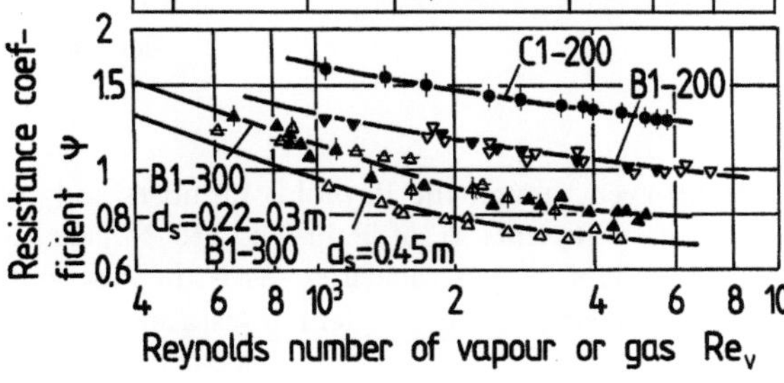

structured packings. In the case of the perforated packings, Mellapak 250 Y and Montz B2-500, the size of the packing surface was not found to have any influence on the resistance coefficient ψ, see Fig. 3-10.

Despite the fact that the specific packing areas of Ralu-Pak 250 YC and Mellapak 250 Y are equal, Figs. 3-10 and 3-11 show that the resistance coefficients are

Figure 3-10. Resistance coefficient ψ as function of Reynolds number of vapour or gas Re_V, valid for perforated packings produced by Montz and Sulzer

Perforated structured packings						
TP	packing	type	system	p_T [mbar]	d_S [m]	H [m]
□	Montz packing, metal sheet	B2-500	chlorobenzene/ ethylbenzene	33	0.22	1.4
■				66.7		
■				133		
●	Mellapak (Sulzer)	250YC	air, 293 K	1000	0.22	1.2
○			chlorobenzene/ ethylbenzene	66.7		

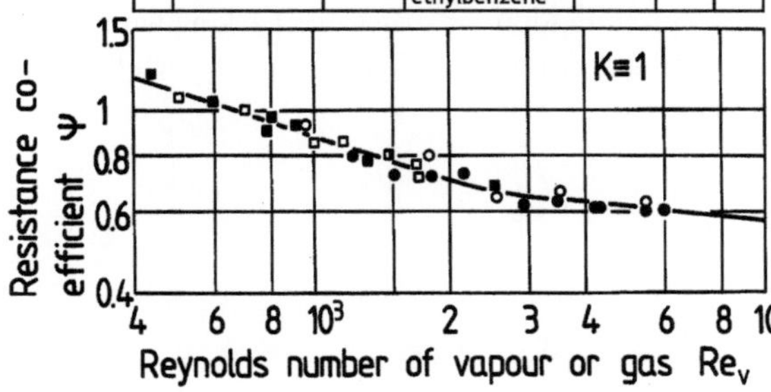

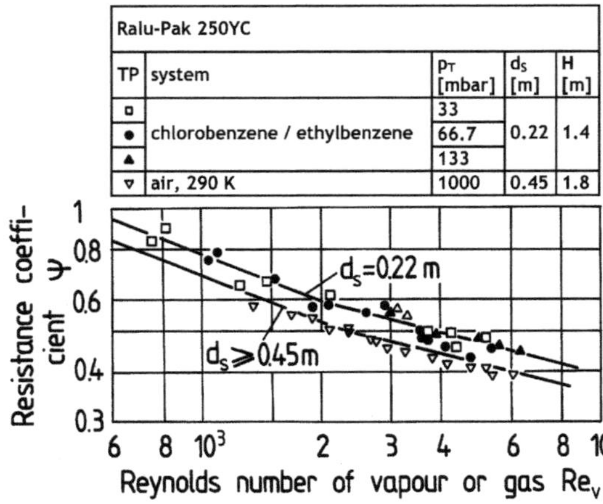

Figure 3-11. Resistance coefficient ψ as function of Reynolds number of vapour or gas Re_V, valid for slit Ralu-Pak 250 YC made of sheet metal, produced by Raschig

different, even though the Reynolds numbers and columns sizes are the same. The resistance coefficients ψ of Ralu-Pak 250 YC are smaller than those of Mellapak 250 Y, for the same gas capacities. As is the case with randomly filled packings, the physical properties of the system investigated have no influence on the resistance coefficient.

Table 3-2 contains a list of the values $K_1 - K_4$ required for calculating the resistance coefficient ψ for structured packings, according to Eq. (3-14), which were taken at columns with different diameters d_S.

3.3
Introducing a Channel Model Based on Partially Perforated Channel Walls

The aim of this work is to determine the law of resistance that applies to classic packing elements as well as modern, lattice packings, which are becoming increasingly popular for practical use, and structured packings as well as stacked packing elements and tube columns, in order to perform the fluid dynamic design of packed columns as accurately as possible.

A standardised description of the pressure drop in packings can be developed in analogy with gas flow in channels [1–9, 11], on the basis of Eq. (3-2).

In a given random or structured packing with a diameter of d_h, acc. to Fig. 3-12 and Eq. (3-2), a pressure drop $\Delta p_{0,l}$ occurs in each flow channel with a length of l, which is given as:

$$\Delta p_{0,l} = \frac{\lambda_l}{2} \cdot \frac{F_V^2}{d_h} \cdot l \tag{3-20}$$

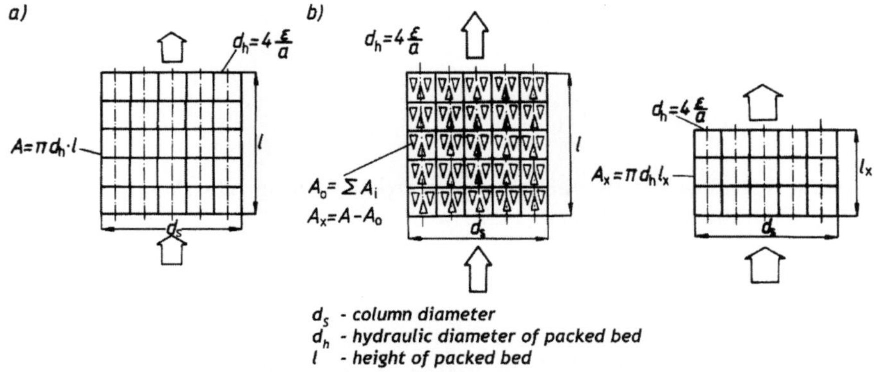

Figure 3-12. Hydraulic model of packed bed as bundle of channels for single-phase flow: (**a**) in channel of the length l, (**b**) in channel of the length l_x for packings with partially perforated wall

Friction forces are generated as a result of gas flow around the channel walls. If the walls are partially perforated, as is the case with Pall or Białecki rings as well as lattice-type packings, the resistance coefficient λ_l in Eq. (3-20) is lower, in comparison to channels with a non-perforated structure.

If one assumes, however, that the same flow model [34, 35] is valid for any type of packing with the same law of resistance, then the decrease in the pressure drop $\Delta p_{0,lx}$ for perforated packing elements with identical channel diameters d_h can only be due to shorter channel lengths l_x.

Furthermore, if the wall surface of a channel is perforated, the length of the perforated channel l_x is shorter, compared to non-perforated packing elements, which results in a pressure drop $\Delta p_{0,lx}$, see Fig. 3-12.

The form factor φ_P represents the proportion of the perforated surface area of a packing element:

$$\varphi_P \equiv \frac{A_0}{A}, \tag{3-21}$$

where A_0 represents the perforated wall surface area and $A = \pi \cdot d \cdot h$ is the total non-perforated wall surface area of a packing element.

For simplification purposes, it also assumed that the influence of deflections within the packing element on the pressure drop is negligible, compared to the pressure drop in channels d_h. Acc. to Fig. 3-12, the ratio of the pressure drops $\Delta p_{0,lx}$ and $\Delta p_{0,l}$ in relation to the gas flow through channels with a diameter of d_h can be equated to the ratio of the equivalent channel lengths l_x/l, assuming the gas capacity factor F_V is the same, Eq. (3-22):

$$\frac{\Delta p_{0,l_x}}{\Delta p_{0,l}} = \frac{l_x}{l}, \tag{3-22}$$

This ratio, in turn, can be equated to the proportion of the non-perforated packing surface area (l-φ)

$$\frac{l_x}{l} = 1 - \varphi_P \quad \rightarrow \quad l_x = l \cdot (1 - \varphi_P) \tag{3-23}$$

Hence, it is possible to derive the following correlation for the ratio $\Delta p_{0,lx}/\Delta p_{0,l}$, based on Fig. 3-12b and Eqs. (3-22) and (3-23):

$$\left(\frac{\Delta p_{0,l_x}}{\Delta p_{0,l}}\right)_{F_V, d_h = const.} = 1 - \varphi_P \tag{3-24}$$

Equations (3-24), (3-2) and (3-8) now lead to Eq. (3-25), describing the pressure drop Δp_0 which occurs during channel flow in a packing with a height H, if the channel walls have a partially perforated or fully closed structure:

$$\frac{\Delta p_0}{H} = \psi_0 \cdot (1 - \varphi_P) \cdot \frac{1 - \varepsilon}{\varepsilon^3} \cdot \frac{F_V^2}{d_p \cdot K} \quad \left[Pam^{-1}\right], \tag{3-25}$$

which results in the following equation:

$$\psi = \psi_0 \cdot (1 - \varphi_P) = \frac{K_A}{Re_v} + K_B, \tag{3-26}$$

where ψ_0 represents the resistance coefficient of flow channels with non-perforated walls and ψ represents the resistance coefficient of flow channels with perforated walls [34, 35].

3.3.1 Determining the Resistance Coefficient for Non-perforated Packing Elements

Most of the recent experimental data which is currently available relates to metal Pall rings, where the individual packing element, Fig. 1-2a, has a partially perforated wall, hence $\varphi_P > 0$.

The evaluation of the experimental data taken by Maćkowiak [1, 2], Krehenwinkel and Knapp [9] as well as Billet [12], using metal Pall rings, led to the following correlation for the product $\psi_0 \cdot (1 - \varphi_P)$ in Eq. (3-26), acc. to Figs. 3-4a and 3-4b:

$$\psi_{Pall\ rings} = \frac{K_A}{Re_V} + K_B = \frac{522.4}{Re_V} + 2.306, \tag{3-27}$$

The constants $K_A = 522.4$ and $K_B = 2.306$ in Eq. (3-27) were determined using the minimisation procedure.

Based on a Pall ring size of d = 0.015–0.080 m, the proportion of the perforated surface area φ is given by the arithmetic mean numerical value of $\varphi_P \approx 0.28$, which, acc. to Eqs. (3-25) and (3-27), results in the following correlation for the resistance coefficient ψ_0 in relation to gas flow through non-perforated packing elements [34, 35]

$$\psi_0 = \frac{725.6}{\mathrm{Re}_V} + 3.203. \tag{3-28}$$

The concurrence between the experimental resistance coefficients for non-perforated packings with $\varphi_P = 0$, such as ceramic Raschig rings, and the calculation acc. to Eq. (3-28) is sufficiently good for practical applications, see Fig. 3-30.

3.3.2 Determining the Pressure Drop in Single-Phase Flow – Final Equation

Based on the value $\varphi_P \approx 0.28$, which is valid for metal Pall rings, Eqs. (3-28) and (3-25) now lead to the final equation for determining the pressure drop in single-phase flow through random and structured packings [34, 35]:

$$\frac{\Delta p_0}{H} = \left(\frac{725.6}{\mathrm{Re}_V} + 3.203\right) \cdot (1 - \varphi_P) \cdot \frac{1-\varepsilon}{\varepsilon^3} \cdot \frac{F_V^2}{d_p \cdot K}; \tag{3-29}$$

To apply this equation, which is based on the evaluation of experimental data and valid in the range of Reynolds numbers $\mathrm{Re}_V \in (200\text{–}20000)$, it is necessary to ascertain the geometric packing data a and ε as well as the proportion of the perforated surface area of the packing wall φ_P. This parameter is a form-specific packing factor, also known as **form factor φ_P.**

The numerical values of the form factors φ_P have been compiled in Tables 6-1a and 6-1b for a number of random packings, including: Pall rings, VSP rings, Hiflow rings, Białecki rings, I-13 rings, Intalox made of metal, Envipac, Hackette, Mc-Pac, R-Pac and other types of packings made of metal, plastic and ceramic.

The same table also contains the numerical values of the form factors $\varphi_{P,\mathrm{exp}}$ for more than 200 types of packings, which were determined by evaluating experimental pressure drop data relating to packings with single-phase flow.

3.3.3 Evaluation of Results

(a) In the case of the tested classic, ring-type packing elements made of metal and plastic, the concurrence between the calculated values $\varphi_{P,\mathrm{calc}}$ and the experimentally derived form factors $\varphi_{P,\mathrm{exp}}$, based on Tables 6-1a and 6-1b, is very good.

For metal Białecki rings, however, the experimentally derived form factors $\varphi_{P,\mathrm{exp}}$ were marginally smaller than the calculated values, as the proportion of the triangular cut-outs in the wall of the individual packing elements, which are bent inwards, provide additional resistance for the gas flow [34, 35].

In the case of modern, lattice-type packings, such as Hiflow, Nor-Pac, VSP rings, PSL rings, R-Pac and Mc-Pac, the concurrence between the calculated values and the experimentally derived form factors $\varphi_{P,exp}$ was also found to be satisfactory, as illustrated by the numerical example.

(b) For some ring-type packing elements, such as metal Cascade Mini Rings (CMR), small 18 mm Ralu rings and 15 mm Hiflow rings or Glitsch Pall rings, the experimentally determined form factors were considerably higher than the calculated values.

These findings can be explained by looking at the results of the pressure drop investigations for 30 mm metal Glitsch Pall rings, acc. to Fig. 3-13, which are referred to in Table 6-1a as 30 mm without collars (oK) and with collars (mK).

Subject to the investigation were two types of identical-sized rings with d = 30 mm, the only difference consisting in the fact that one type of 30 mm Pall ring had an exterior distance collar. This minor form difference led to a considerable difference in the packing density N between both types of packing elements, resulting in a pressure drop reduction. Hence, the resistance coefficients were substantially lower, see Fig. 3-13. Another result of this investigation was the fact that the packing density N was reduced by applying a distance collar to the individual packing ring, thus creating additional void space within the packing. This, in turn, led to a reduction in the pressure drop and an increase in the operating capacity of the packing element.

This was confirmed by studies using CMR Glitsch rings, where the effective form factor $\varphi_{P,exp} = 0.437$ exceeded the calculated value $\varphi_{P,calc} = 0.223$ approximately by a factor of 2 [34, 35].

The same result was obtained by comparing the calculated and experimental form factors for small 15 mm Hiflow rings and 18 mm Ralu rings made of plastic, which also had collars.

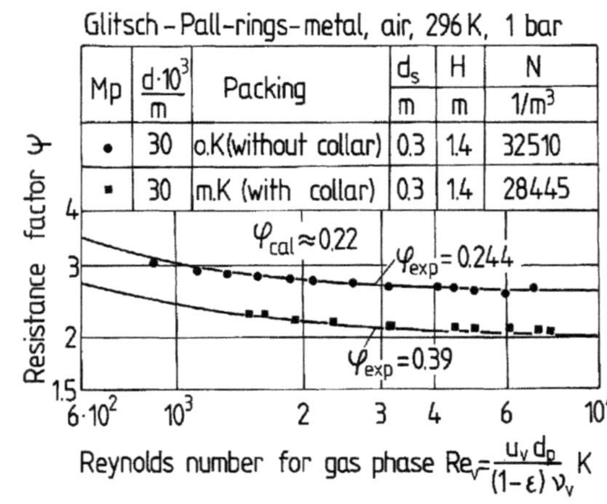

Figure 3-13. Resistance coefficient ψ as function of Reynolds number of vapour or gas Re_V, valid for: (**a**) 30 mm Glitsch Pall ring o.K - without collars: $\varphi_{P,cal} = 0.224$ and $\varphi_{P,exp} = 0.214$, (**b**)·30 mm Glitsch Pall ring m.K – with collars: $\varphi_{P,cal} = 0.214$ and $\varphi_{P,exp} = 0.390$

3.4 Conclusions Chapter 3

1. Random Packings

1. The influence of the system on the resistance coefficient ψ is sufficiently described for practical applications by the kinematic viscosity, expressed in the modified Reynolds number. In addition, it is sufficient to determine the function $\psi = f(\mathrm{Re}_V)$ based on simulation tests with air under ambient conditions, in order to then calculate the pressure drop $\Delta p_0/H$ of the dry packing, using Eq. (3-8), accurately enough for rectification systems under vacuum and normal pressure and for the high pressure range, as well as for any type of absorption process.
2. The analysis of the results presented in this work, in relation to the resistance coefficient ψ for different types of packings, leads to the following conclusion: If the individual packing elements of one type of packing all have the same amount of open wall area, as is the case for metal Pall rings, then the law of resistance $\psi = f(\mathrm{Re}_V)$, applicable to one packing size, is also valid for the other packings dimensions. Otherwise, the correlation $\psi = f(\mathrm{Re}_V)$ would have to be determined separately for each packing size of the same type of packing, as it would be impossible to predict.
3. Based on a simple extended channel model [34, 35], it was possible to develop a description of the dry pressure drop in random and structured packings. It was assumed that a pressure drop only occurs during gas flow through non-perforated flow channels. In the case of perforated packing elements, which are mostly used today, the proportion of deflections and cut-outs in the wall of the individual packing elements φ_P can be calculated, which makes it possible to estimate the channel length.
   Examples of this type of packing are ring-type packing elements, such as Pall rings, Białecki rings, VSP rings, Hiflow rings, PSL rings, I-13 rings, Mc-Pac, SR-Pac, R-Pac, Ralu-Flow and metal Intalox saddles IMTP.
4. The pressure drops $\Delta p_0/H$ of the dry packing, based on experiments with randomly filled as well as stacked packing elements, can be described by the generally valid Eq. (3-25):

$$\frac{\Delta p_0}{H} = \psi_0 \cdot (1 - \varphi_P) \cdot \frac{1-\varepsilon}{\varepsilon^3} \cdot \frac{F_V^2}{d_p \cdot K} \quad \left[Pam^{-1}\right],$$

where ψ_0 represents the resistance coefficient for non-perforated packing elements, acc. to Eq. (3-28)

$$\psi_0 = \frac{725.6}{\mathrm{Re}_V} + 3.203$$

which is a function of the modified Reynolds number Re_V. Hence, φ_P is a form factor, representing the perforated surface area of the wall of an individual packing element.

The technical data of the relevant packing elements, a and ε, and the experimentally determined form factors $\varphi_{P,exp}$ are listed in Tables 6-1a and 6-1b.

In order to classify a packing and calculate the pressure drop $\Delta p_0/H$, acc. to Eq. (3-25), it is necessary to ascertain the geometric packing data a and ε as well as a third variable, which is the packing factor φ_P. To determine the pressure drop $\Delta p_0/H$ for complicated types of random and structured packings as accurately as possible, it is recommended to determine the form factor φ_P by experiment, as shown in Tables 1a, 1b and 1c.

Based on the Eq. (3-25), it is therefore possible to predict the pressure drop of dry packed beds simply by ascertaining the geometric data a and ε as well as the shape of the respective packing element (φ_P-factor) without conducting any experimental studies. This is a significant result, in particular for new designs as well as the modification of existing types of packings.

Finally, it can be concluded that it is possible, based on the simple model structure presented in this book, to predetermine the pressure drop $\Delta p_0/H$ in single-phase flow for given systems and operating conditions purely by ascertaining the geometric data of the packing element, a and ε, and the shape of the packing (φ_P-factor), in accordance with Eq. (3-29). In the case of geometrically simple, ring-type packing elements without any additional exterior distance collars, the form factor φ_P can be estimated with sufficient accuracy for practical applications without performing any experiments, see numerical Example 1.

Furthermore, a comparison between the calculated and experimental form factors shows that by applying collars to the individual packing elements, it is possible to create additional free packing surface area, thus reducing the packing density N. As a result, the experimentally determined form factors φ_P are found to be higher than the calculated values.

Depending on the type and design of the packing, this can lead to a doubling of the form factor values φ_P, as was reported for Glitsch CMR and Pall rings, Ralu rings made of plastic and 15 mm plastic Hiflow rings, see Table 6-1a.

The gas flow for Reynolds numbers $Re_V \geq 2100$ is practically turbulent. In this range, the resistance coefficient ψ is only marginally dependent on the Reynolds number Re_V. In practical applications, columns equipped with larger-diameter packing elements of $d > 0.025$ m are always operated in the turbulent range.

The numerical Examples 3.3 and 3.4 illustrate the correlations presented above.

5. During the course of this work, the resistance coefficients ψ for a number of modern random and structured packings, shown in Fig. 1-2b, were determined using various test systems.

   The numerical values of the resistance coefficients ψ for turbulent gas flow are in the following ranges for packed columns used today:

$\psi \approx 0.1-0.3$	for structured packings, plate or sheet-metal packings, stacked packings and tube columns
$\psi \approx 0.4-1$	for sheet-metal packings (Sulzer packing, Ralu-Pak, Montz-Pack)
$\psi \approx 1$	for random, perforated packing elements (Nor-Pac, Hiflow rings, Envipac, VSP, Hackette, Ralu-Flow) made of metal and plastic
$\psi \approx 1.5$	for ceramic Hiflow rings and metal Mc-Pac

$\psi \approx 2$	for Top-Pack, VSP metal, Intalox of metal and ceramic
$\psi \approx 2.42$	for Pall rings made of metal and plastic
$\psi \approx 2.62$	for Białecki rings made of metal and plastic
$\psi \approx 3.1$	for ceramic Raschig rings

Extensive studies on the pressure drop in single-phase flow in packings have shown that the resistance coefficient ψ in Eq. (3-8) for column diameters of $d_S \leq 0.8 \div 1.0$ m is not only dependent on the shape of the packing and the flow range, but is also a function of the column diameter d_S. This is due to the seal rings built into the individual packing elements, which reduce the free surface area available for gas flow. In the case of larger column diameters of around $d \geq 1$ m, the influence of the seal rings on the resistance coefficient is negligible.

The dry packing form factor φ_P data are listed in Table 6-1a,b.

2. Structured Packings

1. In contrast to random packings, see Fig. 3-2, the pressure drop $\Delta p_0/H$ of dry structured packings is reduced, as the column diameter d_S increases, see Fig. 3-8a–c. This is due to the fact that the influence of seal rings in large columns becomes less significant, and is practically negligible from $d_S \geq 1$ m. The experimental pressure drop data for small columns can be up to 65% higher than for large columns. This is why it is very difficult to predict the pressure drop and the maximum loading capacity of packed columns, based on experiments with small test columns. Instead, it requires a method such as the one described above for Mellapak 250Y, shown in Fig. 3-8d.
2. The introduction of the wall factor K into the Reynolds number Re_V, according to Eqs. (3-10) and (3-25), for determining the pressure drop $\Delta p_0/H$ is not justified. Instead, $K = 1$ applies. The Eq. (3-25) for calculating the pressure drop $\Delta p_0/H$ is therefore as follows for structured packings:

$$\frac{\Delta p_0}{H} = \psi_0 \cdot (1 - \varphi_P) \cdot \frac{1-\varepsilon}{\varepsilon^3} \cdot \frac{F_V{}^2}{d_P} \quad \left[Pam^{-1}\right]$$

where

$$\psi_0 = f(\varphi_P, \mathrm{Re}_V), \quad \mathrm{Re}_V = \frac{u_V \cdot d_P}{(1-\varepsilon) \cdot \nu_V} \quad \text{and} \quad K = 1$$

Packing form factors φ_P are listed in Table 6-1c.
3. The resistance coefficient ψ of structured packings also depends on whether they are perforated, non-perforated or slit. The resistance coefficients ψ of slit packings have been found to be lower than those for perforated packings.

The non-perforated packings have the highest resistance coefficients ψ, see Figs. 3-9 and 3-11. They are comparable to those of modern, randomly filled lattice packings, see Figs. 3-9 and 3-7.

The diagrams, see Figs. 3-2 to 3-11 and 3-14 to 3-37, are based on experiments [1-37], in which the main variables were modified within the following ranges:

$$\begin{array}{rcccl} 0.14 & \leq & \rho_V & \leq 128 & \text{kgm}^{-3} \\ 6.4 & \leq & \eta_V \cdot 10^6 & \leq 18 & \text{kgm}^{-3}\,\text{s}^{-1} \\ 0.66 & \leq & \varepsilon & \leq 0.987 & \text{m}^3\,\text{m}^{-3} \\ 0.15 & \leq & d_S & \leq 1.8 & \text{m for packings} \\ 0.025 & \leq & d_S & \leq 0.058 & \text{m for tube columns} \\ 0.008 & \leq & d & \leq 0.100 & \text{m} \\ 0.65 & \leq & H & \leq 7.0 & \text{m} \\ 0.033 & \leq & p_T & \leq 100 & \text{bar} \\ 150 & < & Re_V & \leq 25000 & [-] \end{array} \tag{3-30}$$

4. Tables 3-1, 3-2 and 6-1a-c contain the constants K_1, K_3 and/or exponents K_2, K_4 and the packing form factors φ_P for calculating the resistance coefficient, according to Eq. (3-14) and (3-26), for the random and structured packings tested.

Numerical Examples

Numerical Example 3.1

The aim is to determine the geometric packing surface a and the void fraction ε for 50 mm metal Pall rings, based on a packing density of $N = 6690\ m^{-3}$. According to Table 6-1a, this packing density differs from the standard values provided by the manufacturer, which are as follows:

$N_0 = 6100\ m^{-3}$
$a_0 = 110\ m^2m^{-3}$
$\varepsilon_0 = 0.952\ m^3m^{-3}$

Solution

As a rule, the standard values a_0, ε_0 and N_0, provided by the manufacturer and referred to by the index '0' in this book, are valid for a ratio of $\varphi = d_S/d \cong 10$.

For packing densities of $N \neq N_0$, e.g. $N = 6690\ m^{-3}$, the surface *a* changes, according to the following equation:

$$a = a_0 . \frac{N}{N_0} = 110 \cdot \frac{6690}{6100} = 120.63 \quad \left[m^2 m^{-3}\right] \tag{3-31}$$

The required void fraction ε of the packing elements can be determined, based on the quantity N and on ε_0 and N_0, provided by the manufacturer, using the transposed

Eq. (3-4). The following applies for $N = N_0$, according to Eq. (3-4):

$$\varepsilon_0 = 1 - \frac{V_1}{V_S} \cdot N_0 \quad \rightarrow \quad 1 - \varepsilon_0 = \frac{V_1}{V_S} \cdot N_0 \tag{3-32}$$

and for $N \neq N_0$:

$$\varepsilon = 1 - \frac{V_1}{V_S} \cdot N \quad \rightarrow \quad 1 - \varepsilon = \frac{V_1}{V_S} \cdot N \tag{3-33}$$

Dividing Eq. (3-33) by (3-31), followed by a simple transposition, leads to the following dependency:

$$\frac{1 - \varepsilon}{1 - \varepsilon_0} = \frac{N}{N_0} \quad \text{gives} \quad \varepsilon = 1 - (1 - \varepsilon_0) \cdot N/N_0 \tag{3-34}$$

As a result the void packing of the packed bed is:

$$\varepsilon = 1 - (1 - 0.952) \cdot \frac{6690}{6100} = 0.947 \quad m^3 m^{-3}$$

Numerical Example 3.2

The aim is to determine the pressure drop of a dry as well as an irrigated Mellapak structured packing, type 250Y, in a column with $d_S = 1.0$ m, for F_V factors of $F_V = 1.6$, 2.3 and 3.75 $Pa^{1/2}$. The constants for determining the resistance coefficient for Mellapak, used in a pilot plant with $d_S = 0.22$ m, are known. The physical properties are applicable to the mixture air/water at 293 K and 1 bar.

Solution

Table 6-1c contains the numerical values for the parameters K_1 to K_4 of Eq. (3-14) for calculating the pressure drop of dry packings in small columns with diameters of $d_S =$ 0.22/0.30 m (see Fig. 3-10). They are as follows:

$$K_1 = 8.19, \quad K_2 = -0.321, \quad K_3 = 1.936, \quad K_4 = -0.133$$

If the calculation of the pressure drop $\Delta p_0/H$ is performed for columns with $d_S = 1.0$ m, the correction of the resistance coefficient is required, according to Eq. (3-17). $\varphi_2 \approx 1$ can be assumed for $d_S = 1.0$ m. Analogous to Eq. (3-18), the constants $K_{1,A}$ and $K_{3,A}$ for $d_S = 1.0$ m, according to Eq. (3-19), can now be estimated as follows:

$$K_{1,A} = 0.794 \cdot 8.19 = 6.50 \quad \text{and/or} \quad K_{3,A} = 1.936 \cdot 0.794 = 1.537$$

The following correlation applies for $Re_V \geq 2100$: $\psi_A = K_{3,A} \cdot Re_V^{-K_4}$.

The geometric data of Mellapak 250Y, based on Table 6-1c, is given as:

$$a = 250\ \ m^2m^{-3} \quad \text{and} \quad \varepsilon = 0.975\ \ m^3m^{-3}$$

For $d_P = 6 \cdot (1 - \varepsilon)/a = 0.006$ m and the Reynolds numbers $Re_V \geq 2100$, calculated using Eq. (3-10), with K = 1, the resistance coefficients ψ_A can be calculated as:

$$(a)\ F_V = 1.6\ \sqrt{Pa} \quad \Rightarrow Re_V = 2335.3 \quad \psi_A = 0.548$$
$$(b)\ F_V = 2.3\ \sqrt{Pa} \quad \Rightarrow Re_V = 3386.6 \quad \psi_A = 0.521$$
$$(c)\ F_V = 3.75\ \sqrt{Pa} \quad \Rightarrow Re_V = 5474.0 \quad \psi_A = 0.489$$

The numerical values for the pressure drop of the dry packing are calculated as follows, based on Eq. (3-8) and for K = 1:

$$\frac{\Delta p_0}{H} = \psi_A \cdot \frac{1-\varepsilon}{\varepsilon^3} \cdot \frac{F_V{}^2}{dp} \quad \Rightarrow$$

$$(a)\ \tfrac{\Delta p_0}{H} = 63.1 \quad Pam^{-1} \qquad \left(\tfrac{\Delta p_0}{H}\right) \exp \cong 60\ Pam^{-1} \qquad [25]$$

$$(b)\ \tfrac{\Delta p_0}{H} = 126 \quad Pam^{-1} \qquad \left(\tfrac{\Delta p_0}{H}\right) \exp \cong 130\ Pam^{-1} \qquad [25]$$

$$(c)\ \tfrac{\Delta p_0}{H} = 309.0\ Pam^{-1} \qquad \left(\tfrac{\Delta p_0}{H}\right) \exp \cong 300\ Pam^{-1} \qquad [25]$$

Conclusion

The concurrence of the experiment with the calculated estimate confirms the model of transferability, which allows experimental results on the fluid dynamics of structured packings to be transferred from small to large-scale plants. This method can also be applied to other structured packings and column diameters.

Numerical Example 3.3

The aim is to determine the pressure drop $\Delta p_0/H$ during air flow through a packing, which consists of metal packing elements Mc-Pac, size 1, material 1.4571 (SS 316Ti) at 1 bar and 293 K in a column with a diameter of $d_S = 0.32$ m [26, 27, 34, 35].

The following values are given:

Geometric packing data:	$a = 185.1\ m^2/m^3$;	$\varepsilon = 0.974\ m^3/m^3$
Form factor:	$\varphi_{P,exp} = 0.532$	acc. to Table 6-1a
	$\varphi_{P,calc} = 0.50$	acc. to [34]
Physical properties:	$\rho_V = 1.22\ kg/m^3$;	$\nu_V = 15.2 \cdot 10^{-6}\ m^2/s$
Experimental results:	for $u_V = 2.39$ m/s =>	$(\Delta p_0/H)_{exp} = 405.2$ Pa/m [26, 27]

Solution

Equation (3-3)	for:	$d_p = 6 \cdot \frac{1-\varepsilon}{a} = 6 \cdot \frac{1-0.974}{185.1} = 0.000843\ m,$
Equation (3-6)	for:	$K = \left(1 + \frac{2}{3} \cdot \frac{1}{1-0.974} \cdot \frac{0.000843}{0.32}\right)^{-1} = 0.9366,$
Equation (3-10)	for	$Re_V \frac{2.389.0.000843}{(1-0.974) \cdot 15.2 \cdot 10^{-6}} \cdot 0.9366 = 4772.9$
and	for	$F_V = u_V \cdot \sqrt{\rho_V} = 2.39 \cdot \sqrt{1.22} = 2.64\sqrt{Pa}$

lead to Eq. (3-29) for $\varphi_{P,exp} = 0.532$ describing the pressure drop of the dry packing $\Delta p_0/H$

$$\frac{\Delta p_0}{H} = \left(\frac{725.6}{4772.9} + 3.203\right) \cdot (1 - 0.532) \cdot \frac{1-0.974}{0.974^3} \cdot \frac{2.64^2}{0.000843 \cdot 0.9366} = 390.0\ \text{Pa/m}$$

The relative error is therefore given as:

$$\delta\left(\frac{\Delta p_0}{H}\right) = \frac{390.0 - 405.2}{405.2} \cdot 100 = -3.75\%$$

in relation to the experimental value $\varphi_{P,exp}$.

Using the calculated value $\varphi_{P,calc} = 0.5$, based on the geometry of the packing Mc-Pac size 1, the pressure drop is found to be $\Delta p_0/H = 416.7$ Pa/m, based on Eq. (3-29). Hence, the relative error is:

$$\delta\left(\frac{\Delta p_0}{H}\right) = \frac{416.7 - 405.2}{405.2} \cdot 100 = +2.6\%$$

Numerical Example 3.4

To determine is the single phase pressure drop $\Delta p_0/H$ during air flow through a packed bed filled with random Mc-Pac, size 2, material 1.4571 (SS 316Ti) at 1 bar and 278 K in a column with a diameter of $d_S = 0.60$ m and packing height $H = 3.0$ m [new Envimac data 2009].

The following values are given for $N = 7899\ 1/m^2$:

Geometric packing data:	$a = 92.94\ m^2/m^3$;	$\varepsilon = 0.982\ m^3/m^3$
Form factor:	$\varphi_{P,exp} = 0.532$	acc. to Table 6-1b
	$\varphi_{P,calc} = 0.50$	acc. [34]
Physical properties:	$\rho_V = 1.262\ kg/m^3$;	$\nu_V = 14.4 \cdot 10^{-6}\ m^2/s$
Experimental results:	for $u_V = 2.356$ m/s =>	$(\Delta p_0/H)_{exp} = 190$ Pa/m [2009]

Equation (3-3)	For:	$d_P = 6 \cdot \frac{1-\varepsilon}{a} = 6 \cdot \frac{1-0.982}{92.94} = 0.001162\ m,$
Equation (3-6)	for:	$K = \left(1 + \frac{2}{3} \cdot \frac{1}{1-0.982} \cdot \frac{0.001162}{0.6}\right)^{-1} = 0.9331,$
Equation (3-10)	for	$Re_V = \frac{2.356 \cdot 0.001162}{(1-0.982) \cdot 14.4 \cdot 10^{-6}} \cdot 0.9331 = 9855.4$
and	for	$F_V = u_V \cdot \sqrt{\rho_V} = 2.356 \cdot \sqrt{1.262} = 2.65\ \sqrt{Pa}$

Solution

lead to Eq. (3-29) for $\varphi_{P,exp} = 0.532$ describing the pressure drop of the dry packing $\Delta p_0/H$

$$\frac{\Delta p_0}{H} = \left(\frac{725.6}{9855.4} + 3.203\right) \cdot (1-0.532) \cdot \frac{1-0.982}{0.982^3} \cdot \frac{2.65^2}{0.001162 \cdot 0.9331} = 188.78\ Pa/m$$

The relative error is therefore given as:

$$\delta\left(\frac{\Delta p_0}{H}\right) = \frac{188.78 - 190.0}{190.0} \cdot 100 = -0.64\%$$

in relation to the experimental value $\varphi_{P,exp}$.

Using the calculated value $\varphi_{P,calc} = 0.5$, based on the geometry of the packing Mc-Pac size 2, the pressure drop is found to be $\Delta p_0/H = 201.69$ Pa/m, based on Eq. (3-29). Hence, the relative error is:

$$\delta\left(\frac{\Delta p_0}{H}\right) = \frac{201.69 - 190}{190} \cdot 100 = +6.2\%$$

Annex Chapter 3

Tables Chapter 3

Table 3-1a. Technical data of packings and parameters for calculating the resistance coefficients, based on Eq. (3-14), for random metal packings

Packing type	Size	d_S [m]	H [m]	N [m^{-3}]	a [m^2m^{-3}]	ε [%]	$Re_V < 2100$		$Re_V \geq 2100$	
							K_1	K_2	K_3	K_4
Pall ring	15 mm	0.22	1.45	253817	407.92	92.63	10	−0.18	3.23	−0.0343
		0.22	1.50	235000	377.70	93.10	10	−0.18	3.23	−0.0343
		0.30	0.80	229225	368.40	93.30	10	−0.18	3.23	−0.0343
		0.50	1.50	224000	360.00	93.60	10	−0.18	3.23	−0.0343
		0.50	1.50	230700	370.30	93.30	10	−0.18	3.23	−0.0343
		0.80	2.00	236200	379.60	93.25	10	−0.18	3.23	−0.0343
	25 mm	0.30	1.46	53894	223.50	95.36	10	−0.18	3.23	−0.0343
		0.154	1.30	55603	232.00	93.73	10	−0.18	3.23	−0.0343
		0.75	3.00	–	215.00	94.20	10	−0.18	3.23	−0.0343
		0.435	1.65	51500	215.00	94.20	10	−0.18	3.23	−0.0343
		0.30	0.80	52388	218.71	94.10	10	−0.18	3.23	−0.0343
		0.435	1.65	53276	222.41	94.00	10	−0.18	3.23	−0.0343
		0.80	2.00	51500	215.00	94.20	10	−0.18	3.23	−0.0343
		0.22	1.31	53200	222.10	94.00	10	−0.18	3.23	−0.0343
		0.30	2.00	51500	215.00	94.20	10	−0.18	3.23	−0.0343
	35 mm	0.30	1.20	19000	145.0	94.8	10	−0.18	3.23	−0.0343
		0.75	3.00	19000	150.0	94.5	10	−0.18	3.23	−0.0343
		0.50	2.90	19000	150.0	94.5	10	−0.18	3.23	−0.0343
		0.50	2.00	19260	147.0	94.7	10	−0.18	3.23	−0.0343
		0.50	3.95	19230	146.8	94.74	10	−0.18	3.23	−0.0343
		0.80	2.00	19650	150.0	94.6	10	−0.18	3.23	−0.0343
		0.50	2.00	19200	146.9	94.7	10	−0.18	3.23	−0.0343
		0.50	3.95	19230	146.9	94.7	10	−0.18	3.23	−0.0343
		0.80	3.75	19320	147.4	94.7	10	−0.18	3.23	−0.0343
	38 mm	0.30	1.46	15572	149.64	95.19	10	−0.18	3.23	−0.0343

Table 3-1a. (continued)

Packing type	Size	d_S [m]	H [m]	N [m^{-3}]	a [m^2m^{-3}]	ε [%]	$Re_V < 2100$		$Re_V \geq 2100$	
							K_1	K_2	K_3	K_4
	50 mm	0.75	3.00	6100	110.0	95.2	10	−0.18	3.23	−0.0343
		0.50	2.00	6100	110.0	95.2	10	−0.18	3.23	−0.0343
		0.50	3.95	6480	116.9	94.9	10	−0.18	3.23	−0.0343
		0.50	1.33	6100	110.0	95.2	10	−0.18	3.23	−0.0343
		0.30	3.70	6690	120.4	94.8	10	−0.18	3.23	−0.0343
		1.20	5.50	–	110.0	95.2	10	−0.18	3.23	−0.0343
	80 mm	0.75	3.0	1600	78.0	96.0	10	−0.18	3.23	−0.0343
		0.80	2.0	1650	80.4	96.0	10	−0.18	3.23	−0.0343
Białecki ring	25 mm	0.15	1.50	55000	238	94.0	10.17	−0.17	4.17	−0.0522
		0.30	1.40	56156	243	94.0	10.17	−0.17	4.17	−0.0522
		0.15	0.83	–	210	94.3	10.17	−0.17	4.17	−0.0522
	35 mm	0.30	1.20	19000	155	96.7	10.17	−0.17	4.17	−0.0522
		0.30	0.74	19000	155	96.7	10.17	−0.17	4.17	−0.0522
		0.15	0.83	–	155	95.0	10.17	−0.17	4.17	−0.0522
	50 mm	0.30	1.40	6300	110.0	97.3	10.17	−0.17	4.17	−0.0522
		0.30	1.35	6000	104.8	96.8	10.17	−0.17	4.17	−0.0522
Raschig ring	15 mm	0.50	1.5	241000	352.33	91.95	36.245	−0.19	12.00	−0.0455
	25 mm	0.50	2.0	52200	223.00	91.90	36.245	−0.19	12.00	−0.0455
	35 mm	0.50	2.0	19300	152.40	92.89	36.245	−0.19	12.00	−0.0455
	50 mm	0.50	2.0	6630	112.20	94.90	36.245	−0.19	12.00	−0.0455
Ralu ring	35 mm	0.75	3.0	–	138.07	96.4	10	−0.18	3.23	−0.0343
	50 mm	0.75	3.0	6730	112.16	97.2	10	−0.18	3.23	−0.0343
VSP ring	size 1	0.30	1.45	33434	199.6	97.20	9.1	−0.18	3.37	−0.05
	size 1	0.45	2.00	31572	188.0	97.64	9.7	−0.18	3.37	−0.05
	size 2	0.45	2.0	7841	104.56	98.0	9.1	−0.18	3.37	−0.05

Table 3-1a. (continued)

Packing type	Size	d_S [m]	H [m]	N [m^{-3}]	a [m^2m^{-3}]	ε [%]	$Re_V < 2100$		$Re_V \geq 2100$	
							K_1	K_2	K_3	K_4
Pall ring (VFF)	58 mm	0.45	2.00	5983	107.9	97.0	8.0	−0.18	2.848	−0.045
Top-Pak	size 2	0.45	2.0	2800	75.0	98.0	8	−0.18	2.85	−0.045
IMTP	#40	0.45	2.0	43400	131.7	97.5	8.57	−0.18	3.17	−0.05
Hiflow ring	28 mm	0.30	1.46	40790	202.9	96.20	4.49	−0.118	2.98	−0.066
		0.45	2.00	36638	182.0	96.55	4.49	−0.188	2.98	−0.066
	58 mm	0.45	2.0	5146	100.76	97.7	2.18	−0.06	2.18	−0.06
PSL ring	50 mm	0.30	1.45	6248	110.74	97.63	10	−0.18	3.23	−0.0343
Glitsch ring	no. 0.5	0.30	1.14	560811	356.8	95.5	6.534	−0.148	4.333	−0.092
CMR 404	no. 1A	0.30	1.4	158467	232.5	97.12	5.4	−0.14	3.241	−0.0733
	no. 1.5A	0.30	1.4	60744	176.37	97.37	5.4	−0.14	3.241	−0.0733
	no. 2	0.45	2.0	32956	150.68	98.87	8.124	−0.18	2.903	−0.0455
	no. 3A	0.45	2.0	9914	100.36	98.746	8.124	−0.18	2.903	−0.0445
Glitsch ring	30 P	0.30	1.36	32509	168.92	95.7	10	−0.17	4.03	−0.05
Mc-Pac	size 1	0.30	3.0	63584	185.1	97.4	5.4	−0.159	2.382	−0.052
	size 2	0.30/1.60	1.4−2.5	7700	90.6	98.2	5.4	−0.159	2.382	−0.052
		0.60	1.5	8500	100.0	98.1	5.4	−0.159	2.382	−0.052
		0.60	3.0	7789	92.2	98.2	5.4	−0.159	2.382	−0.052

Table 3-1b. Technical data of packings and parameters for calculating the resistance coefficients, based on Eq. (3-14), for random packings made of plastic

Packing type	Size	d_S [m]	H [m]	N [m^{-3}]	a [m^2m^{-3}]	ε [%]	$Re_V < 2100$		$Re_V \geq 2100$	
							K_1	K_2	K_3	K_4
Pall ring PP/PVDF	25 mm	0.22	0.91	54500	235.7	88.3	10	−0.18	3.23	−0.0343
		0.30	0.90	55200	239.8	88.0	10	−0.18	3.23	−0.0343
		0.15	1.40	55000	232.7	88.0	10	−0.18	3.23	−0.0343
		0.22	1.40	53050	228.8	88.5	10	−0.18	3.23	−0.0343
		0.22	1.50	55583	239.8	88.6	10	−0.18	3.23	−0.0343
	35 mm	0.30	1.29	16080	142.93	92.2	10	−0.18	3.23	−0.0343
		0.45	1.20	17140	148.00	90.8	10	−0.18	3.23	−0.0343
	50 mm	0.45	2.00	6480	95.4	92.6	10	−0.18	3.23	−0.0343
		0.30	1.33	6765	111.1	91.9	10	−0.18	3.23	−0.0343
		0.75	3.00	6400	110.0	92.0	10	−0.18	3.23	−0.0343
		0.30	1.45	6820	100.24	92.17	10	−0.18	3.23	−0.0343
Białecki ring PP	50 mm	0.30	1.4	6400	115	93.0	10.17	−0.17	4.17	−0.0522
Envipac PP	size 1A*	0.30	1.36	42431	156.20	93.20	2.817	−0.07	2.12	−0.02
	size 1	0.30	1.40	51681	138.02	93.65	6.05	−0.187	2.14	−0.0513
	size 2	0.30	1.45	6339	76.1	96.40	4.792	−0.187	1.6974	−0.0513
		0.45	2.0	6763	81.3	96.13	4.792	−0.187	1.6974	−0.0513
		0.60	3.0	6422	77.1	95.9	4.792	−0.187	1.6974	−0.0513
	size 3	0.30	1.45	1808	57.0	96.2	4.792	−0.187	1.6974	−0.0513
		0.45	2.0	1962	58.8	96.0	4.792	−0.187	1.6974	−0.0513
		0.60	2.0	2018	60.5	95.8	4.792	−0.187	1.6974	−0.0513
		0.60	3.03	2094	66.0	95.5	4.792	−0.187	1.6974	−0.0513
Dtnpac PP	size 1	0.45	2.0	29039	135.0	92.00	5.4	−0.159	2.382	−0.052
		0.45	2.0	24570	127.4	93.35	5.4	−0.159	2.382	−0.052
	size 2	0.30	1.4	10212	112.0	93.75	4.792	−0.187	1.693	−0.0513
		0.60	1.1	9300	100.0	94.3	4.792	−0.187	1.693	−0.0513

Table 3-1b. (continued)

Packing type	Size	d_S [m]	H [m]	N [m^{-3}]	a [m^2m^{-3}]	ε [%]	$Re_V < 2100$		$Re_V \geq 2100$	
							K_1	K_2	K_3	K_4
Hiflow ring	28 mm	0.30	0.9	45500	190.0	92.0	4.615	−0.145	2.258	−0.0418
	50 mm	0.30	1.4	6280	107.94	93.2	3.52	−0.154	1.6	−0.0418
	90 mm	0.45	2.0	1337	69.52	96.25	3.06	−0.0418	1.392	−0.0418
Hiflow-Super	50 mm	0.45	2.00	6196	84.0	94.06	2.61	−0.0843	2.61	−0.0843
PP		0.30	1.44	5927	80.3	94.10	2.538	−0.0843	2.538	−0.0843
Hiflow saddle	50 mm	0.45	2.0	9500	82.6	94.08	2.62	−0.069	2.62	−0.069
Intalox saddle	35 mm	0.30	1.36	27790	176	90.0	8.11	−0.108	8.11	−0.108
	50 mm	0.45	1.21	9875	139.41	89.54	10	−0.18	3.23	−0.0343
		0.45	2.00	7930	111.90	91.6				
Super-Torus saddle	size 2	0.75	3.0	5600	110	94.0	3.206	−0.0755	3.206	−0.0755
Nor–Pac	17 mm	0.30	1.4	247982	309	92.0	7.63	−0.256	1.825	−0.069
	22x27 PVDF	0.30	1.4	49009	248	91.3	10.98	−0.284	1.963	−0.059
	28 mm	0.154	1.3	44960	181.9	93.0	7.63	−0.256	1.825	−0.069
		0.40	1.4	44500	180.0	92.7				
	38 mm	0.30	1.4	18629	125.9	93.80	7.63	−0.256	1.825	−0.069
	50 mm	0.30	2.00	7570	93.30	95.0	7.63	−0.256	1.825	−0.069
		0.30	1.43	7332	90.37	95.2	7.63	−0.256	1.825	−0.069
Ralu ring	38 mm	0.75	3.0	–	150.0	93.45	17.5	−0.321	3.4	−0.107
	50 mm	0.45	2.0	5841	112.73	93.85	17.5	−0.321	3.4	−0.107
		0.75	3.0	5508	106.30	94.20				
Tellerette	50 mm	0.50	–	–	110	93.0	1.65	−0.0418	1.65	−0.0418
		0.45	2.0	9188	95	93.4	1.158	–	1.158	–
VSP ring size 2	58 mm	0.45	2.0	5006	75.9	95.2	1.633	−0.05	1.633	−0.05
Hackette	45 mm	0.45	2.0	12741	1357	92.81	10.98	−0.27	2.22	−0.065
Glitsch CMR	No. 1	0.45	2.00	24150	185.78	93.84	5.4	−0.18	3.241	−0.0733
	No. 1	0.30	1.41	25568	198.69	93.48				
	No. 2	0.45	2.0	6244	128.64	95.21	9.5	−0.18	3.12	−0.0343

Table 3-1c. Technical data of packings and parameters for calculating the resistance coefficients, based on Eq. (3-14), for random packings made of ceramic

Packing type	Size	d_S [m]	H [m]	N [m^{-3}]	a [m^2m^{-3}]	ε [%]	$Re_V < 2100$		$Re_V \geq 2100$	
							K_1	K_2	K_3	K_4
Pall ring ceramic	25 mm	0.50	1.1	43000	205.7	74.8	10	−0.18	3.23	−0.0343
		0.22	1.5	36621	156	75	10	−0.18	3.23	−0.0343
	50 mm	0.45	1.00	0	112.95	19.3	12.68	−0.235	3.81	−0.081
		0.22	1.46	6238	116.2	18.7				
Raschig ring ceramic	8 mm	0.10	1.0	0	550	63.0	34.6	−0.307	6.05	−0.079
	15 mm	0.23	0.63	220000	300.8	67.6	34.6	−0.307	6.05	−0.079
	25 mm	0.30	0.8	48175	185.4	67.85	34.6	−0.307	6.05	−0.079
		0.22	1.0	52377	201.0	65.00	34.6	−0.307	6.05	−0.079
		0.30	1.0	46000	177.0	69.30	34.6	−0.307	6.05	−0.079
	35 mm	0.30	1.2	15890	121.5	77.1	34.6	−0.307	6.05	−0.079
		0.30	1.0	0	133.9	72.3	34.6	−0.307	6.05	−0.079
	50 mm	0.30	0.73	6400	92.89	78.2	34.6	−0.307	6.05	−0.079
Raschig ring glass	25 mm	0.15	1.0	54081	203.2	80.46	17.508	−0.209	5.264	−0.0525
		0.15	1.0	54799	205.9	80.20	17.508	−0.209	5.264	−0.0525
Hiflow ring ceramic	20 mm	0.30	1.25	121314	280.0	76.2	3.285	−0.094	1.6	–
	35 mm	0.45	1.73	17156	110.0	82.6	3.047	−0.065	3.047	−0.065
		0.30	1.36	17383	114.7	82.8	3.047	−0.065	3.047	−0.065
	50 mm	0.30	1.20	5120	89.77	80.86	3.047	−0.065	3.047	−0.065
		0.45	1.45	5000	97.66	81.20	3.047	−0.065	3.047	−0.065
		0.30	0.90	4890	85.65	81.73	3.047	−0.065	3.047	−0.065
	75 mm	0.45	1.86	1904	54.8	86.52	2.4	−0.065	2.4	−0.065
Intalox saddle ceramic	25 mm	0.30	0.87	63890	185	72.4	15.5	−0.255	3.8	−0.071
	38 mm	0.30	1.00	19000	125.8	75.7	15.5	−0.255	3.8	−0.071
		0.385	3.05	20037	133.0	74.3	15.5	−0.255	3.8	−0.071
		0.45	2.0	19625	129.8	75.0	15.5	−0.255	3.8	−0.071
		0.50	1.33	20000	132.96	74.324	15.5	−0.255	3.8	−0.071
50 mm	0.30	1.34	8952	125.8	75.7	15.5	−0.255	3.8	−0.071	
	0.45	1.40	8880	97.7	77.3	15.5	−0.255	3.8	−0.071	
	0.75	3.00	9000	99.0	77.0	15.5	−0.255	3.8	−0.071	
	0.385	3.05	8998	99.0	77.0	15.5	−0.255	3.8	−0.071	
	0.22	1.47	8180	50.0	79.1	15.5	−0.255	3.8	−0.071	

Table 3-2. Technical data of packings and parameters for calculating the resistance coefficients, based on Eq. (3-14), for structured and stacked packings

Packing, material	Type	dS [m]	H [m]	N [m^{-3}]	a [$m^2\ m^{-3}$]	ε [%]	ReV < 2100		ReV ≥ 2100	
							K1	K2	K3	K4
Montz	B1-300	0.22	1.4	–	300	97.2	10.50	−0.321	2.49	−0.133
		0.30	1.5		300	97.2	9.205	−0.321	2.185	−0.133
		0.45	2.0		300	97.2	8.97	−0.321	2.13	−0.133
	B1-200	0.30	1.40	–	200	97.8	4.36	−0.182	3.00	−0.133
		0.30	1.56		200	97.8	4.36	−0.182	3.00	−0.133
		0.22	1.34		200	97.8	4.626	−0.182	3.18	−0.133
	B1-100	0.30	1.4	–	100	98.7	5.835	−0.162	5.835	−0.162
	B2-500X	0.22	1.52	–	500	96.0	3.915	−0.321	0.9293	−0.133
	C1-200	0.30	1.56	–	200	95.4	7.00	−0.215	3.728	−0.133
	A3-500	0.45	2.0	–	500	95	1.21	−0.14	–	–
Mellapak	250Y	0.22	1.25	–	250	96	8.19	−0.321	1.936	−0.133
	250Y PP	0.30	–	–	250	95.5	1.62	−0.13	1	−0.068
Pall rings stacked, ceramic	50 mm	0.40	0.90	8200	153.1	78.4	4.513	−0.18	2.25	−0.089
		0.50	0.90	7500	140.0					
		0.22	1.48	7320	116.2					
Białecki rings stacked, metal	50 mm	0.151	1.70	7821	128.24	96.18	4.773	−0.18	2.552	−0.0966
		0.30	1.45	7654	128.50	96.26	4.675	−0.18	2.47	−0.0966
	25 mm	0.15	1.4	60000	259	92.8	7.588	−0.291	1.738	−0.0966
PSL rings stacked, metal	50 mm	0.28	1.45	7882	139.8	96.8	4.03	−0.153	2.34	−0.082

Table 3-2. (continued)

Packing, material	Type	dS [m]	H [m]	N [m^{-3}]	a [m^2 m^{-3}]	ε [%]	ReV < 2100		ReV ≥ 2100	
							K1	K2	K3	K4
Hiflow rings stacked, ceramic	50 mm	0.45	1.0	5683	102.3	75.8	0.515	−0.075	0.515	−0.075
Ralu-Pak	250YC	0.22	1.42	–	250	96.3	9.72	−0.383	3.52	−0.25
		0.30	1.80		250	96.3	11.5	−0.383	4.00	−0.25
		0.45	1.62		250	96.3	10.09	−0.383	3.656	−0.25
Gempak, sheet-metal packing	202AT	0.22	1.47	–	202	97.0	12.447	−0.321	4.50	−0.188
		0.30	1.47		202	97.0	10.51	−0.321	3.8	−0.188
Impuls packing, ceramic	50 mm	0.40	1.4	7890	105.9	82.4	2.64	−0.115	2.61	−0.115
Waben packing, PP	–	0.30	0.8	–	209.4	81.7	2.749	−0.359	0.543	−0.1471
FX-260 plastic packing	VFF	0.45	2.0	–	250	96.2	1.616	−0.156	0.83	−0.069
Euroform K = 1	50 mm	0.30	1.7	–	110	93.6	17.5	−0.104	1.75	−0.104
Nor Pac PP	size 3	0.30	1.54	–	130	84.48	61.296	−0.4575	1.0274	−0.224

Diagrams for Single-Phase Flow in Packed Columns - Law of Resistance

Figure 3-14. Resistance coefficient ψ as function of Reynolds number of vapour or gas Re_V, valid for 12–50 mm random Białecki rings made of metal – $\varphi_P = 0.208$ for Eq. (3-26)

Białecki ring, metal						
TP	$10^3 \cdot d$ [m]	system	p_T [mbar]	d_S [m]	H [m]	N [1/m³]
▼	12	chlorobenzene/ ethylbenzene	33	0.22	1.4	419570
◆	25	ethylbenzene	35 ÷ 1000	0.15	1.5	55000
▽	25	air, 293 K	1000	0.15	1.5	57800
△	35			0.22	1.3	18200
●	50			0.45	0.64	6700
ɸ	50			0.3	1.35	6280

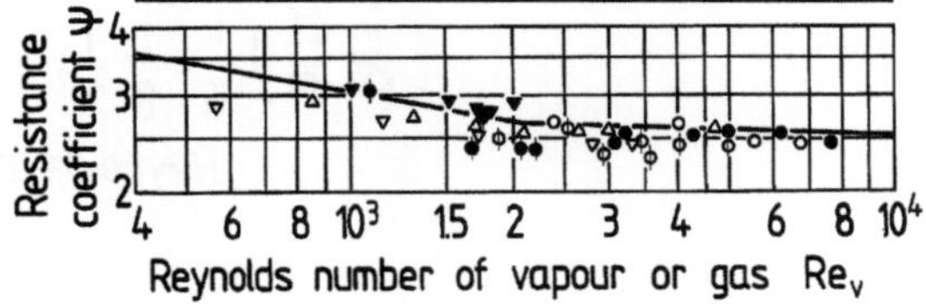

Figure 3-15. Resistance coefficient ψ as function of Reynolds number of vapour or gas Re_V, valid for random VSP rings sizes 1 and 2, made of metal – $\varphi_P = 0.345$ for Eq. (3-26)

TP	system	p_T [mbar]	d_S [m]	H [m]	N [1/m³]
VSP ring size 2, produced by VFF, metal (1.4402)					
●	chlorobenzene/ ethylbenzene	66.7	0.22	1.48	7621
○	air, 293 K	1000	0.45	2.0	7841
VSP ring size 1, produced by VFF, metal (1.4402)					
■	chlorobenzene/ ethylbenzene	66.7	0.22	1.47	30948
□	air, 293 K	1000	0.30	1.46	33434
⌽			0.45	2.0	31572

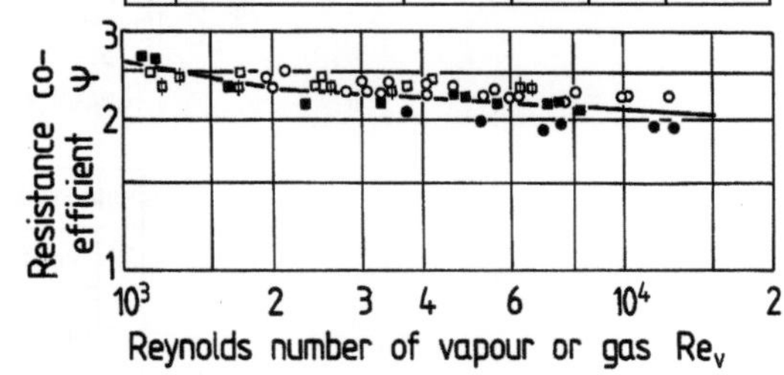

Figure 3-16. Resistance coefficient ψ as function of Reynolds number of vapour or gas Re_V, valid for 28 mm ($\varphi_P = 0.51$) and 58 mm ($\varphi_P = 0.63$) random Hiflow rings made of metal

Hiflow ring, metal

TP	$10^3 \cdot d$ [m]	system	p_T [mbar]	d_S [m]	H [m]	N [1/m³]
■	28	chlorobenzene/ ethylbenzene	66.7	0.22	1.5	38857
□			1000	0.45	2.0	36638
⌽				0.60		38534
●	58	air, 293 K	1000	0.3	1.35	5000
○				0.45	1.45	5075
ϕ				0.45	2.0	5146

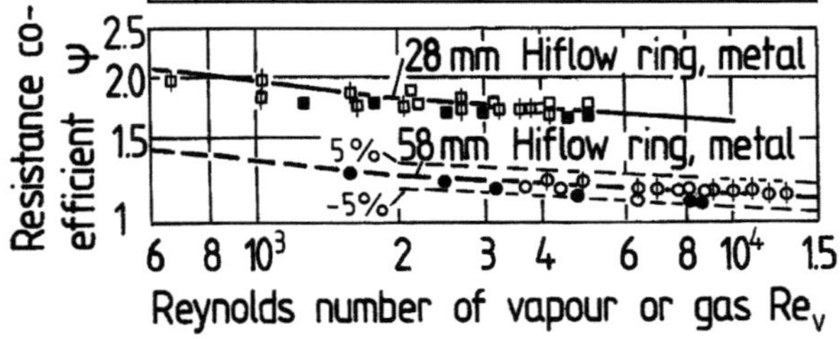

Figure 3-17. Resistance coefficient ψ as function of Reynolds number of vapour or gas Re_V, valid for 28–90 mm random Hiflow rings made of plastic – $\varphi_P = 0.54–0.736$ for Eq. (3-26)

Hiflow ring, PP

TP	system	p_T [mbar]	d_S [m]	H [m]	N [1/m³]
○	air, 288 K	1000	0.3	1.34	42576
ϕ	methanol / water	465	0.22	1.40	47544

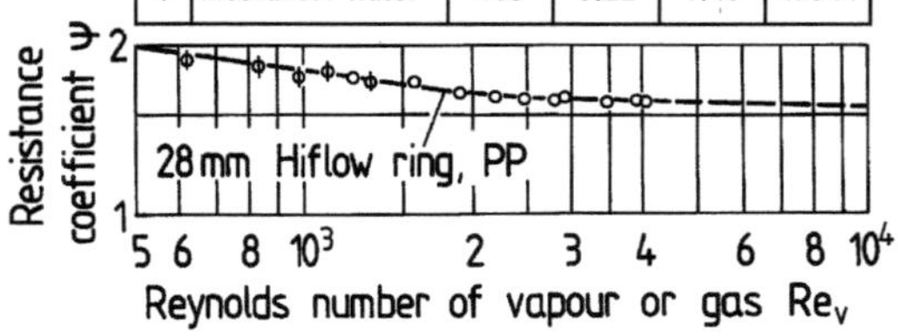

Hiflow ring, PP

TP	system	p_T [mbar]	d_S [m]	H [m]	N [1/m³]	lit.
■	air, 293 K	1000	0.3	1.45	6598	Author
▲		1000	1	1	1328	Kozioł (1984) TU-Wrocław

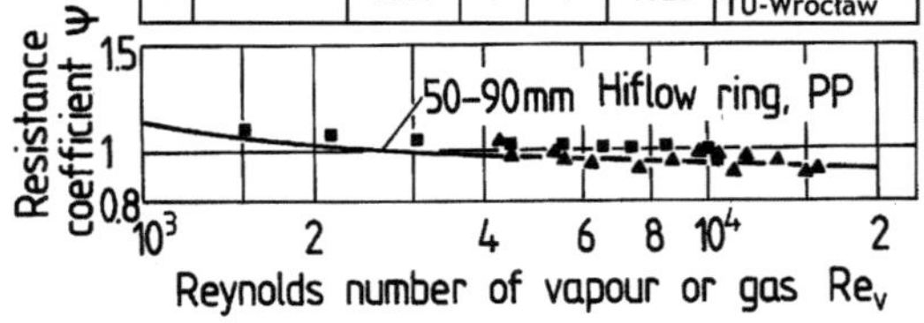

Figure 3-18. Resistance coefficient ψ as function of Reynolds number of vapour or gas Re_V, valid for 53 mm random Hiflow Super rings ($\varphi_P = 0.618$) and Hiflow saddles ($\varphi_P = 0.583$)

50 mm Hiflow saddle, PP					
TP	system	p_T [mbar]	d_S [m]	H [m]	N [1/m³]
■	air, 292,5 K	1000	0.45	2.0	9500

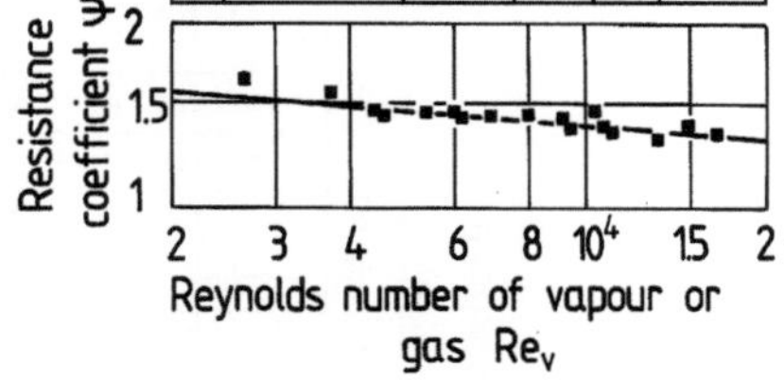

Hiflow ring, PP					
TP	system	p_T [mbar]	d_S [m]	H [m]	N [1/m³]
⊖	air, 291 - 295 K	1000	0.3	1.44	5509
o			0.3	1.35	6050
•			0.45	2.0	6195

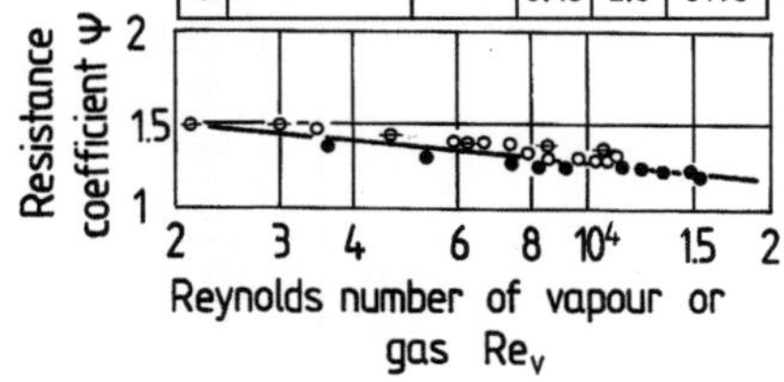

Hiflow ring, ceramic, production year 1988						
TP	d·10³ [m]	system	p_T [mbar]	d_S [m]	H [m]	N [1/m³]
+	20	air, 292 - 296 K	1009	0.3	1.10	90383
•	38		1007	0.45	1.46	13240
▫	50		996	0.45	1.0	5361
▾	75		997	0.45	1.85	1808

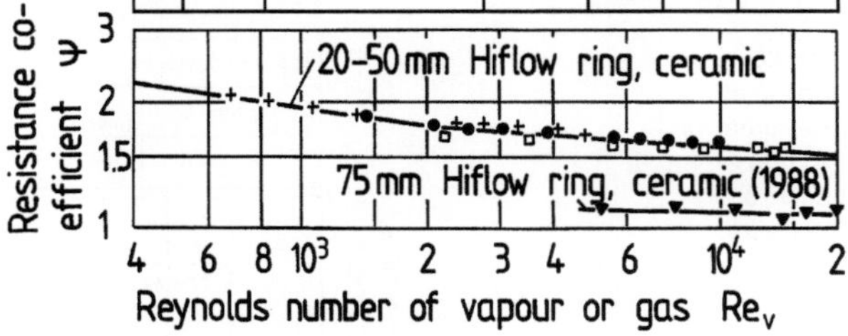

Figure 3-19. Resistance coefficient ψ as function of Reynolds number of vapour or gas Re_V, valid for 20–75 mm random Hiflow rings made of ceramic (1988) – $\varphi_P = 0.47–0.672$ for Eq. (3-26)

28 mm NorPac ring, plastic					
TP	system	p_T [mbar]	d_S [m]	H [m]	N [1/m³]
o	air, 289 K	1000	0.3	1.35	44500
•		1000	0.45	0.93	47900
ϕ	toluene / n-octane	80	0.22	1.45	46800
⊖	chlorobenzene / ethylbenzene	67	0.22	1.45	45580

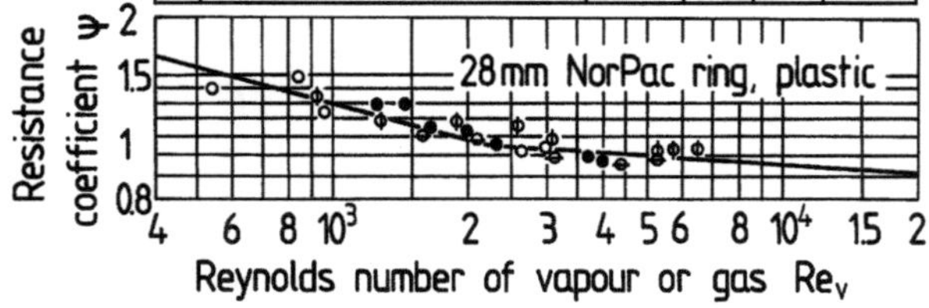

38 and 50 mm NorPac ring, plastic						
TP	d·10³ [m]	system	p_T [mbar]	d_S [m]	H [m]	N [1/m³]
△	38	chlorobenzene/ ethylbenzene	33	0.22	1.45	18477
▲	38	air, 291 K	997	0.3	1.40	18629
⍋	38		997	0.45	1.40	19881
□	50		1000	0.45	1.24	7943
■	50		1000	1.0	1.0	7119
⌽	50		1010	0.3	1.45	7320

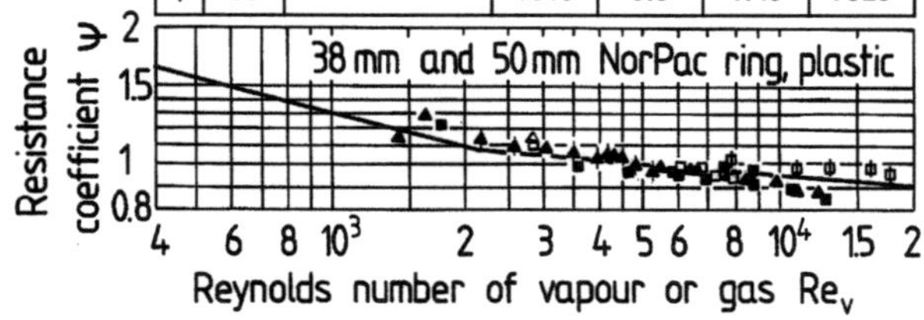

Figure 3-20. Resistance coefficient ψ as function of Reynolds number of vapour or gas Re_V, valid for 28–50 mm random Nor-Pac packing made of plastic – $\varphi_P = 0.694$ for Eq. (3-26)

Figure 3-21. Resistance coefficient ψ as function of Reynolds number of vapour or gas Re_V, valid for 45 mm random Hackette made of plastic – $\varphi_P = 0.665$ for Eq. (3-26)

45 mm Hackette, PP					
TP	system	p_T [mbar]	d_S [m]	H [m]	N [1/m³]
⊖	air, 293 K	1000	0.45	0.87	12252
•				2.0	12741

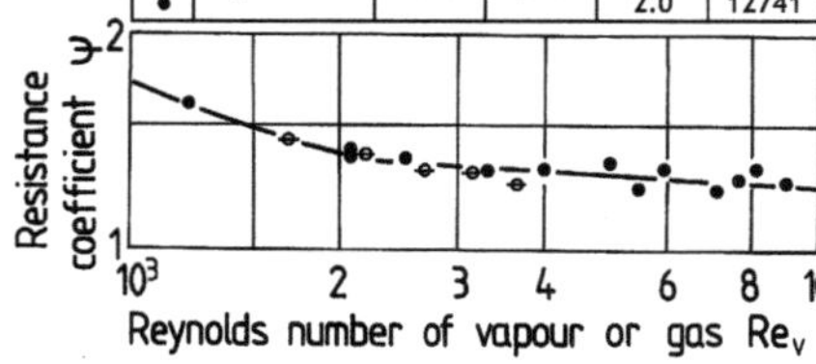

TP	d·10³ [m]	packing material	system	p_T [mbar]	d_S [m]	H [m]	N [1/m³]
○	58	Pall ring, VFF metal	air, 293 K	1000	0.45	2	5983
●	80x60	Top-Pak size 2 metal					2800
△	45	Top-Pak size 1 metal	chlorobenzene/ ethylbenzene	66,7	0.22	1.46	19881
▲	45	Top-Pak size 1 metal	air, 293 K	1000	0.3	1.45	7943

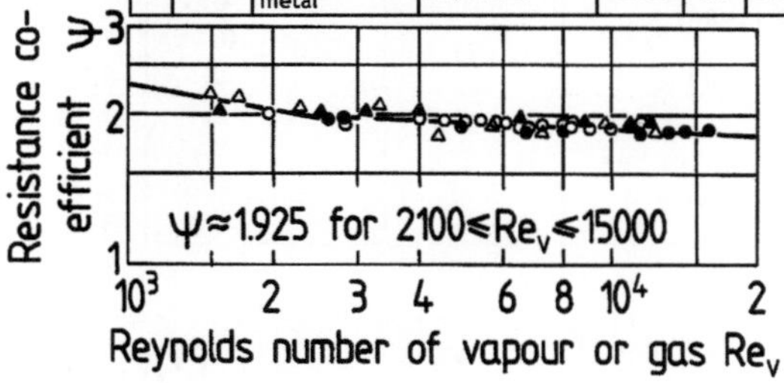

Figure 3-22. Resistance coefficient ψ as function of Reynolds number of vapour or gas Re_V, valid for 58 mm random Pall rings made of metal and Top-Pak sizes 1 and 2 – $\varphi_P = 0.417–0.424$ for Eq. (3-26)

Figure 3-23. Resistance coefficient ψ as function of Reynolds number of vapour or gas Re_V, valid for random metal Mc-Pac rings sizes 1 and 2 – $\varphi_P = 0.532$ for Eq. (3-26)

McPac ring, metal

TP	Type	d·10³ [m]	system	p_T [mbar]	d_S [m]	H [m]	N [1/m³]
○	size 1	30	air, 293 K	1000	0.316	1.5	63584
●					0.345	1.65	70000
□	size 2	65	air, 293 K	1000	0.316	3.0	7285
■					0.6	3.0	7899
■					1.6	2.5	7700
◇			chlorobenzene/ ethylbezene	233	0.6	1.5	8500
◆				300			
◆				532			

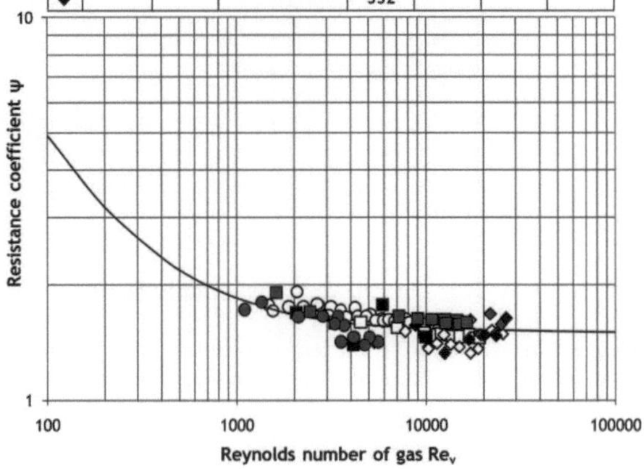

Envipac, PP							
TP	Type	d·10³ [m]	system	p_T [mbar]	d_S [m]	H [m]	N [1/m³]
□	size 1	32	air, 293 K	1000	0.3	0.95	49122
□					0.3	1.39	52681
■					0.32	4.9	53000
○	size 2 (1988)	60			0.3	1.45	6339
●					0.45	1.81	6773
● (grey)	size 2 (2009)				0.6	3.00	6422
△	size 3	80			0.3	1.44	1808
▲					0.45	1.98	1962
▲ (grey)					0.6	2.05	2018
△					0.6	3.03	2094

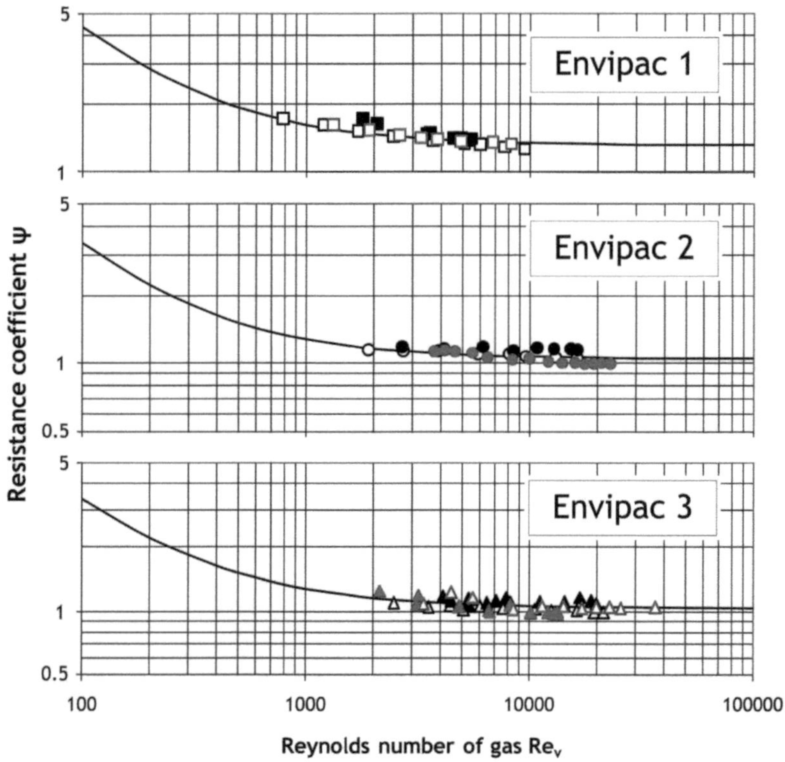

Figure 3-24. Resistance coefficient ψ as function of Reynolds number of vapour or gas Re_V, valid for random Envipac packing sizes 1 ($\varphi_P = 0.611$), 2 and 3 ($\varphi_P = 0.676$), made of plastic

Figure 3-25. Resistance coefficient ψ as function of Reynolds number of vapour or gas Re_V, valid for random Dtnpac packing sizes 1 ($\varphi_P = 0.676$) and 2 ($\varphi_P = 0.60$), made of plastic

Dtnpac, PP							
TP	Type	$d \cdot 10^3$ [m]	system	p_T [mbar]	d_S [m]	H [m]	N [1/m³]
ϕ	size 1	45	air, 293 K	1000	0.45	1.94	24570
o						2.0	29039
□	size 2	70			0.3	1.43	10212
■					0.6	1.10	9300

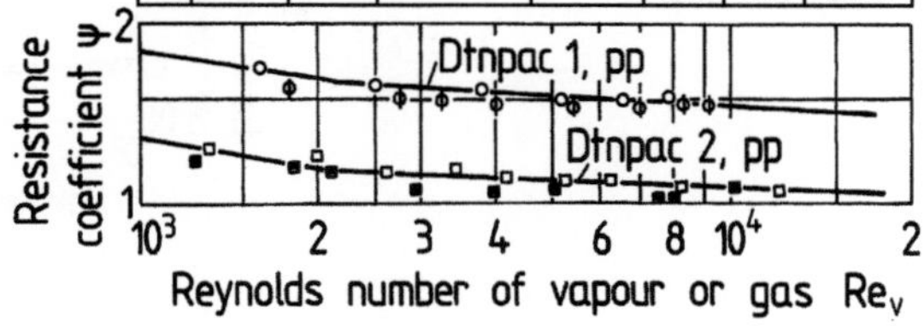

Figure 3-26. Resistance coefficient ψ as function of Reynolds number of vapour or gas Re_V, valid for random R-Pac rings sizes 1 ($\varphi_P = 0.54$) and 2 ($\varphi_P = 0.40$) and SR-Pac rings size 2 ($\varphi_P = 0.60$), made of ceramic

R-Pac, ceramic							
TP	Type	$d \cdot 10^3$ [m]	system	p_T [mbar]	d_S [m]	H [m]	N [1/m³]
◇	size 1	30	air, 293 K	1000	0.316	3.00	56000
□	size 2	55				1.36	18200
■ (grey)			chlorobenzene/ethylbezene	280	0.600	1.50	22000
■				560			

SR-Pac, ceramic							
TP	Type	$d \cdot 10^3$ [m]	system	p_T [mbar]	d_S [m]	H [m]	N [1/m³]
○	size 2	65	air, 293 K	1000	0.316	1.36	3575
● (grey)			chlorobenzene/ethylbezene	266	0.600	1.50	4953
●				505			

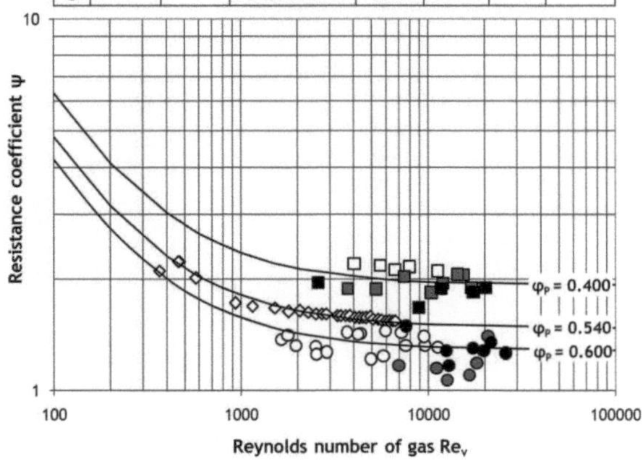

Figure 3-27. Resistance coefficient ψ as function of Reynolds number of vapour or gas Re_V, valid for random Ralu Flow packing size 2, made of plastic – $\varphi_P = 0.604$ for Eq. (3-26)

Ralu-Flow size 2, plastic

TP	$d \cdot 10^3$ [m]	system	p_T [mbar]	d_S [m]	H [m]	N [1/m³]
○	58	air, 293 K	1000	0.3	1.44	4714
●	58			0.6	2.12	4860

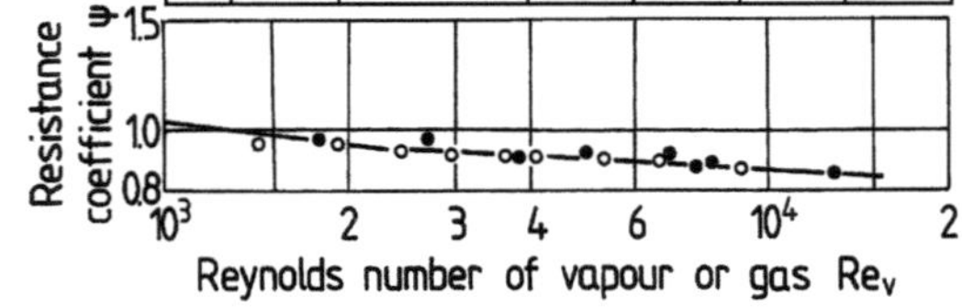

Figure 3-28. Resistance coefficient ψ as function of Reynolds number of vapour or gas Re_V, valid for 38 ($\varphi_P = 0.048$) and 50 mm ($\varphi_P = 0.28$) random Intalox saddles made of plastic

Intalox saddle, PP

TP	$d \cdot 10^3$ [m]	system	p_T [mbar]	d_S [m]	H [m]	N [1/m³]
▽	38	air, 293 K	1000	0.3	1.36	27790
▲	50	air, 296 K		0.3	1.34	8656
△				0.45	2.0	7930

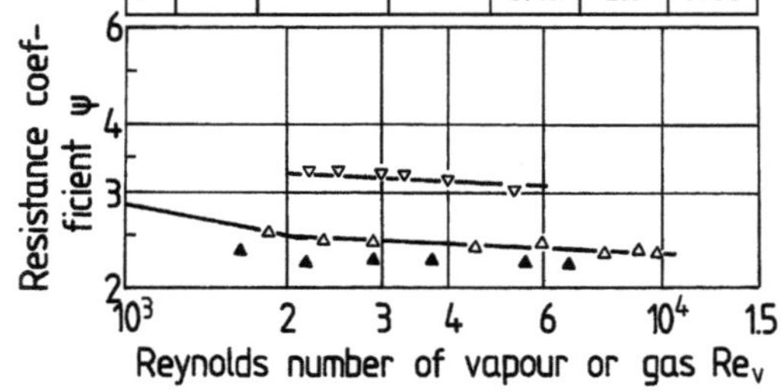

Figure 3-29. Resistance coefficient ψ as function of Reynolds number of vapour or gas Re_V, valid for 25–38 mm random Intalox saddles made of ceramic – $\varphi_P = 0.327$ for Eq. (3-26), experimental data acc. to [A, 30]

Intalox saddle, ceramic, $d_S/d \geq 6$

TP	$d \cdot 10^3$ [m]	system	p_T [mbar]	d_S [m]	H [m]	N [1/m³]
●	25	chlorobenzene/ ethylbenzene	66	0.22	1.47	59415
▲	38					16898
○	25	air, 284 K	1000	0.3	0.87	63890
▲	38	air, 290 K			1.34	18925
△	38	air, 293 K		0.45	2.0	19528

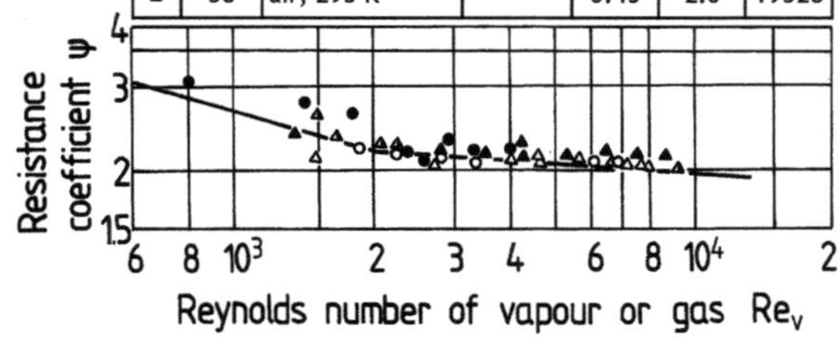

Raschig ring, ceramic							
TP	$d \cdot 10^3$ [m]	system	p_T [mbar]	d_S [m]	H [m]	N [$1/m^3$]	Lit.
▽	15	chlorobenzene / ethylbenzene	66.7	0.22	1.44	220940	Author
♦	25		66.7		1.24	51620	
●	25		133				
	50						
	50						
-o-	25	benzene / toluene	1000	0.3	2	46000	[9]
o	25	air, 293 K	1000	0.3	0.76	48176	Author
△	35			0.5	2	17115	[11]
□	50			0.45	1.4	6400	[6]

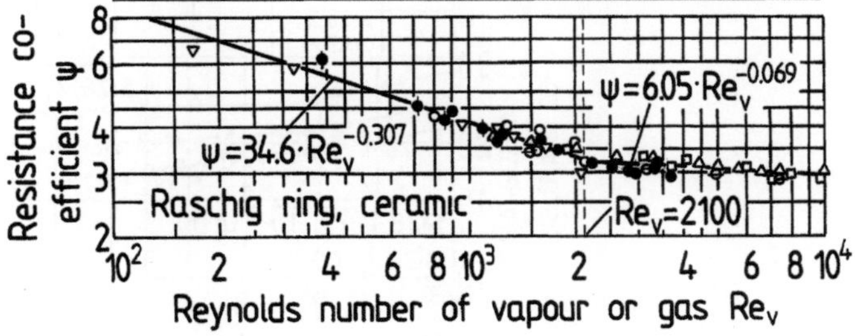

Figure 3-30. Resistance coefficient ψ as function of Reynolds number of vapour or gas Re_V, valid for 25 mm random Raschig rings made of ceramic – $\varphi_P = 0$ for Eq. (3-26), experimental data acc. to [31]

Ralu ring, PP						
TP	$d \cdot 10^3$ [m]	system	p_T [mbar]	d_S [m]	H [m]	N [$1/m^3$]
△	50	air, 293 K	1000	0.3	0.87	5913
▲				0.45	2	5841

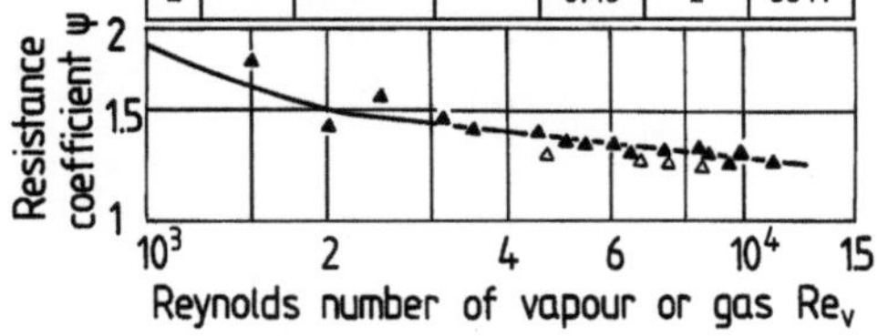

Figure 3-31. Resistance coefficient ψ as function of Reynolds number of vapour or gas Re_V, valid for 50 mm random Ralu rings made of plastic – $\varphi_P = 0.60$ for Eq. (3-26)

Figure 3-32. Resistance coefficient ψ as function of Reynolds number of vapour or gas Re_V, valid for random CMR Glitsch rings made of metal and plastic, various sizes

Glitsch CMR rings, CMR-304 of metal and CMR-PP of plastic

TP	Type	system	p_T [mbar]	d_S [m]	H [m]	N [1/m³]
✗	size 1 A CMR-304	air, 293 K	1000	0.3	1.4	158457
▽	size 1,5 A CMR-304			0.3	1.4	60744
▼ (Turbo)	size 1,5 A CMR-304			0.3	1.4	60310
▲	size 2 A CMR-304			0.45	2.0	32950
□	size 3 A CMR-304			0.45	2.0	9914
●	size 1 A CMR-PP			0.3	1.41	25568
○				0.45	2.0	24150
△	size 2 A CMR-PP			0.45	2.0	6242

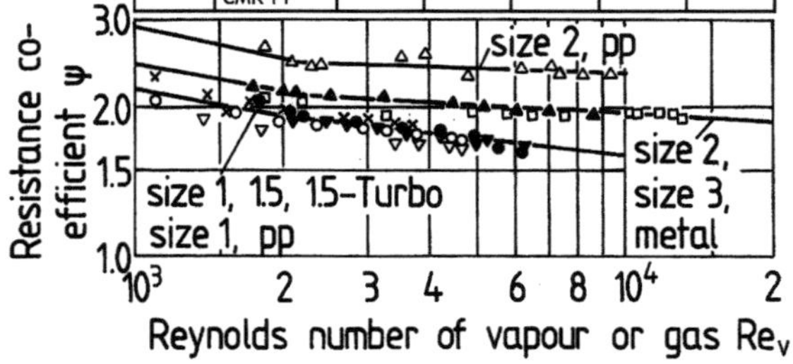

Figure 3-33. Resistance coefficient ψ as function of Reynolds number of vapour or gas Re_V, valid for random Raschig Super rings made of metal and plastic, sizes 0.3-3 – $\varphi_P = 0.43–0.65$ for Eq. (3-26), experimental data acc. to [33]

Raschig Super Ring, metal

TP	Type	d·10³ [m]	system	p_T [mbar]	d_S [m]	H [m]	N [1/m³]
◆	0.3	15	air, 293 K	1000	0.288	1.0	180000
■	0.5	20			0.288	1.0	145000
△	0.7	25			0.288	2.0	46500
✳	1.0	30			0.44	2.0	32000
○	1.5	38			0.288	2.0	13750
●	2.0	50			0.75	3.0	9500
◇	3.0	75			0.44	2.0	4300

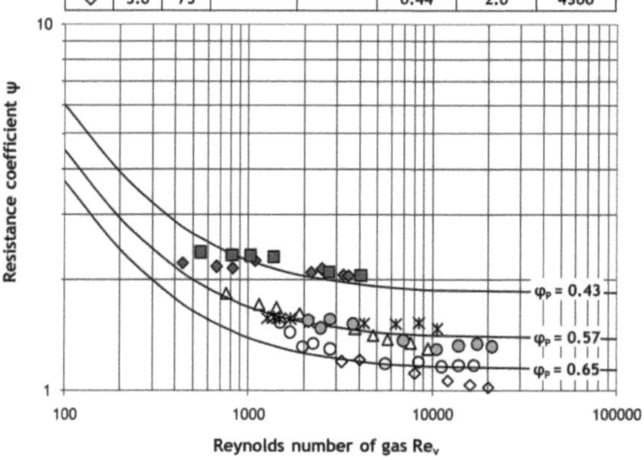

Figure 3-34. Resistance coefficient ψ as function of Reynolds number of vapour or gas Re_V, valid for random Nutter rings made of metal, sizes 0.7–3 – $\varphi_P = 0.326$ for Eq. (3-26), experimental data acc. to [36]

Nutter ring, metal						
TP	Type	system	p_T [mbar]	d_S [m]	H [m]	N [1/m³]
◆	0.7	air, 293 K	1000	0.3	2.25	167400
■	1.0					67100
△	1.5					26800
□	2.0					13600
✱	2.5					8800
●	3.0					4200

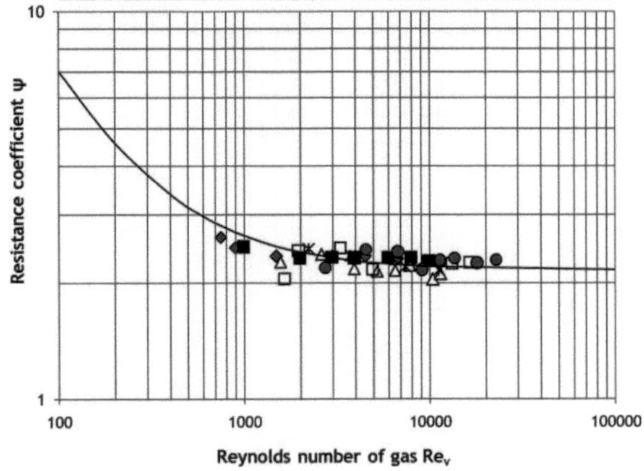

Figure 3-35. Resistance coefficient ψ as function of Reynolds number of vapour or gas Re_V, valid for random Rauschert metal saddle ring made of metal, sizes 25–70 – $\varphi_P = 0.378–0.647$ for Eq. (3-26), experimental data acc. to [32]

Rauschert metal saddle ring, metal					
TP	Type	$d \cdot 10^3$ [m]	system	p_T [mbar]	d_S [m]
■	25-4	25	air, 293 K	1000	0.8
△	50-5	50			
×	60-5	60			
●	40-4	40			
□	70-5	70			
	70-6	70			

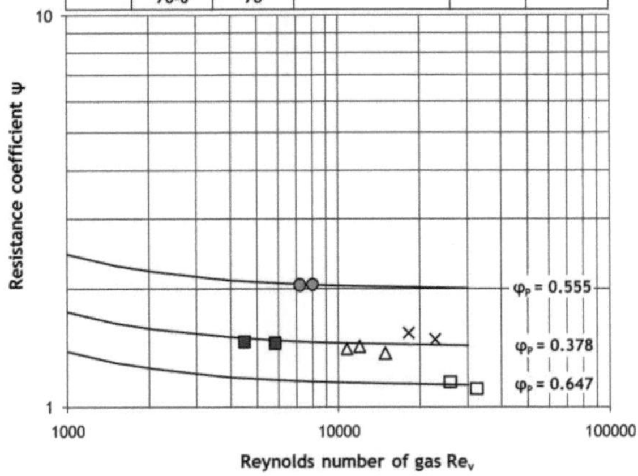

FKP, PP							
TP	Type	a [m²/m³]	system	p_T [mbar]	d_S [m]	H [m]	φ_P [-]
△	327	125	air, 293 K	1000	0.60	3.04	0.856
○	319	150	air, 293 K	1000	0.60	3.05	0.921
●	319	150	air, 293 K	1000	0.60	1.20	0.921

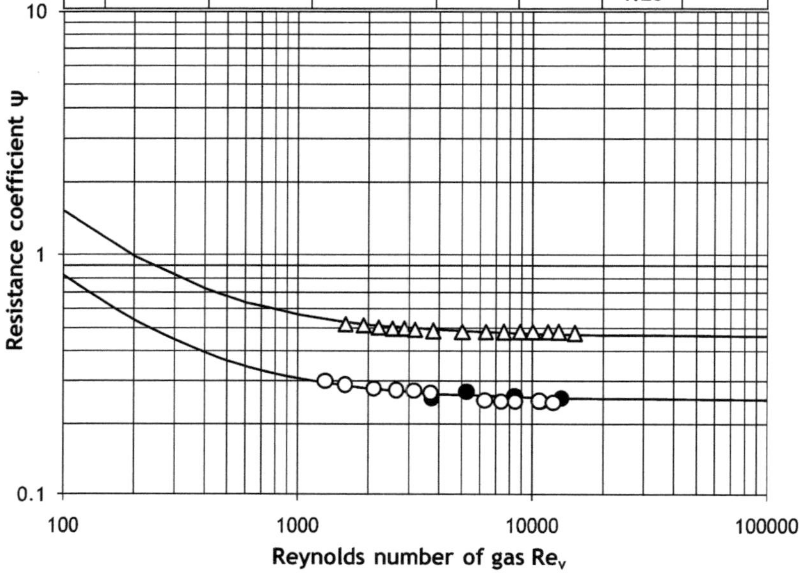

Figure 3-36. Resistance coefficient ψ as function of Reynolds number of vapour or gas Re_V, valid for structured packing GEA-H2 FKP made of plastic, sizes 319–327 – $\varphi_P = 0.856 - 0.921$ for Eq. (3–26)

NET, PP							
TP	Type	a [m²/m³]	system	p_T [mbar]	d_S [m]	H [m]	φ_P [-]
□	38-150	80	air, 293 K	1000	0.60	3.15	0.824
■					0.60	1.80	

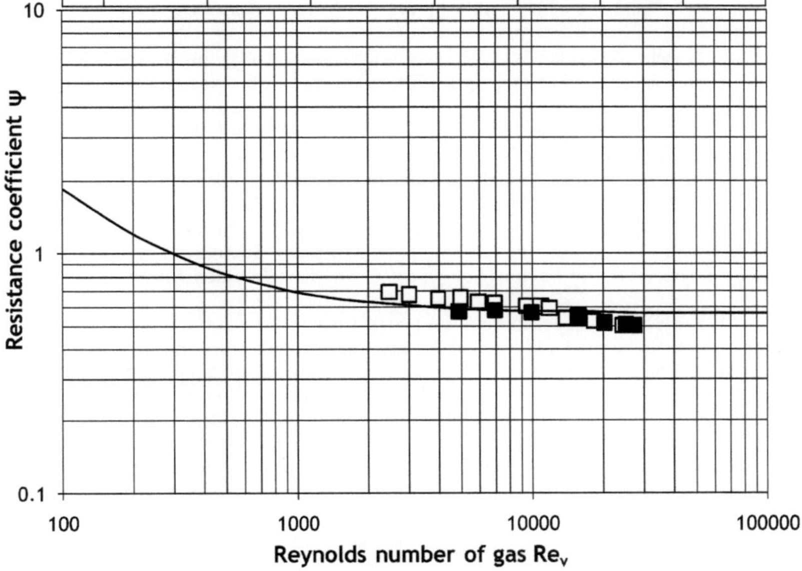

Figure 3-37. Resistance coefficient ψ as function of Reynolds number of vapour or gas Re_V, valid for structured packing GEA-H2 NET made of plastic, sizes 38–150 – $\varphi_P = 0.824$ for Eq. (3–26)

References Chapter 3

1. Reichelt W. Strömung in Füllkörperapparaten bei Gegenstrom einer flüssigen und einer gasförmigen Phase. Verlag Chemie, Weinheim (1974)
2. Reichelt W. Zur Berechnung des Druckverlustes einphasig durchströmter Kugel- und Zylinderschüttungen. Chem.-Ing.-Tech., vol. 44 (1972), p. 1068/1071
3. Brauer H. Grundlagen der Einphasen- und Mehrphasenströmungen. Verlag Sauerländer, Aarau und Frankfurt/M. (1971)
4. Brauer H, Mewes D. Strömungswiderstand sowie Stoff- und Wärmeübergang in ruhenden Füllkörperschichten. Chem.-Ing.-Tech., vol. 44 (1972) no. 1 and 2, p. 93/96
5. Ergun S. Fluid flow through packed columns. Chem. Eng. Progr., vol. 48 (1952) no. 2, p. 89/94
6. Maćkowiak J. Einfluss der Führungsflächen auf die Hydraulik und den Stoffübergang in Füllkörperkolonnen bei der Absorption. Dissertation TU-Wrocław (1975)
7. Billet R, Maćkowiak J. Neues Verfahren zur Auslegung von Füllkörperkolonnen für die Rektifikation. „vt"-verfahrenstechnik, vol. 1–7 (1983) no. 4, p. 203/211
8. Billet R, Maćkowiak J. How to use the absorption data for design and scale-up of packed columns. Fette, Seifen, Anstrichmittel, vol. 86 (1984) no. 9, p. 349/358
9. Schmidt R. Zweiphasenstrom und Stoffaustausch in Schüttungsdichten. VDI-Forschungsheft no. 550, VDI-Verlag, Düsseldorf (1972)
10. Teutsch Th. Druckverlust in Füllkörperschüttungen bei hohen Berieselungsdichten. Dissertation TH München (1962)
11. Kast W. Gesetzmäßigkeiten des Druckverlustes in Füllkörpersäulen. Chem.-Ing.-Tech., vol. 36 (1964) no. 5, p. 464/468

12. Bemer GG, Kalis GAJ. A new method to predict hold-up and pressure drop in packed columns. Trans. I. Chem. Eng., vol. 56 (1978), p. 200/205
13. Billet R. Industrielle Destillation. Verlag Chemie, Weinheim (1973)
14. Billet R, Maćkowiak J. Hochwirksame metallische Packung für Gas- und Dampf/Flüssig-Systeme. Chem.-Ing.-Tech., vol. 57 (1985) no. 11, p. 976/978
15. Billet R, Maćkowiak J. Application of modern Packings in Thermal separation Processes. Paper at Achema 1985 – Frankfurt a.M. on 10 June 1985. as well as Chem. Eng. Technol. vol. 11 (1988), p. 213/227
16. Meier W, Hunkeler R, Stöcker WD. Performance of a new regular tower packing 'Mellapak'. 3rd Int. Symp. on Distillation, London, April 1979. as well as Chem.-Ing.-Tech., vol. 51 (1979) no. 2, p. 119/122
17. Bornhütter K. Stoffaustausch von Füllkörperschüttungen unter Berücksichtigung der Flüssigkeitsströmungsform. Dissertation, TU München (1991)
18. Krehenwinkel H. Experimentelle Untersuchungen der Fluiddynamik und der Stoffübertragung in Füllkörperkolonnen bei Drücken bis 100 bar. Dissertation, TU Berlin, Dezember (1986)
19. Maćkowiak J, Ziołkowski Z. Untersuchungen der Fluiddynamik von Kolonnen mit regellosen Füllkörpern Bauart Białecki-Ring und I-13-Rings (orig. Polish). Scientific Papers of the Institute of Chemical Engineering and Heating Equipment of Wrocław Technical University (1973), no. 21, p. 3–24
20. Ziołkowski Z, Maćkowiak J. Badanie oporów przepływu jednofazowego w kolumnach wypełnionych pierścieniami Białeckiego i I-13 (orig. Polish). Inż. Chem. vol. IV (1974) no. 4, p. 703
21. Ziołkowski Z, Maćkowiak J, Kowalski J. Badanie oporu jednofazowego przepływu dla układanych pierścieni Białeckiego 50 mm (orig. Polish). Inż. Chem. vol. V (1975) no.2, p. 429
22. Maćkowiak J, Suder S. Hydraulika i wymiana masy w kolumnie wypełnionej układanymi pierścieniami Białeckiego (orig. Polish). Inż. Chem. i. Procesowa vol. 8 (1977) no. 3, p. 651–664
23. Maćkowiak J, Suder S. The new tube column with Pall-Rings for counter-current and cucurrent processes. Dissertation Vesprem (Hungary) 1979 and. Chem.-Ing.-Techn., vol. 50 (1978) no. 7, p. 550–551
24. Maćkowiak J. Hydraulische Untersuchungen der geordneten 50-mm keramischen Raschigringe. Internal study – TU-Wrocław (1975)
25. Kuźniar J, Niżański A. Investigations on Mellapak 250Y made of sheet metal (orig. Polish). III- Symposium volume: Rectification, Absorption, Extraction p. 65–74. Szklarska Poręba, 13–15. 09.1999
26. Maćkowiak J. Mc-Pac – ein neuer metallischer Füllkörper für Gas-Flüssigkeitssysteme. Chem.-Ing.-Techn. vol. 73, no. 1+2 (2001) p. 74–79
27. Maćkowiak J. Mc-Pac – Nowe metalowe wypełnienie dha układów gaz-ciecz (orig. Polish). Inż. Chemiczna i Procesowa, vol. 21 (2000) p. 679–689
28. Maćkowiak J, Ługowski Z. Geringe Apparatevolumina und Betriebskosten mit neuen keramischen Füllkörpern - R-Pac und SR-Pac. Verfahrenstechnik, vol. 29 (1995), no. 6, p. 19–22
29. Maćkowiak J, Szust J. Hydraulika i wymiama masy w kolumnach wypełnionych ceramicznymi pierścieniami R-Pac i SR-Pac w układach gaz/ciecz (orig. Polish). Inż. Chem. i Procesowa, vol. 18 (1997), no. 4, p. 675–691
30. Bańczyk L, Woźniak A, Szymkowiak E. Hydraulika kolumny wypełnionej ceramicznymi pierścieniami Białeckiego (orig. Polish). Inż. Chem. i Procesowa, vol. 7 (1977), no. 1, p. 261–274
31. Bańczyk L, Woźniak A, Jarzynowski M, Grobelny A. Hydraulika kolumny wypełnionej ceramicznymi pierścieniami IChN oraz Raschiga (orig. Polish). Inż. Chem. i Procesowa, vol. 9 (1979), no. 1, p. 15–28
32. Linek V, Moucha T, Rejl F. Hydraulic and mass transfer characteristics of packings for absorption and distillation columns. Rauschert-metall-sattel-rings. Trans IChemE, Vol 79, Part A, October 2001
33. Schultes M. Raschig Super-Ring. A new Fourth Generation Packing Offers New Advantages. Trans IChemE, Vol 81, Part A, January 2003
34. Maćkowiak J. Neue Aspekte der Modellierung bei der Fluiddynamik von Packungskolonnen. Chem.-Ing.-Techn., vol. 78 (2006), no.8, p. 1079–1086
35. Maćkowiak J. Extended channel model for prediction of the pressure drop in single-phase flow in packed columns. ChERD, Vol 87 (2009), p. 123–134
36. Metal Random Packing - Superior random packings combined with innovative mass transfer technology. Technical Information by Sulzer Chemtech Ltd., 22.64.06.40 - IV.06 – 50, Switzerland
37. Billet R, Schultes M. Modelling of pressure drop in packed columns. Chem. Eng. Technol. vol. 14 (2004), p. 89/95

Chapter 4

Pressure Drop of Irrigated Random and Structured Packings

4.1 Introduction and Literature Overview

4.1.1 Significance of Pressure Drop for Packed Column Design

It is important to know the total pressure drop Δp of the irrigated packed bed when designing packed columns for gas/liquid systems in counter-current flow of the phases. In absorption as well as desorption processes, the total pressure drop Δp of the packing determines the blower capacity and thus the major part of the operating costs of the process. In rectification, the sum of the top pressure p_T and the total pressure drop Δp gives the bottom pressure p_W, which determines the bottom temperature t_W. The bottom temperature, in turn, determines the effective temperature difference, at a given heating medium temperature, at which the reboiler in the distillation column must be operated.

High pressure drops are likely to result in significant changes in the relative volatility along the column height, in particular in vacuum rectification processes with a high number of theoretical stages [5, 32], which means that the reflux ratios R, required to achieve the specified product purity, are considerably higher than in the case of low pressure drops Δp. Higher reflux ratios R lead to an increase in steam consumption and therefore in operating costs.

The total pressure drop of the packing Δp at a known packing height H, the calculation of which was the topic of studies such as [13–15], is given by the product $\Delta p = (\Delta p/H)\cdot H$, in which the quotient $\Delta p/H$ stands for the pressure drop per 1 m height of the packed bed.

Based on the known pressure drop $\Delta p/H$ and the specific pressure drop $\Delta p/n_t$, Eq. (1-4) can be used to calculate the theoretical separation efficiency of the random or structured packing n_t/H, and therefore the total height of the packed bed H, if the number of theoretical stages n_t is known.

J. Maćkowiak, *Fluid Dynamics of Packed Columns*,
Chemische Technik/Verfahrenstechnik, DOI 10.1007/b98397_4,

This chapter takes a closer look at the calculation of the pressure drop of the irrigated packed bed $\Delta p/H$ for randomly filled and stacked packing elements as well as structured packings.

4.1.2
Literature Overview

Table 4-1 contains a list of some important studies [1–21, 25, 41–48] on the pressure drop of irrigated packings in columns with conventional packing elements, such as Raschig rings, Pall rings and Intalox saddles. The work can be divided into three groups.

The first group includes the graphic methods and can be found in the fundamental work of Sherwood et al. [29], Eckert [9], Bolles and Fair [8], Mersmann [2, 21, 44], Schmidt [7], Schumacher [10], etc. They are based on capacity diagrams, in which the dimensionless pressure drop lines $\Delta p/(\rho_L \cdot g \cdot H)$ are plotted against the empirical flooding point, mostly determined using the test system air/water. The graphic methods allow for a quick estimation of the pressure drop up to the flooding point. Apart from the dimensionless pressure drop $\Delta p/(\rho_L \cdot g \cdot H)$, the capacity diagrams can also be used to plot other constant parameter curves, e.g. relating to the liquid hold-up [2, 21, 55] or the volumetric mass transfer area [7], which is one of their additional advantages [7, 9, 21]. This results in system-related capacity diagrams, such as the ones presented by Schmidt [7], which are relevant for practical applications.

The capacity diagrams, showing the pressure drop parameter [1, 8, 9, 21, 29, 55], are very useful for estimating the pressure drop $\Delta p/H$. Mersmann [21, 55] and Reichelt [1] give an accuracy of ±25% and ±20%, respectively. According to Molzahn and Wolf [25] as well as Blaß and Kurtz [42, 43], the accuracy of the graphic correlations is ±50%.

The second group includes empirical methods, which are based on tests using air/water. Here, the influence of physical properties, operating conditions, such as the specific liquid load u_L, and packing-specific variables on the quotient $\Delta p/\Delta p_0$ in Eq. (3-1) is expressed by the dimensionless numbers Re_L, Re_V, Fr_L and/or empirical power law models. Please refer to Teutsch [3, 16], Weiß et al. [11], Maćkowiak [17], Leva [20], Kast [45], Kolev [46], Beck [47] and Kleinhükelkotten [48] and others, see Table 4-1b Table 4-1c.

Studies by Weiß et al. [11] as well as other authors [3, 25] show that calculations based on these models in relation to irrigated classic packings such as Pall or Raschig rings used for rectification systems, are only in line with experimental data, if the quotients $\Delta p/\Delta p_0$, experimentally determined using the air/water system, are multiplied with the physical property quotients $(\rho_L/\rho_W)^{n_1}$, $(\rho_V/\rho_W)^{n_2}$, $(\eta_L/\eta_W)^{n_3}$ with experimentally derived exponents n_1, n_2, n_3. This improves the accuracy of the experimental results, which can be given with a mean relative error of ±10–25%, [11]. Another disadvantage of this group of methods is the fact that their application is restricted to the range below the loading line. The accuracy of other methods [1, 7, 46–48] in the same range is approximately similar.

Table 4-1a. List of graphical methods for determining the pressure drop of irrigated random packings

Packing	Column d_S [m]	H [m]	System	Validity range	Presentation form of results	Literature
ceramic Raschig rings, Lessing rings, spirals, etc.	–	–	air/water, et al. 1 bar, 293 K	up to flooding point	flood load diagram	Sherwood et al. [29]
Berl saddles, Intalox saddles, metal Pall rings, et al. d = 0.008–0.09 m			air/water, et al. 1 bar, 293 K	up to flooding point	flood load diagram	Eckert [9]
ceramic Raschig rings, Intalox saddles, metal Pall and Raschig rings d = 0.015–0.050 m	up to 1.2	5.5	air/water, et al. 1 bar, 293 K	up to flooding point	flood load diagram	Bolles, Fair [8]
ceramic Raschig rings, Intalox and Berl saddles, d = 0.035 m			air/water, et al. 1 bar, 293 K	up to flooding point	diagrams $\Delta p_0/\left(\sigma_L \cdot g \cdot H\right) = f\left(B_L^*\right)$	Mersmann [2, 21, 44]
ceramic Raschig rings d = 0.008–0.050 m	0.3	2.0	ethanol/water, benzene/water, et al.	up to flooding point	diagrams etc. $u_{V,Fl} = f\left(u_{L,Fl}\right)$	Schmidt [7]
ceramic and metal Raschig rings, Pall rings, saddles et al.			various systems	up to flooding point	diagram: 5 different operating ranges	Schumacher [19], Billet [5], Teutsch [16] and Kast [45]
Raschig rings made of metal and ceramic, Pall rings made of metal d = 0.008–0.025 m	0.040–0.150	1	air/water, et al.	up to flooding point	empirical equation	Reichelt [1]
Pall rings made of metal d = 0.025–0.050 m			various systems	up to flooding point	comparison of data by Billet [5], Teutsch [16], et al. with Eckert's correlation	Molzahn, Wolf [25]

Table 4-1b. List of empirical methods for the prediction of the pressure drop of irrigated random packings

Packing	Column d_S[m]	H [m]	System	Validity range	Presentation form of results	Literature
ceramic **Raschig rings, Pall rings, Intalox- and Berl saddles**	0.29–0.44	1.9	air/water	below loading line, up to u_L < 190 mh^{-1}	$\Delta p/\Delta p_0 = f(K_P)$ diagram	Teutsch [16]
ceramic **Raschig rings, Pall rings, saddles** d = 0.025–0.050 m	0.5	2	air/water	below loading line	$\Delta p/\Delta p_0 = f(Fr_L)$ diagram	Kast [45]
metal and plastic **Białecki rings** and ceramic **Raschig rings** d = 0.025–0.053 m	0.3	from 0.45 to 1.40	air/water	above and below loading line, up to flooding point	$\Delta p/\Delta p_0 = a \cdot \exp(K_P \cdot b)$ $K_P = f(Re_V, Re_L, Fr_L)$ a, b = constant	Maćkowiak [17]
ceramic **Raschig rings, Intalox saddle** d = 0.010–0.050 m	–	–	air/water	below loading line	empirical equation	Leva [20]
ceramic and metal **Pall rings, Raschig rings, saddles** d = 0.010–0.070 m	–	–	air/water	below loading line	$\Delta p/\Delta p_0 = (1-A)^3$ $A = f(Re_L, Fr_L, k)$	Kolev [46]
Raschig and Pall rings, Intalox saddle made of metal and ceramic d = 0.008–0.050 m	up to 1.2	up to 5.5	various rectification systems	–	empirical equation	Beck [47] and Billet [5]
Pall rings, Intalox saddle, Raschig rings, et al.				below loading line	empirical equation	Kleinhückelkotten [48]
Pall and Raschig rings d = 0.025–0.050 m	0.4	2.0	7 different rectification systems	–	empirical equation	Weiß et al. [11]

Table 4-1c. List of methods based on the channel model for the prediction of the pressure drop of irrigated random packings

Packing	Column d_S[m]	H [m]	System	Validity range	Presentation form of results	Literature
16 mm Raschig rings ceramic	0.146	1.40	various systems, 1 bar, $\eta_L = 1-200$ mPas $\rho_L = 810-1331$ kgm^{-3} $\sigma_L = 27-17$ mNm^{-1}	below loading line	$\Delta p/\Delta p_0 =$ $= (1 - C \cdot h_L)^{-5}$ $C = 2.1$	Buchanan [19]
Raschig rings made of ceramic, **Pall rings** made of metal d = 0.008–0.080 m	up to 1.2	5.5	various systems	below loading line	Eq. (4-9), (4-10)	Bemer, Kalis [12] and Billet [5]
Raschig rings made of glass, porcelain, metal and copper d = 0.008–0.025 m	0.040 0.080 0.150	1.0	air/water air/glycol air/silicone oil	below loading line root-mean-square deviation ± 30%	$\Delta p/\Delta p_0 = f(h_L)$	Blaß, Kurtz [42, 43]
Pall rings, Białecki rings, NSW rings, Raschig rings made of metal, ceramic and plastic d = 0.015–0.080 m	0.15–1.20	0.7–4.5	various systems, $\eta_L = 0.2-50$ mPas $\rho_L = 730-1250$ kgm^{-3} $\sigma_L = 14-72$ mNm^{-1}	below loading line and $Re_L > 10$,	Eq. (4-11) constants in Table 4-2	Billet, Maćkowiak [13, 15]

Other methods, found in literature, for predicting the pressure drop in irrigated random packings are based on the model which is applicable to tube flow, Eq. (3-2), [52–54, 56, 64]. They belong to the third group of models, which is discussed in the following.

Deriving the Correlation for the Quotient $\Delta p/\Delta p_0$ Based on the Channel Model

If liquid passes through the packing, the free cross-section of the channels available for gas flow is reduced, see Figs. Fig. 3-1 and Fig. 4-1, and is given as $\pi \cdot (d - 2\delta_L)^2$.

This means that in two-phase flow the effective gas velocity $\bar{u}'_V$ in the channels increases. The pressure drop of the irrigated packing can be expressed using Eq. (3-2):

$$\frac{\Delta p}{H} = \lambda_L \cdot \frac{\bar{u}'^2_V}{2 \cdot (d_h - 2\delta_L)} \cdot \rho_V. \tag{4-1}$$

Dividing Eq. (4-1) by Eq. (3-2) gives the following relation:

$$\frac{\Delta p}{\Delta p_0} = \frac{\lambda_L}{\lambda} \cdot \frac{\bar{u}'^2_V}{(d_h - 2\delta_L)} \cdot \frac{d_h}{\bar{u}'^2_V} \tag{4-2}$$

Following the substitution of the effective gas velocity $\bar{u}'_V$, according to Eq. (4-3), Eq. (4-2) is converted to Eq. (4-4).

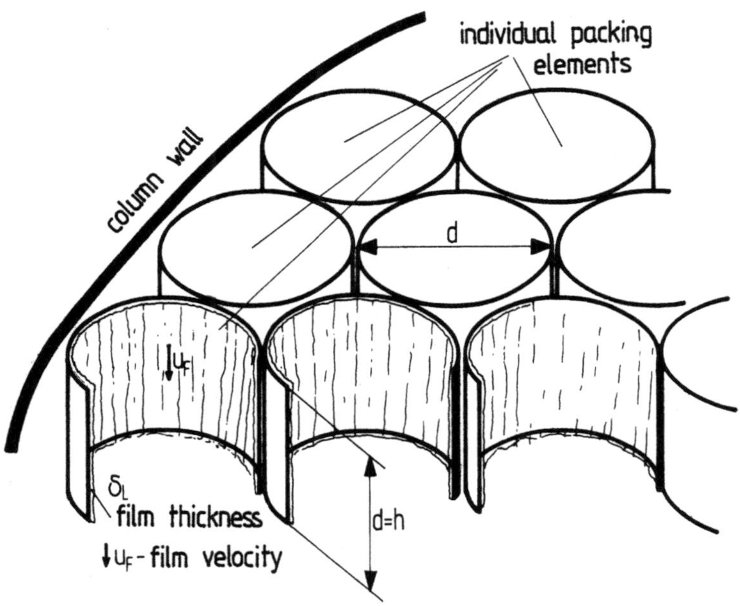

Figure 4-1. Section of packed bed as flow model

$$\bar{u}'_V = \bar{u}_V \cdot \frac{d_h^2}{(d_h - 2\delta_L)^2} \tag{4-3}$$

$$\frac{\Delta p}{\Delta p_0} = \frac{\lambda_L}{\lambda} \cdot \left[\frac{d_h}{(d_h - 2\delta_L)}\right]^5 \tag{4-4}$$

This, in turn, leads to correlation (4-5):

$$\frac{\Delta p}{\Delta p_0} \cong \left[1 - \frac{2\delta_L}{d_h}\right]^{-5} \tag{4-5}$$

In addition, it is possible to describe the mean film thickness δ_L, using correlation (4-6) [19, 12], assuming full wettability of the packing elements.

$$\delta_L = \frac{h_L}{a} \tag{4-6}$$

The pressure drop $\Delta p/\Delta p_0$ is now given as:

$$\frac{\Delta p}{\Delta p_0} \cong \left[1 - \frac{2 \cdot h_L}{a \cdot d_h}\right]^{-5}. \tag{4-7}$$

With $d_h = d$, see Fig. 4-1, and therefore with $d \cdot a = const.$ [19], correlation (4-7) leads to Eq. (4-8):

$$\frac{\Delta p}{\Delta p_0} = (1 - const. \cdot h_L)^{-5}, \tag{4-8}$$

which was derived by Buchanan [19] as early as 1967 and verified by means of experimental data for 16 mm randomly filled ceramic Raschig rings, using various liquids. A numerical value of 2.1 was calculated for the constant in Eq. (4-8). This equation is valid in the range below the loading line.

Buchanan's method [19] was further modified by Bemer and Kalis [12], see Eq. (4-9):

$$\frac{\Delta p}{\Delta p_0} = \left[1 - \frac{h_L}{2 \cdot \varepsilon \cdot \phi \cdot x^{5/3}}\right]^{-5} \tag{4-9}$$

The parameters used in Eq. (4-9) are the efficiency factor $\phi = a'/a$ of the dry packing as well as the contraction coefficient x. Both parameters must be derived by experiments. Bemer and Kalis [12] developed the following equation for calculating the pressure drop $\Delta p/H$ per 1 m for turbulent gas flow:

$$\frac{\Delta p}{H} = 0.29 \cdot \phi^{-2} \cdot F_V^2 \cdot F_P \cdot \left[1 - \frac{h_L}{2 \cdot \varepsilon \cdot \phi \cdot x^{5/3}}\right]^{-5} \quad \left[Pam^{-1}\right]. \tag{4-10}$$

It was verified by means of experimental data for various vapour/liquid systems [5]. In the case of 15–50 mm ceramic Raschig rings, the numerical values for the model parameters ϕ and x were as follows [12]:

$$\phi = 0.6 \quad \text{and} \quad x = 0.435 \tag{4-10a}$$

and in the case of 15–50 mm metal Pall rings [12]:

$$\phi = 0.8 \quad \text{and} \quad x = 0.485 \tag{4-10b}$$

Based on these parameters, the experimental data, primarily derived by Billet [5], is verified with an accuracy of $\delta(\Delta p/H) < \pm 40\%$ for Raschig rings and $\delta(\Delta p/H) < \pm 20\%$ for metal Pall rings. Equation (4-10) is also applicable in the range below the loading line and is based on turbulent gas flow.

Billet's [5] experimental data as well as more recent data [13–15] can be verified more accurately by using the correlation developed by Billet and Maćkowiak [13–15].

$$\frac{\Delta p}{H} = \psi \cdot \frac{1-\varepsilon}{\varepsilon^3} \cdot \frac{F_V^2}{d_P \cdot K} \cdot \left[1 - \frac{h_L}{2 \cdot \varepsilon \cdot \phi \cdot x^{5/3}}\right]^{-5} \quad \left[Pam^{-1}\right]. \tag{4-11}$$

This equation is applicable for gas flow in the transition range and in the turbulent range of the gas phase below the loading line.

Table 4-2 contains a list of the newly determined model parameters ψ, ϕ, x for Eq. (4-11) as well as the constant C for the empirical Eq. (4-12)

$$h_L = C \cdot (Fr_L^0)^{1/3} = C \cdot \left[\frac{u_L^2}{g \cdot d \cdot \varepsilon^2}\right]^{1/3} \quad \left[\mathrm{m^3 m^{-3}}\right] \tag{4-12}$$

used for calculating the liquid hold-up h_L. The values shown in this table refer to various types of packings, including Pall rings made of metal, plastic and ceramic, metal Białecki rings as well as Nor-Pac (NSW rings) and Hiflow rings made of plastic. The experimental values for various systems [5, 13–15] are given more accurately by Eq. (4-11) in the range below the loading line than by using the model parameters ϕ, x, according to Bemer and Kalis [13–15], i.e. $\delta(\Delta p/H) < \pm 10–15\%$ [13–15].

Conclusions– Literature Overview

A number of different models have been published, describing the pressure drop in irrigated random packings for counter-current flow of the phases. Most of these models only cover the complex flow behaviour in the operating range below the loading line, see Table 4-1.

The pressure drop $\Delta p/H$ can be easily estimated using the capacity diagrams [1, 8, 9, 21, 48, 55, 56], see Table 4-1, a number of empirical methods [3, 11, 16, 17, 20, 45 etc.], and/or the pressure drop diagrams such as the one shown in Fig. 2-2a.

Using the methods (4-8), (4-9), (4-11), which are based on the method for tube flow (3-2), it is possible to predict the pressure drop below the loading line for various classic

Table 4-2. List of numerical values for constant C used for predicting the liquid hold-up h_L acc. to Eq. (4-12), as well as the parameter ϕ_m x for Eq. (4-11) for various packings, valid for turbulent liquid and gas flow [13–15]

Packing	Material	$d \cdot 10^3$ [m]	Model parameter for Eq. (4–11)			
			ψ_m	ϕ	x	$C_{Eq. (4\text{–}12)}$
Pall ring	metal	15				
		25				
		35	≈ 2.45	0.81	0.525	1
		50				
		80				
	ceramic	50	≈ 1.95	0.92	0.64	1.15
	plastic	25				
		35	≈ 2.45	0.83	0.57	0.88
		50				
NSW ring (Nor-Pac)	plastic	27				
		38	≈ 1.025	1	0.64	1
		50				
Białecki ring	metal	25				
		35	≈ 2.45	0.79	0.596	1
		50				
Hiflow ring	plastic	50	≈ 1.10	1	0.53	1.068
	metal	–	–	–	–	1.068
	ceramic	–	–	–	–	0.92
Intalox saddle	ceramic	25	–	–	–	1
	plastic	50	–	–	–	0.75
Raschig ring	ceramic	15	–	–	–	1
		25				1
stacked Białecki rings	metal	25	–	–	–	0.88
		35				1

packing elements and some modern types of packing elements. However, they require a few packing-specific constants, which need to be experimentally determined. The calculation of the pressure drop above the loading line and close to the maximum column capacity of approx. 80% of the flooding point is still a difficult task, and there are only a few empirical methods available, e.g. [42, 43] and [17, 55, 56].

This monograph will present a reliable and simple calculation model for determining the pressure drop for any types of systems and column internals throughout the entire operating range up to the flooding point, which can be used to predict the pressure drop of randomly filled and stacked packing elements, tube columns and different types of structured packings and is applicable for any gas/liquid system, for rectification and absorption.

4.2 Liquid Hold-Up

It is important to ascertain the liquid hold-up h_L throughout the entire operating range of a packed column, both for the prediction of the pressure drop and for the constructive design of a packed column. The liquid hold-up at the flooding point determines the constructive design of the column support, while h_L under operating conditions determines the distance between the support and the liquid level at the bottom of the column as well as the position of the gas inlet connection.

4.2.1
Basic Terms

The prediction of the pressure drop in irrigated random packings, acc. to Buchanan [19], Eq. (4-8), Bemer and Kalis [12], Eq. (4-9), and/or Blaß, Kurtz [42, 43] and Maćkowiak [53, 54], requires knowledge of the total liquid hold-up h_L. The work of Gelbe [22], Reichelt [1], Schmidt [7], Buchanan [19], Blass and Kurtz [42, 43] as well as Mersmann, Deixler [44] and Stein [58] contains some comprehensive literature research on the methods for calculating the liquid hold-up in the range below the loading line, and an evaluation of the influence of the physical properties η_L, σ_L, ρ_L, the constructive parameters d_S, d as well as the operating conditions u_L and u_V. The experimental results presented in publications to date mostly refer to thick-walled, classic, ceramic packing elements with $d \leq 0.050$ m and/or, in the case of more recent publications [52–54, 58], to modern random and structured packings. The proportion of the liquid V_{st} retained in pores and void spaces can be substantial in packing elements of this type and size. The amount of liquid V_{st} relating to the packing volume V_S is known as the static liquid hold-up $h_{st} = V_{st}/V_S$. The amount of liquid V_{st} relating to the packing volume V_S, which is part of the liquid flow, makes up the dynamic liquid hold-up, with $h_d = V_d/V_S$. The total amount of liquid $V_L = V_d + V_{st}$ is the total liquid hold-up h_L [22], i.e.:

$$V_L = V_{st} + V_d \rightarrow h_L = \frac{V_L}{V_S} = h_{st} + h_d \quad \left[\mathrm{m^3 m^{-3}}\right]. \tag{4-13}$$

4.2.2
Static Liquid Hold-Up

According to the model by Gelbe [22], the static liquid hold-up h_{st} is dependent on the physical properties and the specific liquid load u_L. If the liquid load u_L tends to zero, the static hold-up reaches its maximum value – adhesion hold-up h_H. The following applies:

$$u_L \rightarrow 0 \quad \Rightarrow \quad h_L = h_{st,\,\max} = h_H. \tag{4-14}$$

As the liquid load u_L increases, the static liquid hold-up decreases and finally reaches zero at the so-called critical $u_{L,crit}$ value, i.e.:

$$u_L = u_{L,crit.} \quad \Rightarrow \quad h_L \cong h_d. \tag{4-15}$$

This means that the total liquid hold-up h_L can be equated with the dynamic liquid hold-up h_D. According to Mersmann and Deixler [44], the static liquid hold-up $h_{L,st}$ prevails in the case of very low, dimensionless liquid load B_L, acc. to Eq. (4-16):

$$B_L = \left[\frac{\eta_L}{\rho_L \cdot g^2}\right]^{1/3} \cdot \frac{u_L}{\varepsilon} \cdot \frac{1-\varepsilon}{\varepsilon \cdot d_P} \tag{4-16}$$

for $B_L < 2 \cdot 10^{-5}$.

Here, the static liquid hold-up reaches the value of the adhesion hold-up h_H. The adhesion hold-up h_H is dependent on the following number [44], see Fig. 4-2,

Figure 4-2. Adhesion liquid hold-up as function of dimensionless number We_L/Fr_L according to Mersmann and Deixler [44]

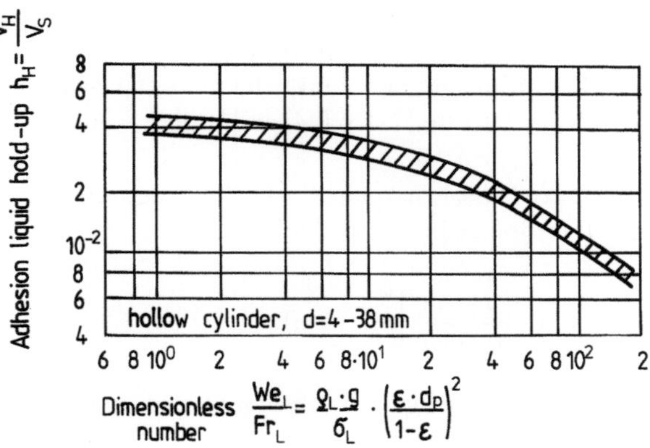

$$\frac{We_L}{Fr_L} = \frac{\rho_L \cdot g}{\sigma_L} \cdot \left[\frac{\varepsilon \cdot d_P}{1-\varepsilon}\right]^2 \tag{4-17}$$

i.e. on the physical properties σ_L, ρ_L and the packing-specific parameters d_P and ε. Figure 4-2 was created based on the experimental results presented by Mersmann and Deixler [45], Gelbe [22] as well as Blaß and Kurtz [42, 43].

The adhesion hold-up values of the packing elements with diameters of $d \geq 0.025-0.090$ m and systems analysed in this book are expected to be very low, i.e. approx. 1% and below, as the dimensionless ratio We_L/Fr_L is in the range of approx. 70–1300.

4.2.3 Dynamic Liquid Load Below the Loading Line

When deriving the methods for calculating the dynamic liquid hold-up, it is assumed that the liquid hold-up is only associated with the trickle film in the random or structured packing. It is also assumed that there are no droplets in the packing and the liquid sprays are also associated with the trickle film.

There are 3 forces acting upon the liquid trickle films in packed columns: gravitational force K_g, viscosity force K_η and resistance force K_ψ [12, 19, 21, 22, 58]. The gravitational force can always be regarded as a driving force. Depending on the two other forces K_ψ and K_η, there are two possible ranges [12, 19, 22]:

(a) The laminar range, in which the viscosity force K_η has a significant effect on the liquid flow,

$$\mathrm{Re}_L = \frac{u_L}{a \cdot \nu_L} < 1 \quad [22] \tag{4-18}$$

and/or 6 [44] and/or 10 [19, 12];

(b) The turbulent range, in which the resistance force K_ψ outweighs the viscosity force, $\mathrm{Re_L} \geq 1$ acc. to [22], 6 [44] and/or $\mathrm{Re_L} \geq 10$ [19, 12].

About (a): Nusselt, as quoted in [19], has derived the following correlation for the mean film thickness δ_L in a vertical tube:

$$\delta_L = \left[\frac{3 \cdot \eta_L \cdot u_L}{\rho_L \cdot g \cdot a}\right]^{1/3} \quad [m]. \tag{4-19}$$

For $h_L = a \cdot \delta_L$, see Eq. (4-6), the liquid hold-up is given as:

$$h_L = \left[\frac{3}{g}\right]^{1/3} \cdot a^{2/3} \cdot (\nu_L \cdot u_L)^{1/3} \quad \left[m^3 m^{-3}\right], \tag{4-20}$$

acc. to Bemer and Kalis [12]. There is no evidence for the applicability of this equation for small-surface, modern packing elements and structured packings, as the comparison with experimental values was only performed using large-surface, ceramic Raschig rings [12]. Buchanan [19] has derived the following Eq. (4-21) for randomly filled Raschig rings:

$$h_L \cong 2.2 \cdot \left[\frac{Fr'_L}{Re'_L}\right]^{1/3} \quad [\mathrm{m^3 m^{-3}}] \tag{4-21}$$

with the dimensionless numbers:

$$Re'_L = \frac{u_L \cdot d}{\nu_L} \tag{4-22}$$

and

$$Fr'_L = \frac{u_L^2}{g \cdot d}. \tag{4-23}$$

The validity of this correlation has been verified for various systems within the wide range of the physical properties $\sigma_L = 29-77\ mNm^{-1}$, $\rho_L = 800-1250\ kgm^{-3}$ and $10^6 \cdot \nu_L = 0.75-310\ m^2s^{-1}$.

About (b): When deriving the equation for calculating the liquid hold-up in the turbulent flow range, $Re_L \geq 1-10$, the losses of energy due to the influence of the viscosity forces on the liquid element in the packing [19, 12] are usually neglected. Based on the balance of forces acting on a fluid element at an inclined plate, Buchanan [19] derived Eq. (4-24):

$$h_L = S' \cdot Fr_L^{1/2} \quad \text{where } S' = const. \tag{4-24}$$

Bemer and Kalis [12] used the balance of forces $K_\psi = K_g$ and correlation (4-25):

$$\frac{1}{2} \cdot \psi_L \cdot u_F^2 \cdot \rho_L = \rho_L \cdot g \cdot \delta_L \quad \text{where } u_F = u_L / h_L \tag{4-25}$$

as well as Eq. (4-6) to derive Eq. (4-26):

$$h_L = \left[\frac{\psi_L}{2g}\right]^{1/3} \cdot a^{1/3} \cdot u_L^{2/3} \quad \left[m^3 \cdot m^{-3}\right] \tag{4-26}$$

In the case of ceramic and metal Raschig rings as well as ceramic Berl saddles, the quotient $(\psi_L/2\,g)^{1/3}$ was given as 0.34 by Bemer and Kalis [12]. The equation for calculating the liquid hold-up h_L is as follows:

$$h_L = 0.34 \cdot a^{1/3} \cdot u_L^{2/3} \quad \left[\mathrm{m}^3 \cdot \mathrm{m}^{-3}\right]. \tag{4-27}$$

It matches the experimental values for the air/water system and the above mentioned types of packings with a relative mean error of ±20% [12].

Buchanan [19] summarised the results of the comprehensive study on the liquid hold-up by using a correlation (4-28), which covers both flow ranges:

$$h_L = 2.2 \cdot \left[\frac{Fr_L'}{Re_L'}\right]^{1/3} + 1.8 \cdot Fr_L'^{1/3} \quad \left[\mathrm{m}^3\mathrm{m}^{-3}\right]. \tag{4-28}$$

The accuracy of this correlation (4-28) matches the one presented by Bemer and Kalis [12], i.e. $\delta(h_L) \leq \pm\, 20\%$.

As is the case with Eq. (4-20), Eq. (4-28) was verified for ceramic Raschig rings of various sizes between 8 and 50 mm.

Figure 4-3 shows a new diagram, created by Mersmann and Deixler [44], which can be used to estimate the entire liquid hold-up h_L, if the dimensionless liquid load B_L, see Eq. (4-16), is known. The diagram is based on experimental data, mostly for ceramic Raschig rings, with a dimension of $d \leq 0.035$ m, taken by Gelbe [22], Blaß and Kurtz [42, 43] as well as Mersmann [21]. The parameter, which is used in Fig. 4-3; is the We_L/Fr_L number, defined in Eq. (4-17).

Figure 4-3. Liquid hold-up h_L as function of dimensionless liquid load B_L according to Mersmann and Deixler [44]

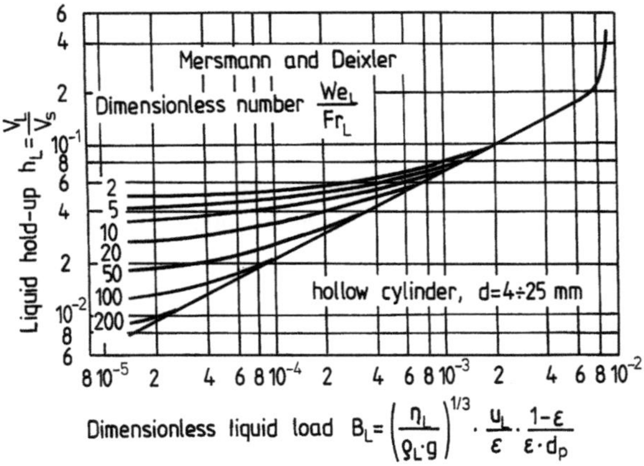

Further models for determining the liquid hold-up can be found in literature [7, 12, 13–15, 20, 21, 22, 42, 42–44, 58]. The work of Gelbe [22] contains some fundamental considerations on the liquid hold-up, and his work was used as a basis for studies by Reichelt [1], Blaß and Kurtz [42] Mersmann and Deixler [44], as well as a publication by Stein [58], in which he presents new models of "straight tubes" and "structured channels", based on trickle film flow.

4.2.4 Analysing the Influence of Various Parameters on Liquid Hold-Up, Based on Literature Data

In the turbulent flow range of the liquid, the liquid load u_L and the packing size d have the biggest influence on the liquid hold-up h_L. In the laminar range, the hold-up is not only influenced by the variables $u_L$and d, but also by the physical properties, viscosity η_L and density ρ_L, of the liquid as well as, to some extent, by the surface tension σ_L [22, 44]. In the range below the loading line, the gas has practically no influence on the liquid hold-up h_L, see e.g. Fig. 2-3.

The influence of the liquid load u_L has been assessed differently in previous studies. For $Re_L \geq 1$, Gelbe [22] found the following correlation:

$$h_L \approx u_L^{5/11} \tag{4-29a}$$

and for $Re_L < 1$:

$$h_L \approx u_L^{1/3}, \tag{4-29b}$$

which is in line with experiments [3, 22, 44] and other models [12, 19].

Leva [20] found the following correlation:

$$h_L \approx u_L^{3/5}. \qquad (4\text{-}29c)$$

In the range $Re_L > 10$, Bemer and Kalis [12] found that:

$$h_L \approx u_L^{2/3} \qquad (4\text{-}29d)$$

According to Brauer [3] and Buchanan [19]:

$$h_L \approx u_L^{0.5-0.6} \qquad (4\text{-}29e)$$

The work of Billet and Maćkowiak [13–15] has confirmed the correlation (4-27) $h_L \approx u_L^{2/3}$, found by Bemer and Kalis [12], also in the case of modern types of packings. In addition, Gelbe [22] observed that the exponent $n = 5/11$ in Eq. (4-29a) is only a usable mean value for $Re_L > 1$. For higher Reynolds numbers Re_L, the numerical value of the exponent would be expected to increase, i.e. $h_L \approx u_L^{\ n}$, where $n > 5/11$. Experiments performed by various authors [22, 12, 44, 13–15] have shown that turbulent liquid flow starts at $Re_L \geq (1 \ldots 10)$.

Based on experimental studies of the liquid hold-up h_L in the range below the loading line, data found in literature lead to the following correlations:

(a) $h_L \approx u_L^{1/2-2/3}$ for $Re_L \geq$ 1–10
and
$h_L \approx u_L^{1/3}$ for $Re_L <$ 1–10
(b) In the turbulent flow range $Re_L \geq$ 1–10, the influence of the physical properties ρ_L, η_L on the liquid hold-up can be practically neglected.
(c) For $Re_L \leq$ 1–10 (1), the liquid hold-up h_L is dependent on the viscosity of the liquid, hence: $h_L \approx \nu_L^{\ 1/3}$.
(d) As the size of the packing elements increases, the liquid hold-up drops, i.e. $h_L \approx d^{-1/3}$ and/or $h_L \approx a^{1/3}$ for $Re_L \geq$ 1–10 (1) and $h_L \approx a^{2/3}$ for $Re_L <$ 1–10 (1).

The investigations of this work therefore focus on the turbulent range, which is relevant for practical applications at vacuum operation as well as for the separation of mixtures under normal and high pressure. In addition, it is important to determine the numerical value of the Reynolds number Re_L, at which turbulent liquid flow in modern packing elements as well as structured packings can be expected. A knowledge of the liquid hold-up is also important for determining the residence time of the liquid in the packed column, Eq. (4-30), e.g. for the separation of thermally unstable mixtures under vacuum as well as for absorption processes with long chemical reactions.

$$\tau = \frac{H \cdot h_L}{u_L} \quad [s] \qquad (4\text{-}30)$$

Literature research has shown that there are well-founded theoretical and empirical methods, which can be used to describe the liquid hold-up in packed columns. In the case of modern, small-surface packing elements and large-surface structured packings, however, there is still a lack of experimental proof with regards to the validity of the models presented in literature.

4.2.5 Test Method, Systems and Packing Elements

The entire liquid hold-up $h_L = V_L/V_S$ was determined by using the test systems air/water and/or air/organic liquid under ambient conditions, following the mass transfer and/or pressure drop tests by simultaneously shutting off the liquid inlet and outlet. The test plants are shown in Figs. 6-1, 6-2, 6-3 and 6-4. The gas supply was also shut off, in order to speed up the dripping of the liquid. Within a maximum of 10–15 min, the volume V_L of the dripping liquid was measured. During this time, even at very low liquid loads of around 3–5 $m^3\ m^{-2}h^{-1}$, approx. 95% of the held-up liquid had been accumulating. This was the case with all larger packing elements, $d \geq 25$–90 mm, with a high void fraction ε. The experimental error was reduced in the range, in which the liquid hold-up is influenced by the gas velocity, i.e. experiments above the loading line, and also in the case of higher liquid loads u_L. The subsequent dripping of liquid from the liquid distributor was prevented by means of a specially constructed tube distributor. The dynamic and static liquid hold-ups were not analysed separately, due to experimental reasons, and particularly as the pressure drop $\Delta p/H$ is determined based on the total liquid hold-up h_L.

During the course of this work, the specific liquid loads were varied from 3 to 80 $m^3 m^{-2} h^{-1}$ in columns with diameters ranging from 150 to 600 mm and packing heights H between 0.8 and 4.3 m. The ratio d_S/d was $d_S/d \geq 5$. The experiments were carried out using pre-flooded random and structured packings, i.e. randomly filled packing elements with a nominal dimension of 15–90 mm, made of ceramic, plastic and metal, as well as Montz sheet metal packings B1-100, B1-200, B1-300, Mellapak 250Y produced by Sulzer, Ralu-Pak 250 YC by Raschig and Gempak 2A T304 by Glitsch as well as other packings made of plastic, see Table 4-3. The evaluation was performed on the basis of literature data given by Bornhütter [55], Suess and Spiegel [57], taken at plants with a column diameter of $d_S = 1$m.

4.2.6 Experimental Results

Some of the experimental results are known from publications, e.g. [13–15, 23]. In those studies, the total liquid hold-up h_L was experimentally determined using the systems air/water, air/ethylene glycol, air/aqueous ethylene glycol solution and air/methanol, in a column with diameters of 0.150 m and/or 0.22 m and a packing height of $H \cong 1.5$ m. The more recent results, which were obtained using mostly small-surface, perforated packing elements in larger-diameter columns, will be discussed in the following.

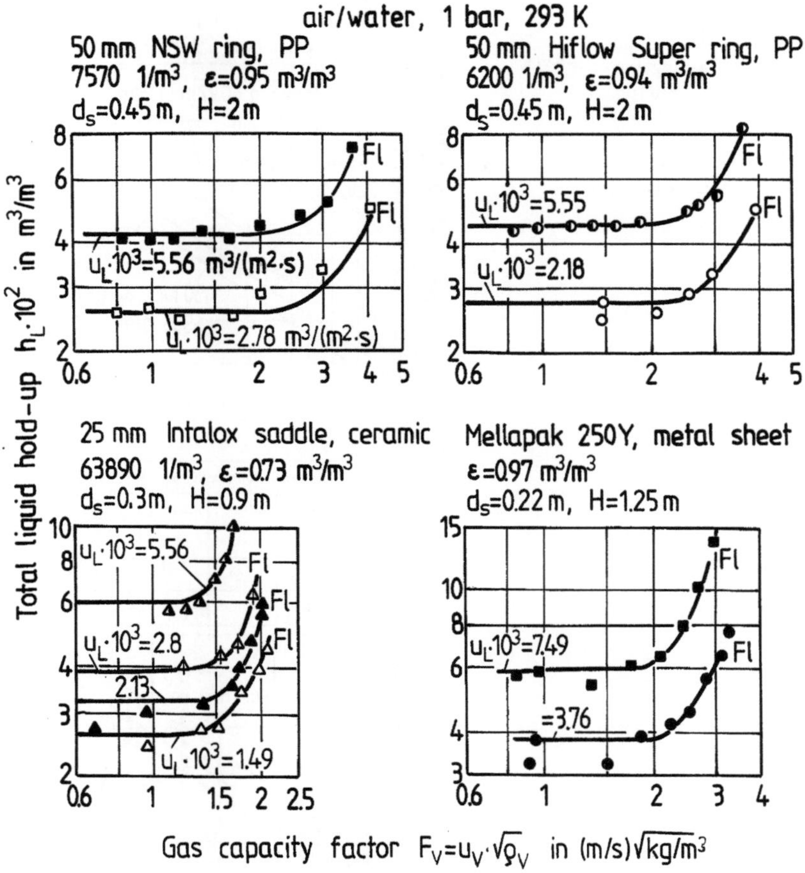

Figure 4-4. Total liquid hold-up as function of gas capacity factor F_V for various specific liquid loads u_L, valid for random 25 mm Intalox saddles, 50 mm NSW rings, Hiflow rings Super and Mellapak 250Y made of sheet metal. System: air/water under normal conditions

4.2.6.1
Liquid Hold-Up Below the Loading Line

Typical curves showing the total liquid hold-up h_L can be seen in Fig. 2-2b for 25 mm metal Białecki rings and Fig. 4-4a–d for selected types of packings with different gas capacity factors F_V, and the specific liquid load u_L as the parameter. The influence of the liquid load u_L on the liquid hold-up of randomly filled Pall rings is shown in Fig. 4-5a–c, the influence of the packing size in Fig. 4-5a,b and the influence of the material in Fig. 4-5c.

Figure 4-6a and b show the dependency $h_L = f(u_L)$ for structured packings and stacked packing elements. The experimental data of Fig. 4-6a is applicable to structured

Figure 4-5. Liquid hold-up h_L as function of specific liquid load u_L below loading line, valid for (a) metal Pall rings, d = 0.015–0.050 m (b) Pall rings made of plastic, d = 0.025–0.050 m (c) 50 mm Pall rings made of various materials

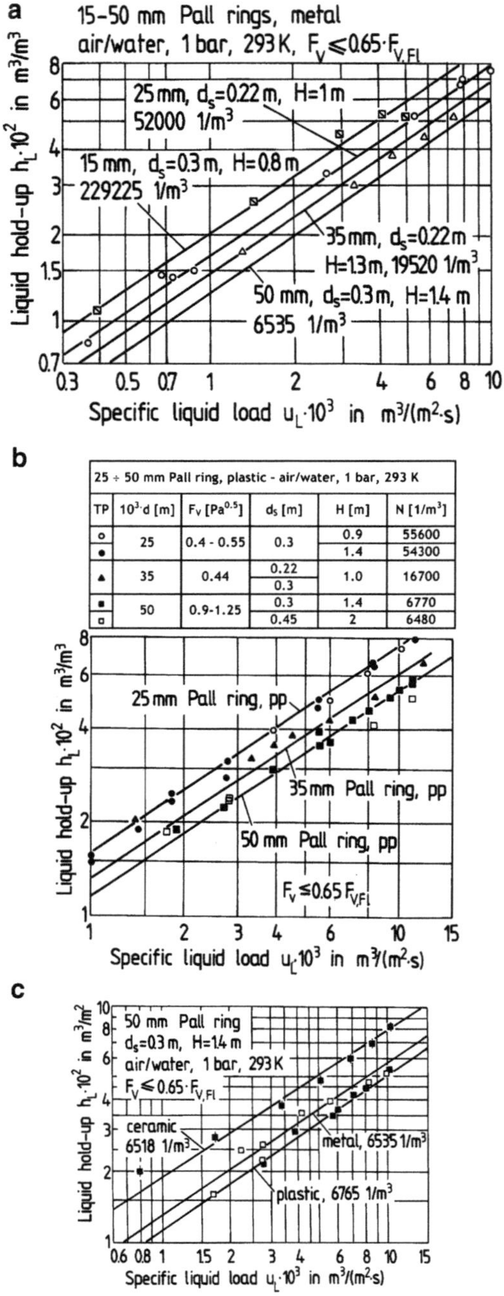

25 ÷ 50 mm Pall ring, plastic - air/water, 1 bar, 293 K

TP	$10^3 \cdot d$ [m]	F_V [$Pa^{0.5}$]	d_S [m]	H [m]	N [1/m³]
○	25	0.4 - 0.55	0.3	0.9	55600
●				1.4	54300
▲	35	0.44	0.22	1.0	16700
			0.3		
■	50	0.9-1.25	0.3	1.4	6770
□			0.45	2	6480

a

air/water, 1 bar, 288 K, $F_V \le 0,65 \cdot F_{V,Fl}$					
TP	packing	material	F_V [$Pa^{0.5}$]	d_S [m]	H [m]
○	Montz B1-300	sheet metal	1.50	0.22	1.42
⌽			1.50	0.30	1.42
⊕			1.14	0.45	1.07
▲	MontzB1-200	sheet metal	<1.33	0.22	1.34
♣				0.30	1.56
■	Montz B1-100	sheet metal	1.05	0.3	1.34
▽	Montz C1-200	PP new	0.55	0.3	1.42
▼		PP old	1.50		
●	Mellapak 250 Y	sheet metal	0.94	0.22	1.25
⏀	Gempak 2 AT 304	sheet metal	≤1.23	0.30	1.47

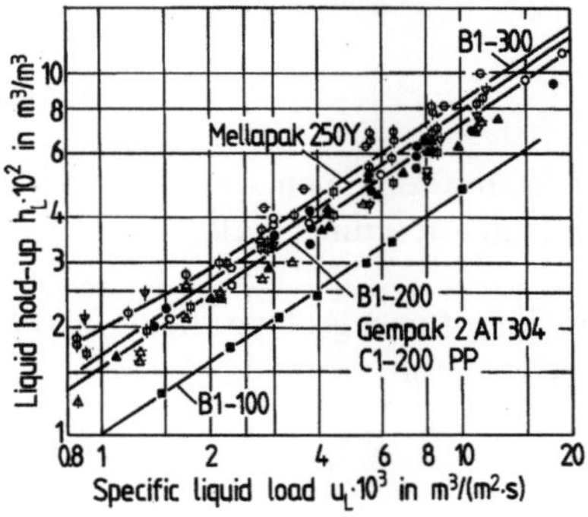

b

air/water, 1 bar, 293 K, $F_V \le 0,65 \cdot F_{V,Fl}$							
TP	packing	material	a [m^2/m^3]	ε [m^3/m^3]	d_S [m]	H [m]	Lit.
◇	25 BR-M-S stacked 25 mm Białecki rings	metal N = 63400 1/m³	275	0.933	0.15	1.5	[A]
▽	50 BR-M-S stacked 50 mm Białecki rings	metal N = 7654 1/m³	128.5	0.963	0.30	1.5	[A]
●	Mellapak 250 X	sheet metal	250	0.97	1.0	3.5	[57]
△	Montz B1-500	sheet metal	500	0.97	0.45	1.7	[A]
□	Montz A3-BX	metal gauze	500	0.95	0.45	2.0	[A]

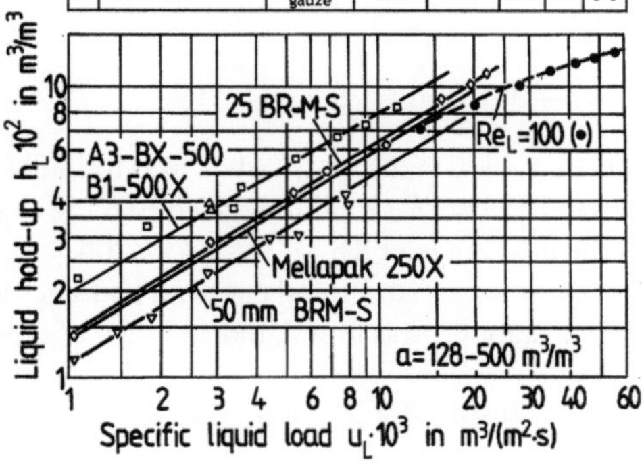

Figure 4-6. Liquid hold-up h_L as function of specific liquid load u_L below loading line, valid for (**a**) Montz packings B1-100Y, B1-200Y, B1-300Y made of sheet metal and C1-200Y made of PP, Mellapak 250Y made of sheet metal and Gempak 202AT made of sheet metal, $d_S = 0.22-0.45$ m and (**b**) valid for 25–50 mm stacked Białecki rings made of metal, Mellapak 250X, Montz B2-500X and Montz A3-500

type Y packings with a flow channel angle of 45°, such as Mellapak 250Y made of sheet metal, Gempak 2AT304 and Montz packings B1-100, B1-200, B1-300 made of sheet metal and C1-200 of plastic. The data shown in Fig. 4-6b is applicable to stacked packing elements and structured packings with a flow channel angle of 30° to the column axis.

A database was created for the purpose of evaluating the large amount of experimental data. It includes approx. 1100 test points for randomly filled and stacked packing elements as well as structured packings made of sheet metal and plastic. Table 4-3, which can be found in the annex of this chapter, contains more detail on the number of experimental series and points, the tested packing elements and systems as well as information on the mean deviations $\delta(h_L)$ of the experimentally derived liquid hold-ups $h_{L,exp}$ from the $h_{L,calc}$ values, which were calculated using three selected methods. The individual methods are defined as $h_{L,calc1}$, $h_{L,calc2}$ and $h_{L,calc3}$ and will be presented in the following. They were selected, as they are suitable for determining the liquid hold-up for randomly filled and stacked packing elements as well as for structured packings.

$h_{L,calc1}$

Here, the modified correlation (4-20) by Bemer and Kalis [12] was used to calculate the h_L value for $Re_L < 2$, using a correction factor of 3/4, see Chap. 2. This numerical value was the result of an evaluation of the available experimental data in the database.

Equation (4-32) was used to determine the liquid hold h_L for turbulent liquid flow $Re_L >$ 2. This equation is obtained by substituting the Froude number of the liquid, according to Eq. (4-31), into Eq. (4-26).

$$Fr_L = \frac{u_L^2 \cdot a}{g} \tag{4-31}$$

$$h_L = C_P \cdot Fr_L^{1/3} \quad [m^3 m^{-3}] \tag{4-32}$$

where

$$C_P = 0.57 \quad \text{for } 2 \leq Re_L \leq 200 \tag{4-33a}$$

The numerical value of $C_P = 0.57$ was found by evaluating experimental data, contained in the database, for randomly filled packing elements and structured packings of type Y, in the flow range of $Re_L \geq 2$ to $Re_L \approx 200$.

In the case of structured packings and stacked packing elements of type X, the following constant C_P was found, independent of type and material:

$$C_P = 0.465 \quad \text{for } 2 \leq Re_L \leq 100. \tag{4-33b}$$

In order to ascertain the operating range, the relative column load $(F_V/F_{V,Fl})_{uL=const}$ was determined for each test point. The gas capacity factor at the flooding point $F_{V,Fl}$ was derived by iteration, using the method (2-67), and the liquid hold-up in the loading range was determined based on the method presented in Chap. 4.2.6.2.

$h_{L,calc2}$

The liquid hold-up h_L for $Re_L < 5$ and $Re_L > 5$ was calculated based on:

$$h_L = 2.2 \cdot B_L^{1/2} \quad [m^3 m^{-3}] \tag{4-34}$$

as presented by Mersmann and Deixler [44].

$h_{L,calc3}$

Here, the liquid hold-up was calculated using the methods by Bemer and Kalis [12], see Eqs. (4-26) and (4-28).

Discussion of Experimental Results – Turbulent Liquid Flow

Figure 4-7 shows a comparison between the experimentally derived liquid hold-ups $h_{L,exp}$ and the h_L values calculated by Bemer and Kalis [12], according to Eq. (4-26). The comparison, which is valid for $Re_L < 10$ and $F_V/F_{V,Fl} \leq 0.65$, shows a systematic deviation of 60 to −10% for the majority of test points. A good conformance between calculated and experimental values was only found for ceramic Raschig rings and ceramic Intalox

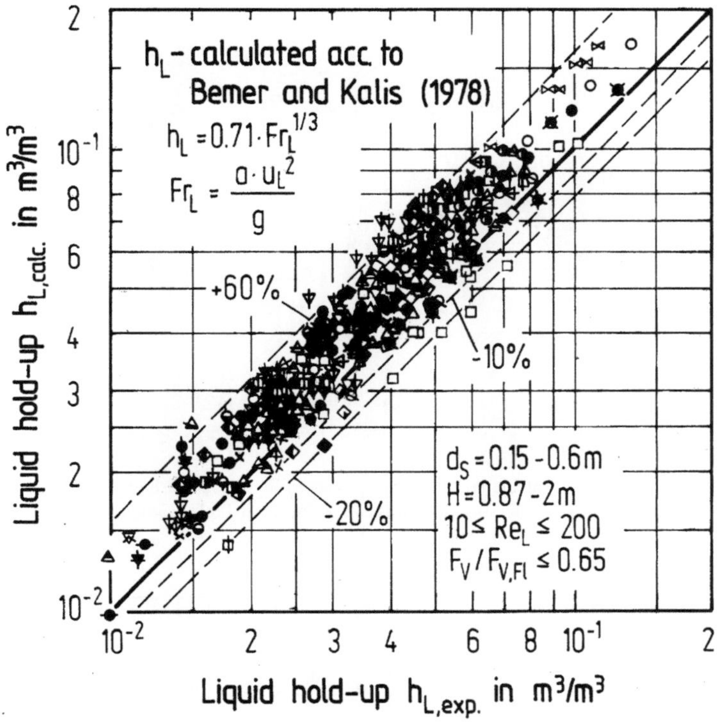

Figure 4-7. Comparison of calculated values $h_{L,calc}$ acc. to Eq. (4-26) by Bemer and Kalis [12] with experimentally determined values $h_{L,exp}$, valid for $Re_L \geq 10$ and $F_V/F_{V,Fl} < 0.65$ as well as for various packings. Table 4-5 relating to Figs. 4-7, 4-8 and 4-9 incl. symbols can be found in the annex of this chapter

saddles. The method used in Eq. (4-26) was originally developed for these packing elements.

Figure 4-8a shows a comparison between the same liquid hold-up data $h_{L,exp}$ and the values calculated using Eq. (4-32), with $C_P = 0.57$, in the range of $Re_L \geq 2$. Here, approx. 75% of around 650 test points deviate from the graph within $\delta(h_L) \leq \pm 20\%$, and 85% within $\delta(h_L) \leq \pm 25\%$. The mean error $\delta(h_L)$ in the determination of the liquid hold-up h_L, according to correlation (4-32), with $C_P = 0.57$, is $\delta(h_L) \cong 15\%$.

Figure 4-8b shows a comparison between the experimental data $h_{L,exp}$ for type X structured packings and for stacked 25–50 mm metal Białecki rings and the liquid hold-up calculated using method (4-32), with a packing constant of $C_P = 0.465$, acc. to Eq. (4-33b). The experimental data is spread around the graph in the flow range of $Re_L = 2$ to $Re_L \approx 100$ with a relative deviation of $\delta(h_L) \leq \pm 15\%$.

An accuracy, similar to the one relating to the calculation of the liquid hold-up using Eq. (4-32a) with $C_P = 0.57$, is achieved when the experimental liquid hold-up data $h_{L,exp}$ is evaluated using Eq. (4-34), acc. to Mersmann and Deixler [44], see Fig. 4-9.

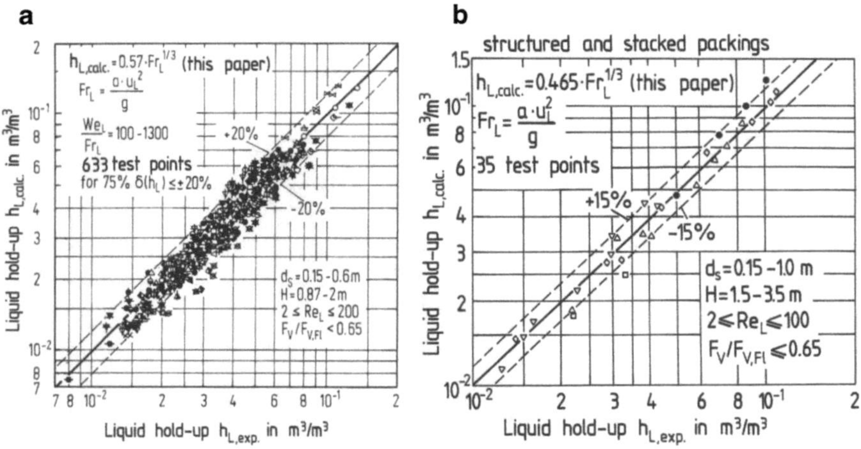

Figure 4-8. Comparison of calculated values $h_{L,calc}$ acc. to Eq. (4-32) with experimentally determined values $h_{L,exp}$, valid for various packings. Symbols can be found in Table 4-5 in the annex to this chapter

Figure 4-9. Comparison of liquid hold-up $h_{L,calc}$ calculated acc. to Eq. (4-34) with experimentally determined values $h_{L,exp}$, valid for $Re_L \geq 5$ and $F_V/F_{V,Fl} < 0.65$, for various packings. Symbols can be found in Table 4-5 in the annex to this chapter

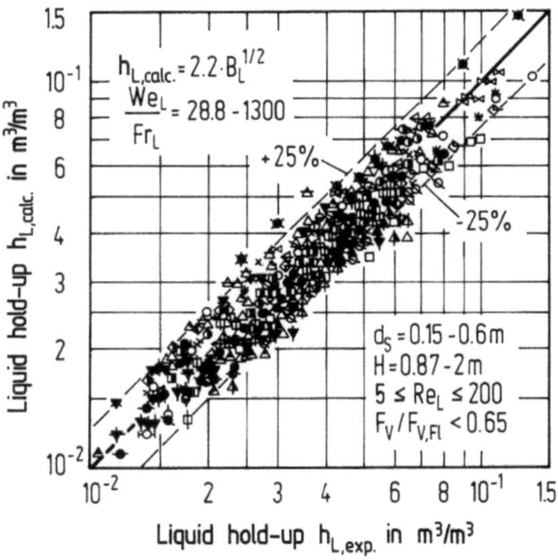

The deviation of the measured data from the graph is $\delta(h_L) \leq \pm 25\%$, with a mean error $\delta(h_L)$ of $\delta(h_L) \cong$ in the determination of the liquid hold-up h_L Table 4-5.
A more detailed analysis of the evaluation results of the experimental data shown in Fig. 4-9, using Mersmann and Deixler's [44] calculation, acc. to (4-34), also leads to the conclusion that the h_L values for higher, dimensionless liquid loads $B_L > 2 \cdot 10^{-4}$ are too small. In the case of turbulent liquid flow, $Re_L \geq 5$, the exponent $n = 1/2$ in Eq. (4-34)

Table 4-5. Symbols specification for Figs. 4-7, 4-8a,b and 4-9

TP	d · 10³ [m]	Packing	Material	System no.	ε [m³/m³]	d_S [m]	H [m]	Literature
○ ◑	25 35	**Białecki rings** random	metal	1, 2, 3	0.940 0.947	0.154 0.220	1.5 1.3	[A]
✱	25	**Raschig rings** random	ceramic	1	0.650	0.220	1.0	
❁	25	**Raschig rings** random		1	0.676	0.226	0.676	[21]
✹	25	**Raschig rings** random	glass	5	0.820	0.15	1.0	[36, 37]
⧌ ✱ ⌹	25 35 50	**Pall rings**	plastic	1, 4	0.880- 0.926	0.22- 0.45	0.91- 2.0	[A]
× ⊕	38 75	**Hiflow rings** random	ceramic	1	0.834 0.865	0.30 0.45	1.34 2.00	
● ●	28 58	**Hiflow rings**	metal	1	0.962 0.970	0.3-0.6 0.45	1.46- 2.0 2.0	
□ ◨ ◧	25 35 58	**Pall rings** random		1, 2, 3	0.950 0.950 0.970	0.30 0.30 0.45	1.46 1.46 2.00	
▼ ▼	32 (size 1) 50 (size 2)	**VSP rings** random		1, 4	0.976 0.980	0.30- 0.45	1.46 2.00	
⦰ ф -⊖-	31.5 (size 1) 50 (size 2) 80 (size 3)	**Envipac** random	plastic (PP)	1, 4	0.930 0.964 0.959	0.30- 0.45	1.36- 2.0	
△ △ ▲	28 50 90	**Hiflow rings** random	plastic	1	0.920 0.935 0.965	0.30 0.30 0.45	1.4 1.4 2.0	
△ △	25 38	**Intalox saddles** random	ceramic	1	0.730 0.757	0.30	0.9 1.4	
◆	50	**NSW rings**	plastic	1, 4	0.950	0.30- 0.45	1.45- 2.0	
△	50	**Ralu rings**		1	0.940	0.30- 0.45	1.45- 2.0	
⬖ ⬗	28 38	**NSW rings** random		1, 4	0.920 0.932	0.30	1.4	
⋈ -◑-	25 35	**Białecki rings stacked**	metal	1, 2, 3	0.928 0.945	0.15 0.22	1.5 1.0	
▽	60 (size 2)	**VSP rings**	plastic	1	0.954	0.30- 0.45	1.40- 2.0	
●	80	**Top-Pac**	metal	1	0.978	0.45	2.0	
⊢⊖⊣	40	**Intalox #40**		1	0.940	0.45	2.0	
▲	53	**Hiflow Super**	plastic	1, 4	0.940	0.30- 0.45	1.4-2.0	
▼ ⧗	70 (size 2) 45 (size 1)	**Dtnpac**		1	0.938 0.920	0.30 0.45	1.4 2.0	

is obviously too small. The experimental data, plotted in Fig. 4-9, can be described slightly better in the range of $B_L > 2 \cdot 10^{-4}$, using the modified method presented in this work, as shown in Eq. (4-35), and correlated with a mean relative error of $\delta(h_L) < 14\%$.

$$h_L = 4.39 \cdot B_L^{0.575} \qquad \left[m^3 m^{-3}\right] \tag{4-35}$$

Plotting the experimental values from Fig. 4-6 for type Y structured packings in the form of $h_L/Fr_L^{\,1/3} = f(u_L)$, as shown in Fig. 4-10, also leads to a constant C_P, whose numerical value for $Re_L \geq 2$ is given as:

$C_P = 0.57 \pm 20\ \%$

air/water, 1 bar, 288 K, $F_V \leq 0.65 \cdot F_{V,Fl}$, $Re_L \geq 2$

TP	packing	material	F_V [$Pa^{0.5}$]	d_S [m]	H [m]
○	Montz B1-300	sheet metal	1.50	0.22	1.42
⌽			1.50	0.30	1.42
⊖			1.14	0.45	1.07
▲	MontzB1-200	sheet metal	<1.33	0.22	1.34
▲				0.30	1.56
■	Montz B1-100	sheet metal	1.05	0.3	1.34
▽	Montz C1-200	PP new	0.55	0.3	1.42
▼		PP old	1.50		
●	Mellapak 250 Y	sheet metal	0.94	0.22	1.25
⌽	Gempak 2 AT 304	sheet metal	≤1.23	0.30	1.47

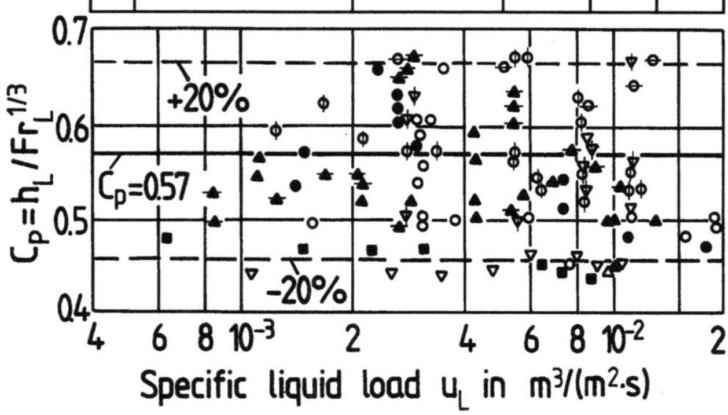

Figure 4-10. Dependency of value C_P for Eq. (4-32) for Montz-packings of various sizes, Sulzer packing type Mellapak 250 Y and Gempak 202 AT made of sheet metal on specific liquid load u_L

This can be used to calculate the liquid hold-up of different types of structured packings, made of various materials. Hence, based on method (4-32) and $C_P = 0.57$, it is possible to determine the liquid hold-up h_L for randomly filled and stacked packing elements as well as for type Y structured packings sufficiently enough for practical applications.

Figure 4-6 shows the influence of the operating time on the liquid hold-up, based on experiments using Montz packing C1-200. C-200 "new", in Fig. 4-10, relates to the experimental values of new packings, whereas the term "old" relates to packings with a longer operating time of approx. 2 weeks. After this operating time, the liquid hold-up increases by approx. 25% and then remains constant. According to Fig. 4-6, no difference was found to exist between the liquid hold-up of the plastic packing C1-200 Y and that of the sheet-metal packing B1-200 Y with the same geometric packing surface area.

In addition, the column diameter was found to have a minor influence on the liquid hold-up h_L of packings with $d_S < 0.3$ m. The experimental values for sheet-metal packings B1-300Y in the column with a diameter of $d_S = 0.22$ m are below the values found for columns with $d_S = 0.3$ m and/or 0.45 m. In the case of the sheet-metal packing B1-100Y with a packing surface area of 100 m^2m^{-3}, the experimental values are approx. 20% below the values calculated acc. to Eq. (4-32). It follows from this that the column size was too small for small-surface packings.

Discussion of Experimental Results – Laminar Liquid Flow

The evaluation of experimental liquid hold-up data leads to the following results:

In the case of laminar liquid flow with Reynolds numbers of $Re_L < 5$, the liquid hold-up h_L can be described reasonably well using Mersmann and Deixler's [44] method, with 85% of the experimental values verified with an accuracy of $\delta(h_L) \leq \pm 20\%$, Fig. 4-11a. Figure 4-11b shows deviations of 0– +50%, in which the $h_{L,exp}$ values are compared to the values calculated acc. to Bemer and Kalis [12], see Eq. (4-20). Hence, the use of a correction factor of 3/4 in Eq. (4-20) has proved to be useful, as it leads to a considerably better conformance between the calculated and experimental values, see Table 4-3, with the exception of some experimental data taken at extremely low liquid loads, which cannot be sufficiently described by any of the methods available.

Finally, it can be noted that the liquid hold-up h_L for any types of random packing elements and type Y structured packings can be calculated using Mersmann and Deixler's [44] model as well as the modified method by Bemer and Kalis [12] with an accuracy of ±20–25%.

In the correlations presented by Bemer and Kalis [12], it is necessary to correct the constant C_P. For $Re_L < 2$, the correction factor in Eq. (4-20) is 0.75. For $Re_L \geq 2$, $C_P = 0.57$ in Eq. (4-32), for packings with a flow channel angle of 45°, and $C_P = 0.465$

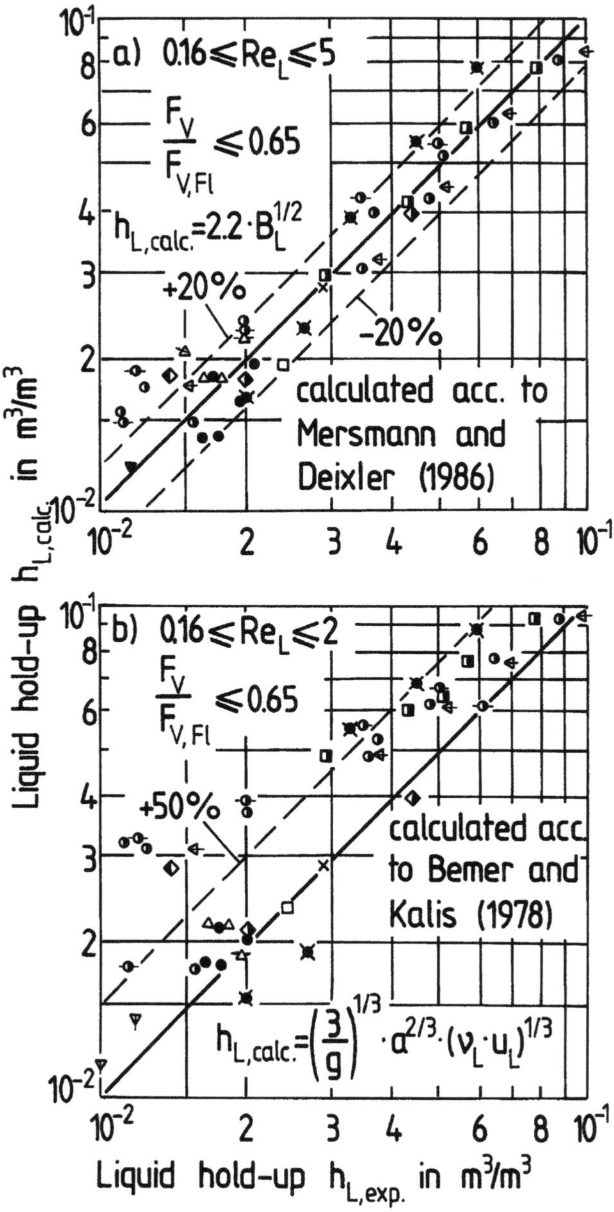

Figure 4-11. Comparison of calculated values $h_{L,calc}$ with experimentally determined values $h_{L,exp}$ (**a**) acc. to Eq. (4-34) by Mersmann and Deixler [44], (**b**) acc. to Bemer and Kalis [12], Eq. (4-20). Symbols can be found in Table 4-6 in the annex to this chapter

in Eq. (4-32) for type X structured packings and stacked packing elements with a flow channel angle of 30°. This applies to the following range of operating and property parameters:

$$0.15 \le \mathbf{Re_L} \le 200 - \text{type Y}$$
$$2 \le \mathbf{Re_L} \le 100 - \text{type X}$$
$$28.8 \le \mathbf{We_L / Fr_L} \le 1300$$
$$58 \le \mathbf{a} \le 500\ m^2 m^{-3}$$
$$0.57 \le \boldsymbol{\varepsilon} \le 0.987\ m^3 m^{-3}$$
$$0.15 \le \mathbf{d_S} \le 1\ m$$
$$0.676 \le \mathbf{H} \le 4.3\ m$$
$$932 \le \boldsymbol{\rho_L} \le 1110\ kgm^{-3}$$
$$21 \le \boldsymbol{\sigma_L} \le 72.4\ mNm^{-1}$$
$$0.58 \le \boldsymbol{\eta_L} \le 14.3\ mPas$$
$$\mathbf{F_V / F_{V,Fl}} \le 0.65$$
$$6 \le \mathbf{d_S / d}$$

Table 4-6. Symbol specification for Fig. 4-11

Symbol	$d \cdot 10^3$ [m]	Packing	System no.[1]	a [m^2/m^3]	ε [m^3/m^3]	We_L/Fr_L [-]	d_S [m]	H [m]
□	25	Pall rings metal	1	233.5	0.854	88.61	0.30	1.46
◨	35		1, 2	~180	0.948	197-388	0.22-0.30	1.3-1.4
⧋	25	Pall rings plastic	1	235	0.883	68.6	0.22	0.91
◀+	35		1, 2, 3, 4	~180	0.910	184-350	0.22	0.70
◧	35	Białecki rings metal	1, 2, 3	~160	0.947	161-386	0.22	1.25-1.30
-◑-	35	Białecki rings metal, stacked	1, 2, 3	~170	1.945	152-201	0.22	1.00
●	28	Hiflow rings metal	1	~190	1.963	110-140	0.3-0.6	1.4-2.0
△	28	Hiflow rings PP	1	198.5	0.916	103	0.30	1.34
×	20	Hiflow rings ceramic	1	309.9	0.854	28.8	0.30	1.25
◈	28	Nor-Pac PVDF	1	193.5	0.922	110.4	0.30	1.45
◆	38		1, 4	159.5	0.932	166-173	0.30	1.40
▼	45	Dtnpac size 1 PP	1	135	0.921	225.5	0.45	2.00
⧍	25	Intalox saddles ceramic	1	185	0.724	74.5	0.30	0.87
✹		Raschig rings ceramic, glass	1,5	180	0.676	50.7	0.154-0.22	0.676-1.0
				203	0.820	247		

[1] system no.:

No.	System	η_L [mPas]	ρ_L [kgm^{-3}]
1	air / water	1.0	998.2
2	air / ethylene glycol	14.3	1106.9
3	air / aqueous glycol solution	8.2	1095.5
4	air / 4% NaOH solution	1.5	1040.0
5	air / silicone oil	9.6	932.0

4.2.6.2
Liquid Hold-Up $h_{L,S}$ Above the Loading Line and Below the Flooding Point

According to Fig. 2-2b and/or 4-4, the liquid hold-up h_L above 65% of the flooding point increases at a constant liquid load u_L and reaches the maximum $h_{L,Fl}$ value of

$$h_{L,Fl} \approx 1.8 \cdot h_L \text{ to } 2.0 \cdot h_L \quad \left[m^3 m^{-3}\right]$$

at the flooding point.

Figure 4-12 shows the correlation between the liquid hold-up h_L and the specific liquid load u_L, with the gas capacity factor F_V as the parameter. Below the loading line, the liquid hold-up h_L increases with u_L to the power of 2/3, whereas above the loading line, the influence of the liquid load on the liquid hold-up h_L is bigger, $h_L \approx u_L{}^n$, with an increasing exponent n, n > 2/3.

Figure 4-12 also includes the phase flow ratios for the respective liquid hold-ups at the flooding point $h_{L,Fl}$. As the phase flow ratio increases λ_0, the ratio between $h_{L,Fl}$ and the liquid hold-up h_L, which is on the line $h_L \approx u_L{}^{2/3}$, for $u_L = u_{L,Fl}$, and which is characteristic of the range below the loading line, decreases.

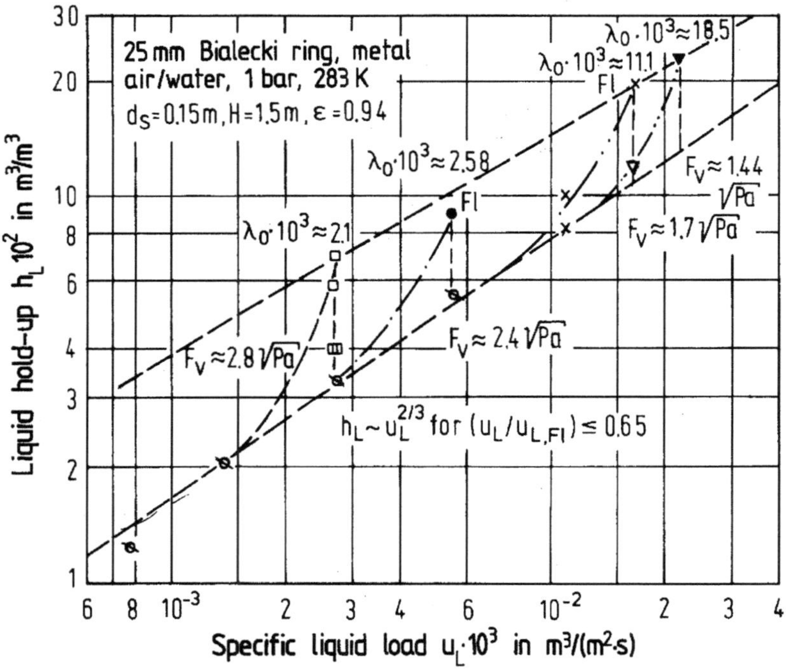

Figure 4-12. Influence of liquid load u_L on liquid hold-up h_L in the range up to flooding point with factor F_V as parameter. Figure is prepared on the basis of experimental data in the Fig. 2-3, symbols: see Fig. 2-3

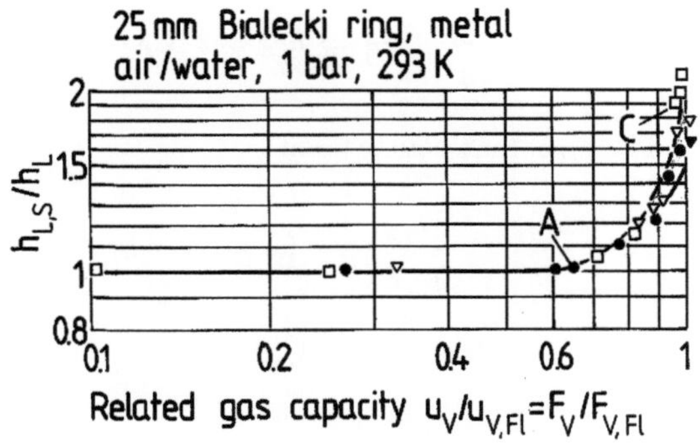

Figure 4-13. Relative liquid hold-up $h_{L,S}/h_L$ as function of relative gas capacity $F_V/F_{V,Fl}$, valid for 25 mm metal Białecki rings; symbols: see Fig. 2-3. Parameter: specific liquid load. A– loading line, C – flooding point

Figure 4-13 can be used to correlate the experimental liquid hold-up data above the loading line. Here, the liquid hold-up $h_{L,S}/h_L$ is presented as a function of the relative gas velocity $F_V/F_{V,Fl}$. The figure was created using the experimental data of Fig. 2-2b.

Figure 4-13 shows that, in the operating range of $u_L = 2.78 \cdot 10^{-3}$ to $22.2 \cdot 10^{-3}$ ms^{-1}, the liquid hold-up $h_{L,S}/h_L$ also increases with the relative column load $F_V/F_{V,Fl}$ and achieves its maximum value at the flooding point. This is higher, by a factor of 1.8 to 2, than in the loading range for $F_V < 0.65\, F_{V,Fl}$, i.e.: $h_{L,Fl} \approx 1.8 \cdot h_L$ to $2.0 \cdot h_L$. Figure 4-13 also shows that the liquid load u_L has a small influence on the liquid hold-up $h_{L,S}/h_L$, see also Eq. (2-47).

In the operating range above the loading line and below the flooding point, the liquid hold-up $h_{L,S}$ can be described using Eq. (4-36):

$$z = f(k) = \left\{ \begin{array}{ll} z_{Fl} - (z_{Fl} - z_0) \cdot \sqrt{1 - \left[\frac{k-k_0}{1-k_0}\right]^2} & \text{for } 0.65 < \frac{F_V}{F_{V,Fl}} < 1 \\ z_0 & \text{for } \frac{F_V}{F_{V,Fl}} \leq 0.65 \end{array} \right\} \tag{4-36}$$

Here, the following correlations have been introduced: $z = \left(h_{L,S}/h_L\right)_{u_L=const.}$ and $k = \left(F_V/F_{V,Fl}\right)_{u_L=const}$. The function in Eq. (4-36) belongs to the so-called C^1 class and is differentiable. In addition, the left and right-hand limit of its derivation equals zero.

The function in Eq. (4.36) has three parameters – k_0, z_{Fl} and z_0 -, which have been determined experimentally. The boundary conditions can be determined based on Fig. 4-13. They are as follows:

$$z_0 = 1, \quad \left(h_{L,S}/h_L\right) = 1$$

and

$$k_0 = 0.65.$$

Based on the additional descriptions for the ratio $h_{L,S}/h_L$

$$z = \frac{h_{L,S}}{h_L} = \frac{h_{L,S}}{C_P \cdot Fr_L^{1/3}} \Rightarrow h_{L,S} = z \cdot C_P \cdot Fr_L^{1/3} \quad \text{for } Re_L \geq 2 \tag{4-37}$$

and for z_{Fl} at the flooding point

$$z_{Fl} = \frac{h_{L,Fl}}{h_L} = \frac{h_{L,Fl}}{C_P \cdot Fr_L^{1/3}} \quad \text{for } u_L = const. \tag{4-38}$$

function (4-36) leads to Eq. (4-39) for $0.65 < F_V/F_{V,Fl} < 1$:

$$z = z_{Fl} - (z_{Fl} - 1) \cdot \sqrt{1 - \left[\frac{k - 0.65}{0.35}\right]^2}. \tag{4-39}$$

Hence, it is possible to determine the variable $z = h_{L,S}/h_L$, if the relative gas capacity $F_V/F_{V,Fl}$, the specific liquid load u_L, the packing surface area a and the void fraction ε are known, using Eqs. (2-47), (2-21) and (4-38). Equation (4-37) is used to define the parameter z, followed by $h_{L,S}$.

At the end of Sect. 4.2, the application of Eq. (4-39) for determining the liquid hold-up $h_{L,S}$ in the loading range is illustrated by means of a numerical example.

The data in Fig. 4-4 and Table 4-3 can also be determined sufficiently accurately for practical applications for $d_S/d > 6$, using Eqs. (4-39), (4-44) and (4-37), with a relative error of $\delta(h_{L,S}) \pm 20\%$.

4.2.7 Conclusions Section 4.2

The evaluation of the large amount of experimental liquid hold-up data leads to the conclusion that turbulent liquid flow can be expected in packed columns at $Re_L \geq 2$, whilst laminar liquid flow is expected at $Re_L < 2$. This corresponds approximately to the model presented by Gelbe [22].

1. The evaluation of experimental data from literature and the author's own database led to the following results.
   The modified Eq. (4-32), developed by Bemer and Kalis [12]

   $$h_L = C_P \cdot Fr_L^{1/3} \quad \text{where } C_P = 0.57 \pm 20\ \% \quad \left[m^3 m^{-3}\right]$$

   satisfactorily allows the determination of the liquid hold-up in the operating range below the loading line for any types of randomly filled packing elements with $d = 0.015 - 0.090$ m and different type Y structured packings with $a = 100 - 500\ m^2m^{-3}$. Equation (4-32) applies to the turbulent range, $Re_L \in (2–200)$. More than 80% of all 1000 experimental data items contained in the database can be verified with a relative error of less than ±20%.

The constant $C_P = 0.57$ for Eq. (4-32) has been re-determined by means of experiments carried out during the course of this work.

The liquid hold-up of stacked packing elements and type X structured packings is lower, compared to random packing elements and type Y structured packings, even if the specific packing surface area is identical. Based on the packing constant $C_P= 0.465$ for Eq. (4-32), the experimental data for packings with a geometric packing surface area of 128 to 500 m^2m^{-3} in the flow range of $Re_L \in (2–100)$ can be verified with a relative error of $\pm 15\%$.
In the case of higher Reynolds numbers, $Re_L > 100$, the experimental data shown in Fig. 4-6b can be described with the following correlation:

$$h_L = C_{P,0} \cdot Fr_L^{1/6} \quad \left[m^3 m^{-3}\right]. \tag{4-40}$$

where $C_{P,0} = 0.2$.

The practical advantage of applying Eq. (4-32) lies in the fact that the geometric packing surface area a is the only variable required to calculate the liquid hold-up. It is possible to determine the liquid hold-up h_L in packed columns more accurately using Eq. (4-32), i.e. with a relative error of $\delta(h_L) \leq \pm\, 10-15\%$, if a packing-specific constant $C_{Pi} \neq C_P$ is introduced, see e.g. Eq. (4-12) and Table 4-2.

2. In the case of laminar liquid flow, $Re_L < 2$, the model developed by Mersmann and Deixler [44], known from literature, as shown in Eq. (4-34):

$$h_L = 2.2 \cdot B_L^{1/2} \quad \left[\mathrm{m^3 m^{-3}}\right],$$

describes the experimental values sufficiently accurately for practical applications, i.e. with an accuracy of $\delta(h_L) \leq \pm 20\%$. A comparable accuracy in relation to the evaluated experimental data of Table 4-3 can be achieved using Eq. (4-20) and a correction factor of 3/4. The equation developed in this work for calculating the liquid hold-up for $Re_L \in (0.15–2)$ is therefore as follows:

$$h_L = \frac{3}{4} \cdot \left[\frac{3}{g}\right]^{1/3} \cdot a^{2/3} \cdot (\nu_L \cdot u_L)^{1/3} \quad \left[\mathrm{m^3 m^{-3}}\right].$$

3. The evaluated experimental hold-up data can also be verified for $Re_L \in (0.1–200)$, using Mersmann's [44] model, below the loading line with a relative error of less than $\pm 25\%$, by means of Eq. (4-34).

$$h_L = 2.2 \cdot B_L^{1/2} \quad \left[\mathrm{m^3 m^{-3}}\right]$$

The We_L/Fr_L number was varied between 28.8 and 1300 in the experiments. In this range, the number was found to have no influence on the liquid hold-up h_L.

4. The experimental determination of the liquid hold-up of random or structured packings of polypropylene (PP) is particularly time-consuming, as the final wettability of the packing surface is only reached after a certain operating period [39].
   Plastic packing elements made of PVDF or those pre-treated with inorganic acid or alkaline solution do not have these properties.
5. If the liquid hold-up h_L is available for a given specific liquid load u_L, it is possible to determine the liquid hold-up $h_{L,S}$ above the loading line for the same u_L value, if the relative gas capacity $F_V/F_{V,Fl} > 0.65$ is known, using the new Eq. (4-39).

Numerical Examples for Section 4.2 – Liquid Hold-Up

Example 1

The aim is to determine the liquid hold-up in the loading range for randomly filled 25 mm Białecki rings made of metal for the test system air/water at ambient conditions.

The column is operated under the following operating conditions: $u_L = 11.1 \cdot 10^{-3}$ ms^{-1} and $u_V = 1.5$ ms^{-1}.

Solution

Based on the numerical Example 2.2 and Fig. 2-2b, the gas velocity at the flooding point $u_{V,Fl}$ is given as:

$$u_{V,Fl} = 1.776 \text{ ms}^{-1}.$$

The numerical value of the relative gas capacity $F_V/F_{V,Fl}$ is:

$$F_V/F_{V,Fl} = u_V/u_{V,Fl} = 1.5/1.776 = 0.845.$$

The numerical value of z_{Fl} is given as:

$$z_{Fl} = h_{L,Fl}/h_L = 0.15/0.08 = 1.875.$$

Acc. to Eq. (4-39), we obtain:

$$z = h_{L,S}/h_L = 1.17.$$

Equation (4-32) is now used to calculate the h_L value below the loading line, based on a geometric packing surface area of $a = 238$ m^2 m^{-3}, which is applicable to 25 mm metal Białecki rings:

$$h_L = 0.57 \cdot \left[\frac{0.0111^2 \cdot 238}{9.81}\right]^{1/3} = 8.22 \cdot 10^{-2} \text{ m}^3\text{m}^{-3}.$$

Based on Fig. 2-2b, the experimental value $h_{L,exp}$ relating to the flow range below the loading line, $F_V/F_{V,Fl} < 0.65$, is given as:

$$h_{L,\,exp} = 0.08\ m^3 m^{-3}.$$

Now Eq. (4-37) is used for the flow range above the loading line, $F_V/F_{V,Fl} \geq 0.65$. The $h_{L,S}$ value is as follows:

$$h_{L,S} = z \cdot h_L = 1.17 \cdot 0.0822 = 0.096\ \mathrm{m^3\,m^{-3}}.$$

Acc. to Fig. 2-2b, we find:

$$h_{L,S,\,exp} = 0.093\ \mathrm{m^3 m^{-3}}.$$

The relative deviation of the experimentally derived values is therefore $\delta(h_L) = +2.75\%$ for the flow range below the loading line and $\delta(h_{L,S}) = +3.2\%$ for the loading range.

4.3 Model for Determining the Pressured Drop of Irrigated Random and Structured Packings, Based on the Known Resistance Coefficient ψ for Single-Phase Flow and the Dimensionless Pressure Drop $\Delta p/\Delta p_0$

4.3.1 Deriving the Model

In order to determine the pressure drop $\Delta p/\Delta p_0$, acc. to Eq. (4-7) or (4-9), it is necessary to know the specific geometric data of the packing element a and ε as well as the packing-specific constants x and ϕ and the total liquid hold-up h_L. The latter can be determined by using one of the models found in literature [12, 44], as quoted in Sect. 4.2.

To extend the validity of the models (4-7) and (4-9) to the operating range above the loading line, it is necessary to know the liquid hold-up $h_{L,S}$ as well as the contraction coefficient x at different operating conditions. Both variables $h_{L,S}$ and x are load-dependent in the operating range. Here, the pressure drop $\Delta p/\Delta p_0$ can be determined more easily throughout the entire operating range up to flooding point by converting Eq. (4-7).

The starting point for this are correlations (4-7) and (4-32), initially for the turbulent flow range $Re_L > 2$. The individual steps are shown in Eqs. (4-41), (4-42), (4-43), (4-44), (4-45), (4-46), (4-47), (4-48), (4-49) and (4-50). Equation (4-7) leads to the following correlation, which is valid for the range below the loading line:

$$\frac{\Delta p}{\Delta p_0} \cong \left[1 - \frac{2 \cdot h_L}{a \cdot d_h}\right]^{-5} \overset{d_h = 4 \cdot \frac{\varepsilon}{a}}{\Rightarrow} \quad \frac{\Delta p}{\Delta p_0} \approx \left[1 - \frac{h_L}{2\varepsilon}\right]^{-5}$$

$$\Rightarrow \frac{\Delta p}{\Delta p_0} = \left[1 - const \cdot \frac{h_L}{\varepsilon}\right]^{-5} \tag{4-41}$$

Equations (4-41) and (4-32) lead to correlation (4-42) for $Re_L \geq 2$ and $F_V/F_{V,Fl} \leq 0.65$:

$$\frac{\Delta p}{\Delta p_0} = \left[1 - const. \cdot \frac{C_P}{\varepsilon} \cdot Fr_L^{1/3}\right]^{-5} = \left[1 - C_{B,0} \cdot \frac{Fr_L^{1/3}}{\varepsilon}\right]^{-5}$$
$$= \left[1 - C_{B,0} \cdot \frac{u_L^{2/3} \cdot a^{1/3}}{g^{1/3} \cdot \varepsilon}\right]^{-5} \tag{4-42}$$

where $C_{B,0}$ = const. C_P is a constant. As $g^{1/3}$ is also a constant, both constants can be combined using the variable C_B.

$$C_B = \frac{C_{B,0}}{g^{1/3}} \quad \left[s^{2/3} m^{-1/3}\right] \tag{4-43}$$

The dimensionless pressure drop can now be described by correlation (4-44):

$$\frac{\Delta p}{\Delta p_0} = \left[1 - C_B \cdot \frac{a^{1/3} \cdot u_L^{2/3}}{\varepsilon}\right]^{-5}. \tag{4-44}$$

Analogously, it is possible to derive the correlation for determining the quotient $\Delta p/\Delta p_0$ for laminar liquid flow below the loading line, $F_V/F_{V,Fl} \leq 0.65$. Here, it is sufficient to substitute Eq. (4-20) into Eq. (4-40). This leads to correlation (4-45):

$$\frac{\Delta p}{\Delta p_0} = \left[1 - C_{C,0} \cdot \left(\frac{3}{g}\right)^{1/3} \cdot \frac{a^{2/3}}{\varepsilon} \cdot (\nu_L \cdot u_L)^{1/3}\right]^{-5} \tag{4-45}$$

where $C_{C,0}$ is a dimensionless constant, which is given by multiplying $(3/g)^{1/3}$ with the parameter C_C, acc. to Eq. (4-46):

$$C_C = C_{C,0} \cdot \left(\frac{3}{g}\right)^{1/3} \quad \left[s^{2/3} m^{-1/3}\right] \tag{4-46}$$

The dimensionless constant $C_{C,0}$, used for calculating the dimensionless pressure drop $\Delta p/\Delta p_0$, acc. to Eq. (4-45), has the following numerical value, acc. to Eq. (4-46):

$$C_{C,0} \equiv 1.$$

Hence, Eq. (4-45) for calculating the dimensionless pressure drop of irrigated random packings now leads to Eq. (4-47):

$$\frac{\Delta p}{\Delta p_0} = \left(1 - \left(\frac{3}{g}\right)^{1/3} \cdot \frac{a^{2/3}}{\varepsilon} \cdot (u_L \cdot \nu_L)^{1/3}\right)^{-5}. \tag{4-47}$$

For turbulent liquid flow $Re_L \geq 2$, Eqs. (3-8), (4-41) and (4-42) result in Eq. (4-48) for determining the pressure drop of irrigated packings:

$$\frac{\Delta p}{H} = \psi \cdot \frac{1-\varepsilon}{\varepsilon^3} \cdot \frac{F_V^2}{d_p \cdot K} \cdot \left[1 - C_{B,0} \cdot \frac{Fr_L^{1/3}}{\varepsilon}\right]^{-5} \quad \left[\mathrm{Pam}^{-1}\right] \tag{4-48}$$

and/or Eq. (4-49)

$$\frac{\Delta p}{H} = \psi \cdot \frac{1-\varepsilon}{\varepsilon^3} \cdot \frac{F_V^2}{d_P \cdot K} \left[1 - C_B \cdot \frac{a^{1/3}}{\varepsilon} \cdot u_L^{2/3}\right]^{-5} \quad \left[\mathrm{Pam}^{-1}\right]. \tag{4-49}$$

For laminar liquid flow, $Re_L < 2$, the following equation for calculating the irrigated pressure drop can be derived (4-50):

$$\frac{\Delta p}{H} = \psi \cdot \frac{1-\varepsilon}{\varepsilon^3} \cdot \frac{F_V^2}{d_P \cdot K} \left[1 - C_C \cdot \frac{a^{2/3}}{\varepsilon} \cdot (\nu_L \cdot u_L)^{1/3}\right]^{-5} \quad \left[Pam^{-1}\right], \tag{4-50}$$

with $C_C = 0.674$ as a result of combining Eqs. (3-8) and (4-45).

Equations (4-48), (4-49) and (4-50) only include two parameters which need to be determined empirically: the resistance coefficient for single-phase flow ψ and the constant C_B for two-phase flow for $Re_L \geq 2$, as well as ψ and C_C for $Re_L < 2$, which must be determined using the experimental pressure drop data $\Delta p_0/H$ of the dry packing and $\Delta p/H$ of the irrigated packing. By using Eqs. (4-48) and (4-50), presented above, the liquid hold-up h_L is therefore no longer required in order to determine the pressure drop $\Delta p/H$. In the operating range below the loading line, the variables C_B and C_C are expected to have constant numerical values, regardless of the type of random or structured packing. These can be determined by substituting into Eq. (4-51) the experimental pressure drop data in single-phase flow $\Delta p_0/H$ and that of the irrigated packing $\Delta p/H$ for a given packing element with a known void fraction ε, packing surface area a and operating conditions u_L and F_V:

$$C_B = \left[\frac{1 - \left[\left[\frac{\Delta p_0}{H}\right] \cdot \left[\frac{H}{\Delta p}\right]\right]^{1/5}}{u_L^{2/3}}\right] \cdot \frac{\varepsilon}{a^{1/3}} \quad \left[\mathrm{s}^{2/3}\mathrm{m}^{-1/3}\right] \tag{4-51}$$

This equation was the result of a simple conversion of Eq. (4-44), for $Re_L \geq 2$.

4.3.2 Comparing Calculated and Experimental Values for Laminar Liquid Flow, $Re_L < 2$

The applicability of Eq. (4-50) for determining the pressure drop of irrigated random packings at laminar liquid flow for Reynolds numbers in the range $0.1 < Re_L < 2$, was verified based on experimental pressure drop data for 25 mm metal Pall rings, taken by

TP	packing	$10^3 \cdot d$ [m]	d_S [m]	H [m]	system	$u_L \cdot 10^3$ [m/s]	$\eta_L \cdot 10^3$ mPa·s	t_L [°C]	lit.
○	Pall ring, metal	25	0.435	1.65	air / glycol	5.55 - 13	18	32	[5]
●					air / turbine oil	5.4 - 11	44	42	
ɸ					air / engine oil	7 - 12.7	51.5	48	
⊖						5.3 - 10.3	91	33	
Δ		35	0.22	1.3	air / glycol	0.3 - 9.4	18	25	[A]
▼	Białecki ring, metal								

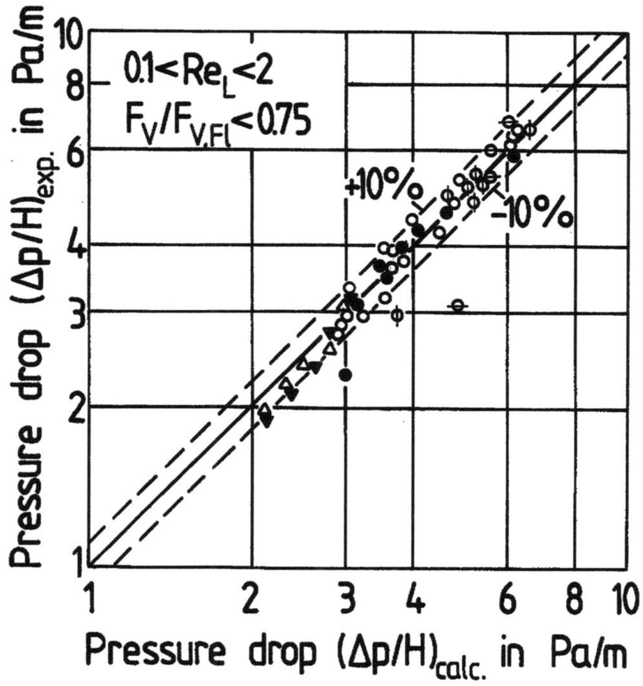

Figure 4-14. Comparison of pressure drop Δp/H calculated acc. to Eq. (4-50) with experimentally determined pressure drop for irrigated packed bed, valid for laminar liquid flow $Re_L < 2$

Billet [5], using the test systems air/water, air/glycol, air/machine oil, air/turbine oil, η_L from 1 to 90 mPas, as well as based on new experimental data for 17 mm plastic Hiflow rings, 15 mm Pall rings made of plastic and metal, 17 mm plastic Nor-Pac, 35 mm metal Pall and Białecki rings using the test system air/ethylene glycol. Figure 4-14 shows the comparison between the experimental pressure drop data $(\Delta p/H)_{exp}$ and the $(\Delta p/H)_{calc}$ values, calculated using Eq. (4-50). 95% of the test points used for the evaluation can be verified for $C_C \cong 0.674$, using Eq. (4-50), with a relative error of ±10%. The resistance coefficient ψ was calculated using Eq. (3-14) as well as the constants and exponents listed in Tables 6-1a–c, depending on the flow range.

4.3.3
Determining the Parameter C_B for Turbulent Liquid Flow

For practical reasons, key importance has been attached to the turbulent flow range, $Re_L \geq 2$, as this is the range mostly used for operating packed columns containing large packing elements in rectification processes under normal pressure and vacuum conditions and/or in pressure rectification and in absorption processes at moderate liquid loads.

Acc. to Fig. 4-15 and Eq. (4-51), the variable C_B is proportional to the factor C_P in Eq. (4-32). Hence, the variable C_B and the factor C_P are expected to be equally dependent on the operating conditions. If the variable C_B is now plotted against the relative gas capacity $F_V/F_{V,Fl}$, Fig. 4-15, the C_B value only increases, once 65% of the gas capacity factor at flooding point (loading line) has been exceeded. Above the loading line, the relative gas capacity $F_V/F_{V,Fl}$ and the specific liquid load u_L are both relevant, due to the fact that the exponent for the specific liquid load in Eq. (4-50) is higher than 2/3 and changes with the operating conditions.

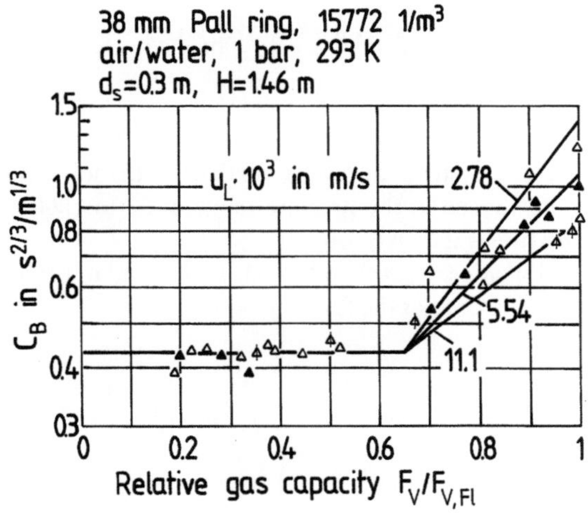

Figure 4-15. Value C_B of random 38 mm metal Pall rings depending on relative gas capacity $F_V/F_{V,Fl}$. Parameter: specific liquid load u_L

4.3.3.1
Determining the Parameter $C_{B,Fl}$ for Operating Conditions at Flooding Point

The correlation between the parameter C_B and the relative gas capacity factor $F_V/F_{V,Fl}$ and specific liquid load u_L, acc. to Fig. 4-15, leads to the following correlation at the flooding point:

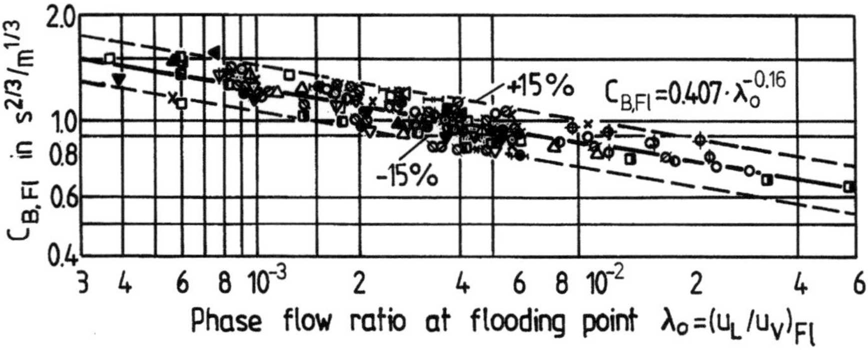

Figure 4-16. Dependency of value $C_{B,Fl}$ on phase flow ratio at the flooding point λ_0, for various systems and types of packings. Symbols can be found in Table 4-7 in the annex to this chapter

$$C_{B,Fl} = f(\lambda_0) \tag{4-52}$$

and for the liquid hold-up $h^0_{L,Fl} = f(\lambda_0)$, see Eq. (2-47).

Figure 4-16 shows an empirical correlation, $C_{B,Fl} = f(\lambda_{0,Fl})$ for a number of random and structured packings. The test points shown here can be correlated for the range $\lambda_0 = (0.3-60)\cdot 10^{-3}$, using the following Eq. (4-53):

$$C_{B,Fl} = 0.407 \cdot \lambda_0^{-0.16} \quad \left[s^{2/3}m^{-1/3}\right] \quad \textit{for} \quad \lambda_{0,Fl} = (0.3 \div 60)\cdot 10^{-3} \tag{4-53}$$

Equation (4-53) was developed on the basis of approx. 200 experimental values for the test systems air/water, methanol/ethanol, chlorobenzene/ethylbenzene, ethylbenzene/styrene in the pressure range of 33 to 1000 mbar. Their validity for any type of mixture is shown below.

Equations (4-48) and (4-53) finally lead to the following equation for calculating the pressure drop at the flooding point:

$$\left[\frac{\Delta p}{H}\right]_{Fl} = \psi_{Fl} \cdot \frac{1-\varepsilon}{\varepsilon^3} \cdot \frac{F^2_{V,Fl}}{d_P \cdot K} \cdot \left[1 - \frac{0.407}{\varepsilon} \cdot \lambda_0^{-0.16} \cdot a^{1/3} \cdot u_L^{2/3}\right]^{-5} \quad \left[\text{Pam}^{-1}\right]. \tag{4-54}$$

4.3.3.2 Determining the Parameter C_B Below the Loading Line

The evaluation of the experimental pressure drop data $\Delta p/H$ and $\Delta p_0/H$ for various test systems – air/water, chlorobenzene/ethylbenzene, ethylbenzene/styrene, methanol/ethanol, toluene/n-octane etc. – acc. to Eq. (4-51), led to the following mean C_B value for the random and type Y structured packings listed in Tables 6-1a–c:

$$C_B \approx 0.40 \pm 12\ \% \quad \left[s^{2/3}m^{-1/3}\right] \tag{4-55}$$

Table 4-7. Table for Fig. 4-16

TP	$d \cdot 10^3$ [m]	Packing	Material	System no.	ε [m^3/m^3]	d_S [m]	H [m]	Literature
○ ◑	25 35	**Białecki rings** random	metal	1, 2, 3	0.940 0.947	0.154 0.220	1.5 1.3	[A]
✱	25	**Raschig rings** random	ceramic	1	0.650	0.220	1.0	
❂	25	**Raschig rings** random		1	0.676	0.226	0.676	[21]
●	25	**Raschig rings** random	glass	5	0.820	0.15	1.0	[36, 37]
⧋ ✦ ⊞	25 35 50	**Pall rings**	plastic	1, 4	0.880- 0.926	0.22- 0.45	0.91- 2.0	[A]
× ⊕	38 75	**Hiflow rings** random	ceramic	1	0.834 0.865	0.30 0.45	1.34 2.00	
● ●	28 58	**Hiflow rings**	metal	1	0.962 0.970	0.3-0.6 0.45	1.46- 2.0 2.0	
□ ◨ ◧	25 35 58	**Pall rings** random		1, 2, 3	0.950 0.950 0.970	0.30 0.30 0.45	1.46 1.46 2.00	
▼ ▼	32 (size 1) 50 (size 2)	**VSP rings** random		1, 4	0.976 0.980	0.30- 0.45	1.46 2.00	
⌀ ⏀ ⊖	31.5 (size 1) 50 (size 2) 80 (size 3)	**Envipac** random	plastic (PP)	1, 4	0.930 0.964 0.959	0.30- 0.45	1.36- 2.0	
△ ⧋ ▲	28 50 90	**Hiflow rings** random	plastic	1	0.920 0.935 0.965	0.30 0.30 0.45	1.4 1.4 2.0	
⧋ ⧋	25 38	**Intalox saddle** random	ceramic	1	0.730 0.757	0.30	0.9 1.4	
◆	50	**NSW rings**	plastic	1, 4	0.950	0.30- 0.45	1.45- 2.0	
⧋	50	**Ralu rings**		1	0.940	0.30- 0.45	1.45- 2.0	
◈ ◈	28 38	**NSW rings** random		1, 4	0.920 0.932	0.30	1.4	
⋈ ◑	25 35	**Białecki rings stacked**	metal	1, 2, 3	0.928 0.945	0.15 0.22	1.5 1.0	
▽	60 (size 2)	**VSP rings**	plastic	1	0.954	0.30- 0.45	1.40- 2.0	
●	80	**Top-Pac**	metal	1	0.978	0.45	2.0	
⊖	40	**Intalox #40**		1	0.940	0.45	2.0	
▲	53	**Hiflow-Super**	plastic	1, 4	0.940	0.30- 0.45	1.4-2.0	
✦ ✖	70 (size 2) 45 (size 1)	**Dtnpac**		1	0.938 0.920	0.30 0.45	1.4 2.0	

and/or the dimensionless constant $C_{B,0}$:

$$C_{B,0} = 0.8562. \tag{4-56}$$

Based on Eqs. (4-48), (4-55) and (4-56), the pressure drop of irrigated random and structured packings $\Delta p/H$ is therefore given as:

$$\frac{\Delta p}{H} = \psi \cdot \frac{1-\varepsilon}{\varepsilon^3} \cdot \frac{F_V^2}{d_P \cdot K} \cdot \left[1 - \frac{0.8562}{\varepsilon} \cdot Fr_L^{1/3}\right]^{-5} \quad \left[\text{Pam}^{-1}\right], \tag{4-57}$$

which, following a simple conversion, leads to Eq. (4-58):

$$\frac{\Delta p}{H} = \psi \cdot \frac{1-\varepsilon}{\varepsilon^3} \cdot \frac{F_V^2}{d_P \cdot K} \cdot \left[1 - \frac{0.4}{\varepsilon} \cdot a^{1/3} \cdot u_L^{2/3}\right]^{-5} \quad \left[\text{Pam}^{-1}\right]. \tag{4-58}$$

4.3.3.3
Determining the Parameter $C_{B,S}$ for the Operating Range Above the Loading Line and Below the Flooding Point to Calculate the Pressure Drop, acc. to Eq. (4-48)

Figure 4-15 shows the correlation between the parameter $C_{B,S}$ above the loading line and the specific liquid load u_L as well as the relative column load $F_V/F_{V,Fl}$. The values $C_B > 0.4$ for $F_V/F_{V,Fl} > 0.65$ indicate that the influence of the specific liquid load u_L and of the quotient $F_V/F_{V,Fl}$ on $\Delta p/\Delta p_0$ is bigger than predicted by Eq. (4-51). Analogous to Eq. (4-36), the correlation between the C_B value and the operating conditions was based on Eq. (4-59). For the operating range above the loading line, $0.65 < F_V/F_{V,Fl} < 1$, the evaluation of all available data items contained in the database led to the following correlation:

$$C_{B,S} = C_{B,Fl} - \left(C_{B,Fl} - C_B\right) \cdot \left[1 - \left(\frac{\left(F_V/F_{V,Fl}\right) - 0.65}{0.35}\right)^{6/5}\right]^{5/6} \quad \left[s^{2/3}m^{-1/3}\right]. \tag{4-59}$$

For random and structured packings of type Y, the C_B value is given as $C_B = 0.40$, acc. to Eq. (4-55), and the $C_{B,Fl}$ value is given by correlation (4-53).

4.3.4
Comparing Calculated and Experimental Values for Turbulent Liquid Flow

Figure 4-17 shows the relative deviation $\delta(\Delta p/H)$ of the experimental pressure drop data up to the flooding point $(\Delta p/H)_{exp}$ from the values calculated acc. to Eqs. (4-48) and (4-50) for 15 to 50 mm randomly filled Pall rings made of metal, plastic and ceramic. The experimental values are applicable for the test systems air/water and air/4% NaOH solution as well as for various rectification systems, acc. to Table 2-2. The symbols used in Fig. 4-17 are each shown with a number that links them to the respective mixture in Table 2-2. Figures 4-17, 4-18, 4-19, 4-20, 4-21, 4-22, 4-23, 4-24, 4-25, 4-26 and 4-27show the spread of the experimental pressure drop data throughout the entire operating range

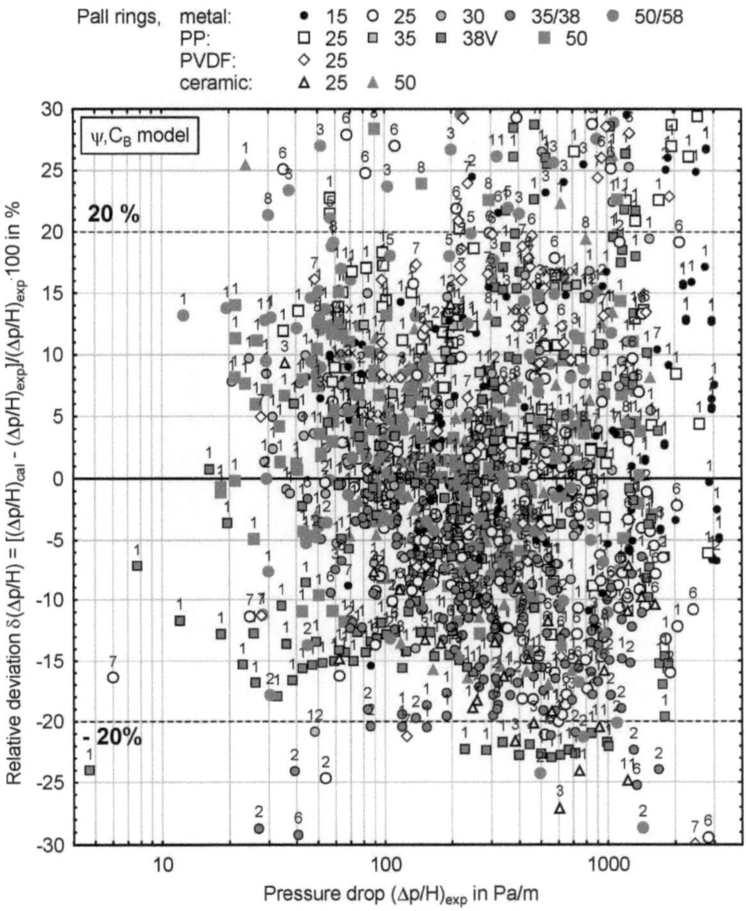

Figure 4-17. Relative deviation $\delta(\Delta p/H)$ of experimental data for determination of pressure drop $(\Delta p/H)_{exp}$ up to flooding point acc. to Eq. (4-49) and (4-54), valid for 15–50 mm metal Pall rings made of metal, plastic and ceramic. No. of system see Table 2-2. Test conditions see Table 4-4

for additional random and structured packings. A total of 10500 experimental data items were evaluated, which are presented in Figs. 4-17, 4-18, 4-19, 4-20, 4-21, 4-22, 4-23, 4-24, 4-25, 4-26 and 4-27. Tables 4-4a–e contain information on the number of test points for the individual packing elements as well as on the test conditions and the mean relative errors

1 $\bar{\delta}_1\left(\Delta p/H\right)$ below the loading line; $F_V/F_{V,Fl} \leq 0.65$
2 $\bar{\delta}_2\left(\Delta p/H\right)$ above the loading line up to the flooding point; $0.65 < F_V/F_{V,Fl} < 1$
3 $\bar{\delta}\left(\Delta p/H\right)$ mean relative error throughout the entire operating range including flooding point, $F_V = F_{V,Fl}$.

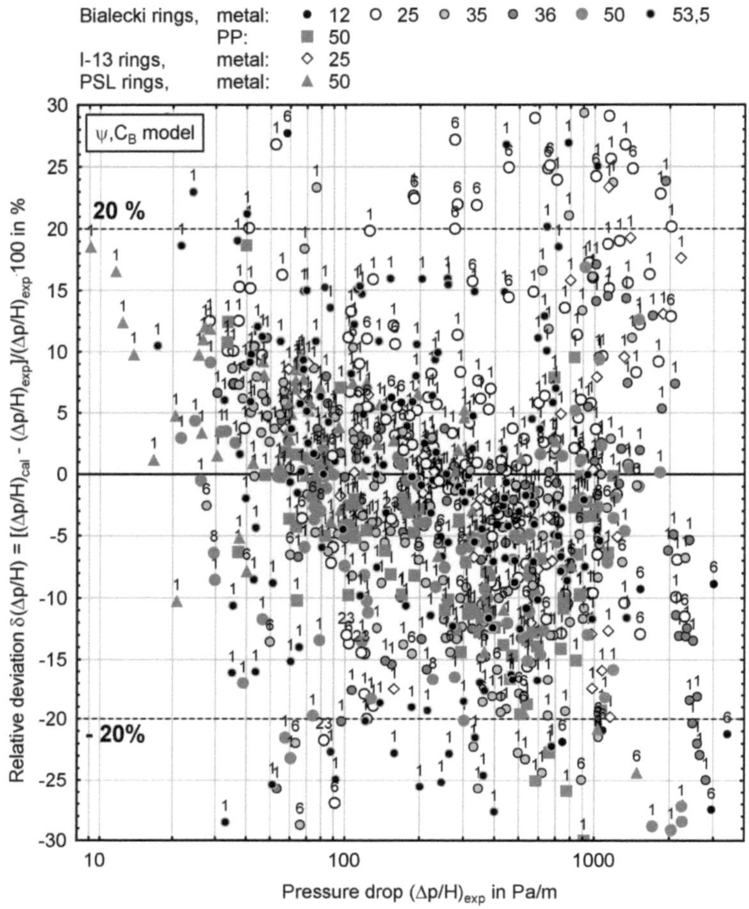

Figure 4-18. Relative deviation $\delta(\Delta p/H)$ of measuring data for determination of pressure drop $(\Delta p/H)_{exp}$ up to flooding point acc. to Eqs. (4-49) and (4-54), valid for random Białecki rings made of metal and plastic and for PSL rings and I-13 rings made of metal. No. of system see Table 2-2. Test conditions see Table 4-4

For the purpose of evaluating the experimental data for vapour/liquid systems held in the database, a computer programme called FDPAK was created for Windows, with which the changes in the substance flows and properties along the column can be taken into account. The programme is described in more detail in Sect. 6.8.

The pressure drops $(\Delta p/H)_{calc}$ were determined using the correlations for the respective ranges, i.e.:

1 Equation (4-49) with $C_B = 0.4 \Rightarrow$ Eq. (4-55) for the operating range below the loading line $F_V/F_{V,Fl} \leq 0.65$,

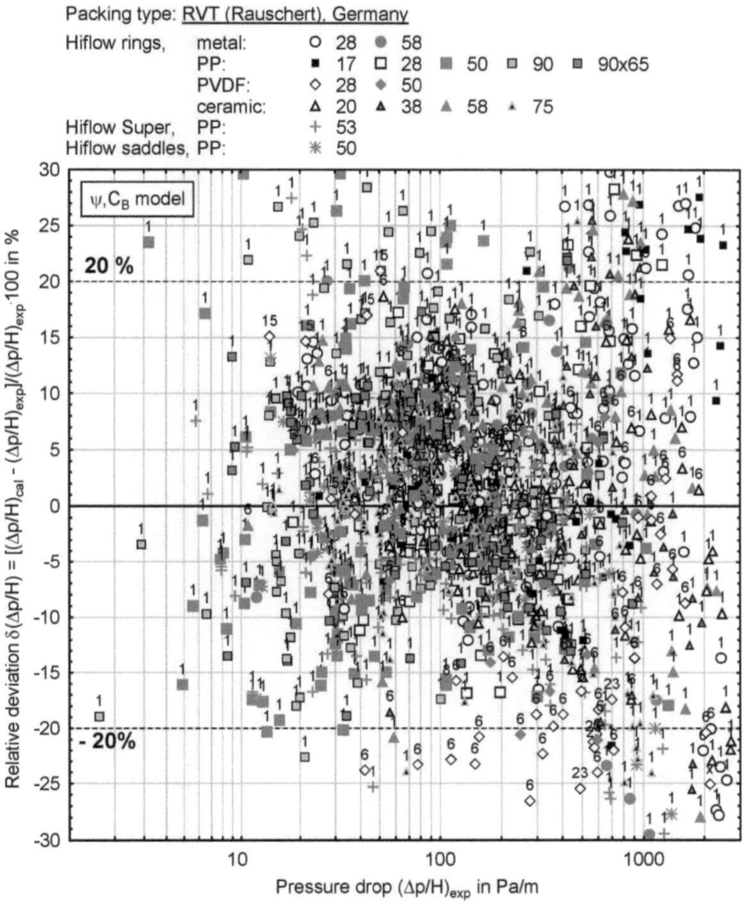

Figure 4-19. Relative deviation $\delta(\Delta p/H)$ of experimental data for determination of pressure drop $(\Delta p/H)_{exp}$ up to flooding point acc. to Eq. (4-49) and (4-54), valid for random packing of type Hiflow by RVT (Rauschert) made of metal, plastic and ceramic. No. of system see Table 2-2. Test conditions see Table 4-4

2 Equation (4-49) with C_B, acc. to Eq. (4-59), for the operating range above the loading line and below the flooding point,

3 Equation (4-49) with $C_{B,Fl}$, acc. to Eq. (4-53), for the flooding point, $F_V/F_{V,Fl} = 1 \pm 0.05$

Figures 4-17, 4-18, 4-19, 4-20, 4-21, 4-22, 4-23, 4-24, 4-25, 4-26 and 4-27 show that the experimental values for the range below the flooding point are practically spread within an error range of ± 20%.

Deviations of ± 30% mostly occur at higher relative gas capacities above 95% of the flood load, which is partly due to the inaccurate determination of the gas velocity at

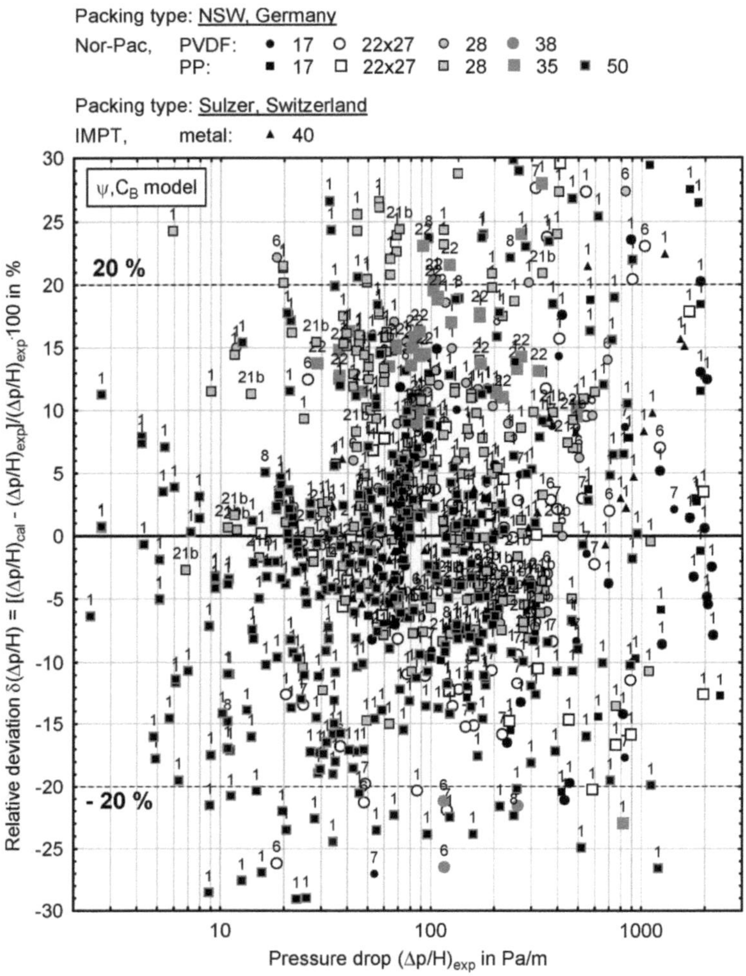

Figure 4-20. Relative deviation $\delta(\Delta p/H)$ of experimental data for determination of pressure drop $(\Delta p/H)_{exp}$ up to flooding point acc. to Eq. (4-49) and (4-54), valid for Nor-Pac packing type NSW made of plastic and metal saddles by Sulzer. No. of system see Table 2-2. Test conditions see Table 4-4

the flooding point $u_{V,Fl}$ and the pressure drop of the dry random or structured packing $\Delta p_0/H$. This applies to around 2% of the test points.

Based on the example of Pall rings, the relative error $\bar{\delta}\left(\Delta p/H\right)$ for the 873 experimental values listed in Table 4-4a throughout the entire operating range, including the pressure drops at the flooding point, is given as:

$$\bar{\delta}\left(\frac{\Delta p}{H}\right) = \frac{1}{n_1} \cdot \sum_{i=1}^{n_1} |\delta_i| \approx 8.14\ \% \tag{4-60}$$

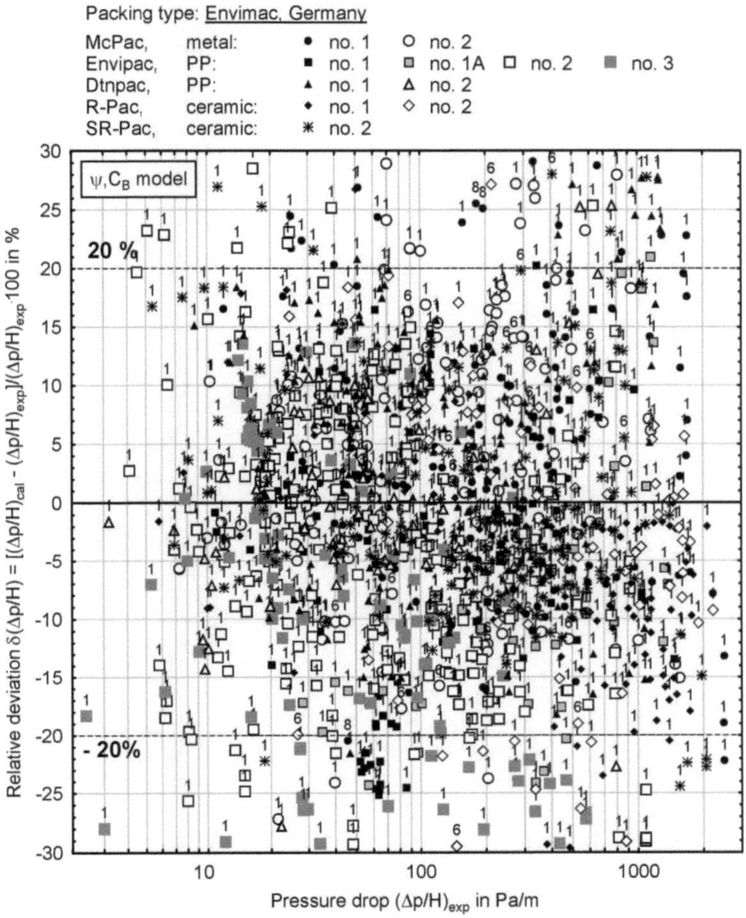

Figure 4-21. Relative deviation $\delta(\Delta p/H)$ of experimental data for determination of pressure drop $(\Delta p/H)_{exp}$ up to flooding point acc. to Eq. (4-49) and (4-54), valid for Envipac, Dtnpac, Mc-Pac, R-Pac, SR-Pac by ENVIMAC made of metal, plastic and ceramic. No. of system see Table 2-2. Test conditions see 4-4

with

$$\bar{\delta}\left(\frac{\Delta p}{H}\right)_1 \approx 6.82\ \% \tag{4-61}$$

for the operating range of $F_V/F_{V,Fl} \leq 0.65$, and

$$\bar{\delta}\left(\frac{\Delta p}{H}\right)_2 \approx 10.47\ \% \tag{4-62}$$

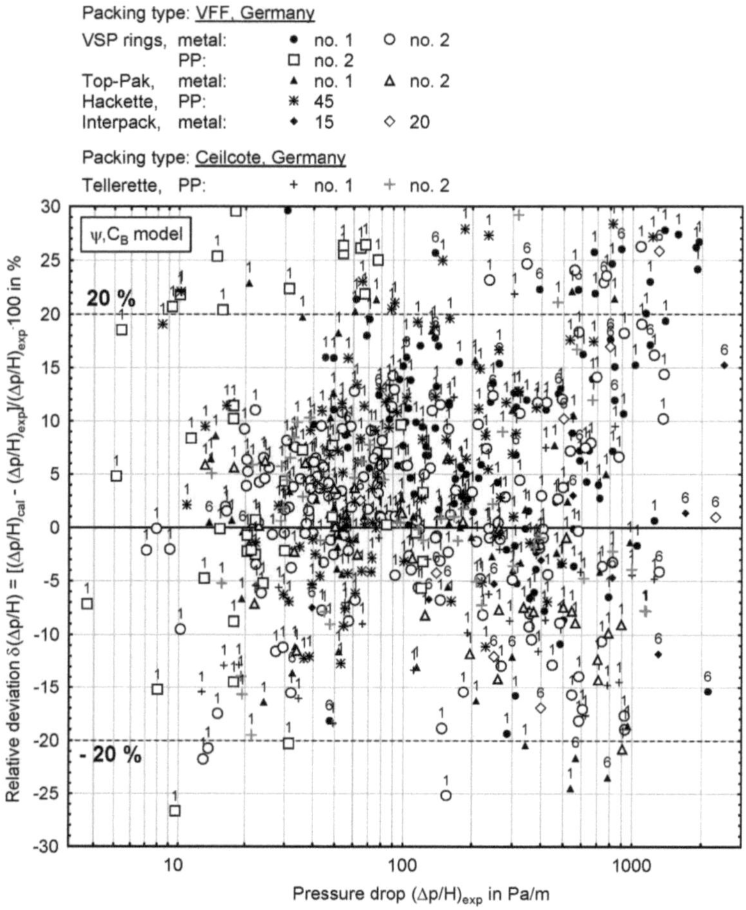

Figure 4-22. Relative deviation $\delta(\Delta p/H)$ of experimental data for determination of pressure drop $(\Delta p/H)_{exp}$ up to flooding point acc. to Eq. (4-49) and (4-54), valid for VSP rings, Top-Pak and Hackette by VFF made of metal and plastic and for plastic Tellerette by Ceilcote. No. of system see Table 2-2. Test conditions see 4-4

for the operating range of $0.65 < F_V/F_{V,Fl} \leq 1$, where n_l is the number of test points. Following thorough evaluation of the experimental data for the various random and structured packings, it can be noted that the mean error $\bar{\delta}\left(\Delta p/H\right)$ for the range below the loading line, up to approx. 80% of the flooding point, remains constant and only increases above this point, see Figs. 4-17, 4-18, 4-19, 4-20, 4-21, 4-22, 4-23, 4-24, 4-25, 4-26 and 4-27.

The mean relative deviations $\bar{\delta}\left(\Delta p/H\right)$ are listed in Tables 4-4a–e at the end of this chapter. They apply to the following random and structured packings made of metal, plastic and ceramic etc.:

as well as to various structured packings:

Montz, Gempak 2AT202,Ralu-Pack 250 YC, Mellapak 250 Y of sheet metal, plastic and gauze tube column with regularly stacked 25—50 mm Białecki rings etc.

8 ÷ 50 mm	Raschig rings of metal and ceramic
15 ÷ 58 mm	Pall rings of metal, ceramic and plastic
12 ÷ 50 mm	Białecki rings of metal and plastic, PSL rings and I-13 rings of metal
17 ÷ 90 mm	Hiflow rings of metal, ceramic and plastic
50 mm	Hiflow saddles of plastic
17 ÷ 50 mm	Nor-Pac (NSW rings) of plastic
15 ÷ 50 mm	Intalox saddles of ceramic and plastic
50 mm	Ralu rings and Ralu-Flow of plastic
25 ÷ 50 mm	stacked Raschig rings, PSL rings, Hiflow rings, Pall rings and Białecki rings of metal and plastic
VSP rings	sizes 1 and 2 of metal and plastic, Hackette, Top-Pak and Interpak
R-Pac	sizes 1 and 2 of ceramic
SR-Pac	size 2 of ceramic
Envipac	sizes 1, 2 and 3 of plastic
Dtnpac	sizes 1 and 2 of plastic etc.
Mc-Pac	sizes 1 and 2 of metal
Top-Pak	sizes 1 and 2 of metal
Ralu Flow	no. 2 of plastic
Glitsch, CMR rings	sizes 0.5A, 1.5A and 2A of metal and plastic
Hackette, Tellerette	sizes 1 and 2 of plastic
Intalox super saddles	sizes 2 and 3 of plastic

The concurrence between the calculated and experimental values, in particular in the range below the loading line, in which columns are preferably operated, is found to be good. Acc. to Tables 4-4a–e, the mean relative error is approx. 8% for all types of randomly filled packing elements and around 11.2% for all types of structured packings. The above-mentioned method of determining the pressure drop in packed columns for the respective ranges has therefore proved to be useful. It should be noted that it was possible to describe the comprehensive data material of approx. 10500 experimental pressure drop items sufficiently accurately, using a single correlation, i.e. Eq. (4-48), for the flow range below the loading line, which is relevant for practical applications, and for any types of packing elements. It should also be pointed out that the uncertainties in the determination of the gas velocity at the flooding point and the pressure drop of the irrigated random or structured packing $\Delta p_0/H$ were taken into account in the calculation of the pressure drop $\Delta p/H$ above the loading line and at the flooding point.

Based on the system of Eqs. (4-48), (4-53), (4-57), (4-58) and (4-59), it is possible to determine the pressure drop of irrigated packings $\Delta p/H$ sufficiently accurately for absorption processes as well as for rectification in the vacuum and normal pressure range up to 30 bar, as the geometric and physical influencing factors can be properly ascertained. In the case of turbulent liquid flow, $Re_L \geq 2$, and for larger packing elements, the wettability of the packing material and the surface tension and viscosity of the trickling liquid has practically no influence on the pressure drop $\Delta p/H$. Larger errors can be expected when determining the pressure drop of structured packings above the loading line, which are caused by the installation of downcomers in columns with smaller diameters. The influence of the downcomers decreases, as the column diameter gets bigger.

The physical properties of the systems, as shown in Table 2-2, and the operating parameters and geometric data, used for the evaluation, vary within the following ranges:

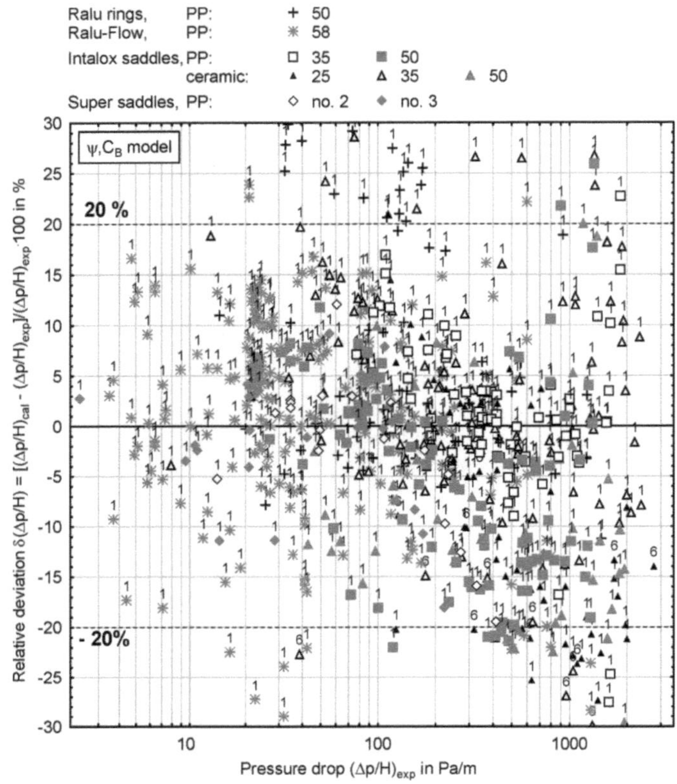

Figure 4-23. Relative deviation $\delta(\Delta p/H)$ of experimental data for determination of pressure drop $(\Delta p/H)_{exp}$ up to flooding point acc. to Eq. (4-49) and (4-54), valid for saddles made of ceramic and plastic, plastic Ralu rings and Ralu-Flow by Raschig. No. of system see Table 2-2. Test conditions see Table 4-4

$$
\begin{array}{rcll}
2 & \leq & \mathbf{Re_L} & \leq 200 \text{ (and/or up to } Re_{L,max}\text{, acc . to Table 4-4a-e} \\
0.15 & \leq & \mathbf{F_V/F_{V,Fl}} & \leq 1 \\
54 & \leq & \mathbf{a} & \leq 550\,m^2\,m^{-3} \\
0.008 & \leq & \mathbf{d} & \leq 0.09\,m \\
0.15 & \leq & \mathbf{d_S} & \leq 1.8\,m \\
0.6 & \leq & \mathbf{H} & \leq 7\,m \\
660 & \leq & \rho_L & \leq 1260\,kgm^{-3} \\
0.03 & \leq & \rho_V & \leq 3.6\,kgm^{-3} \\
14 & \leq & \sigma_L & \leq 72.5\,nNm^{-1} \\
0.2 & \leq & \eta_L \cdot \mathbf{10^3} & \leq 8\,kgm^{-1}\,s^{-1} \\
6.5 & \leq & \eta_V \cdot \mathbf{10^6} & \leq 18.2\,kgm^{-1}\,s^{-1} \\
0 & \leq & \lambda_0 & \leq 0.06 \\
2 & \leq & \Delta P/H & \leq 4000\,Pam^{-1} \\
5 & \leq & ds/d &
\end{array}
\qquad (4.63)
$$

4.3.5 Conclusions Section 4.3

The correlations for determining the pressure drop Δp/H of irrigated random and structured packings, presented in Sect. 4.3, lead to the following conclusions:

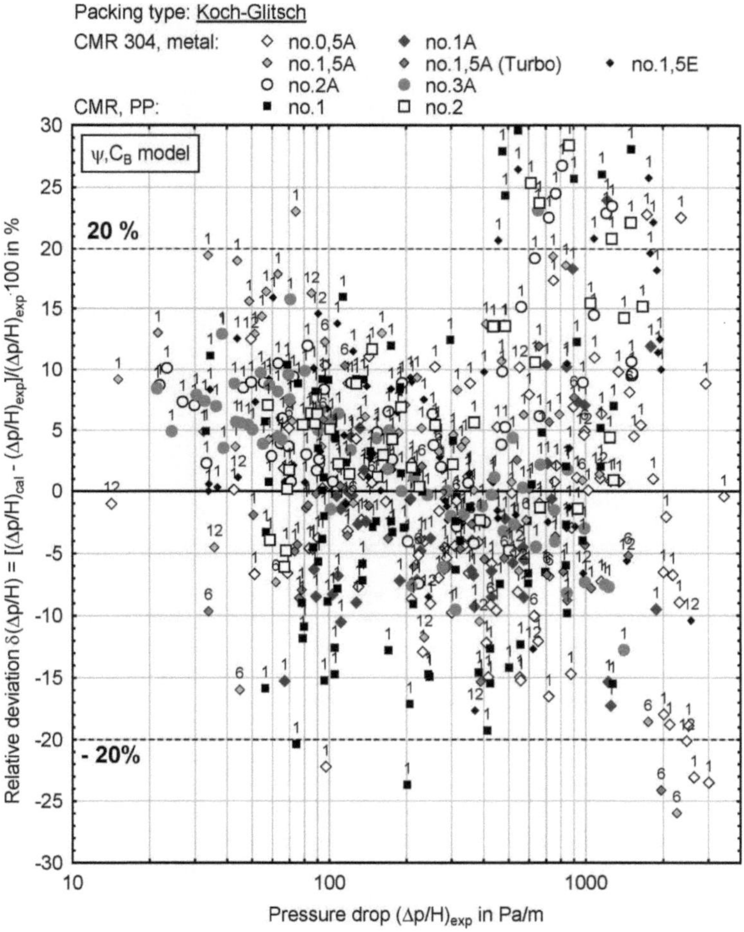

Figure 4-24. Relative deviation $\delta(\Delta p/H)$ of experimental data for determination of pressure drop $(\Delta p/H)_{exp}$ up to flooding point acc. to Eq. (4-49) and (4-54), valid for CMR rings by Koch-Glitsch made of metal and plastic. No. of system see Table 2-2. Test conditions see Table 4-4.

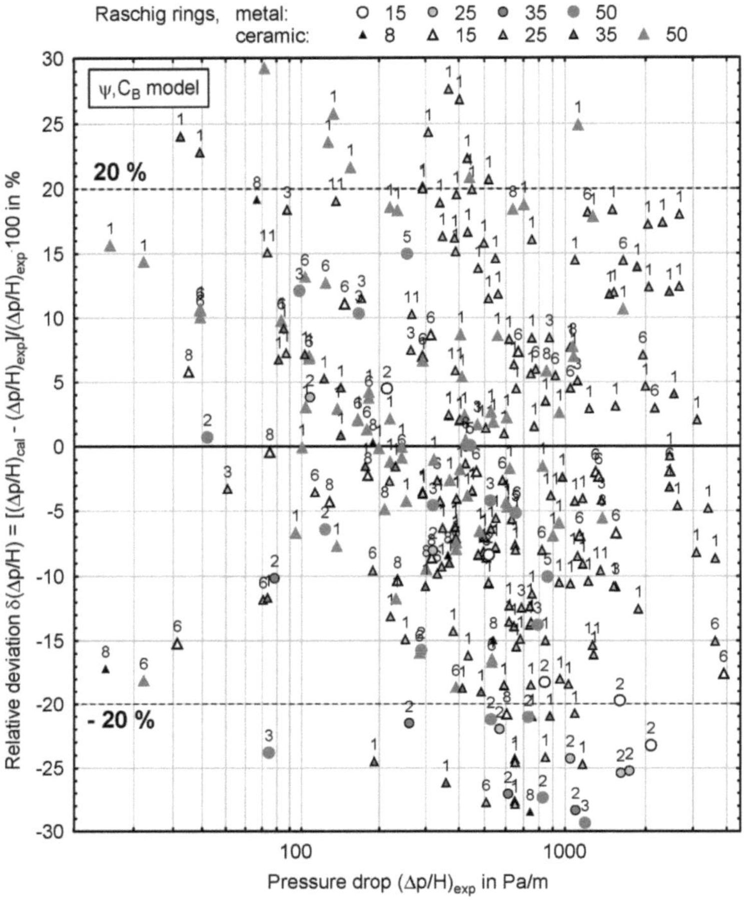

Figure 4-25. Relative deviation $\delta(\Delta p/H)$ of experimental data for determination of pressure drop $(\Delta p/H)_{exp}$ up to flooding point acc. to Eq. (4-49) and (4-54), valid for Raschig rings made of metal and ceramic. No. of system see Table 2-2. Test conditions see Table 4-4

1. In this work, correlation (4-7), which was developed by Buchanan [19], was converted and simplified in such as way that the quotient $\Delta p/\Delta p_0$ can be determined without having to ascertain the liquid hold-up h_L or the packing-specific constants. Instead of a determining a number of packing-specific constants, it is sufficient to determine just one universal, dimensionless variable $C_{B,0}$, based on the experimental pressure drop data $\Delta p/H$ and $\Delta p_0/H$ for any types of random and structured packings, acc. to Eqs. (4-55) and (4-59) in the turbulent flow range of the liquid. The evaluation of the experimental data during the course of this work has shown that the turbulent flow range of the liquid starts at $Re_L \geq 2$ [44]. The dimensionless constant $C_{B,0}$ for the operating range below the loading line was given as $C_{B,0} = 0.8562$, Eq. (4-56).

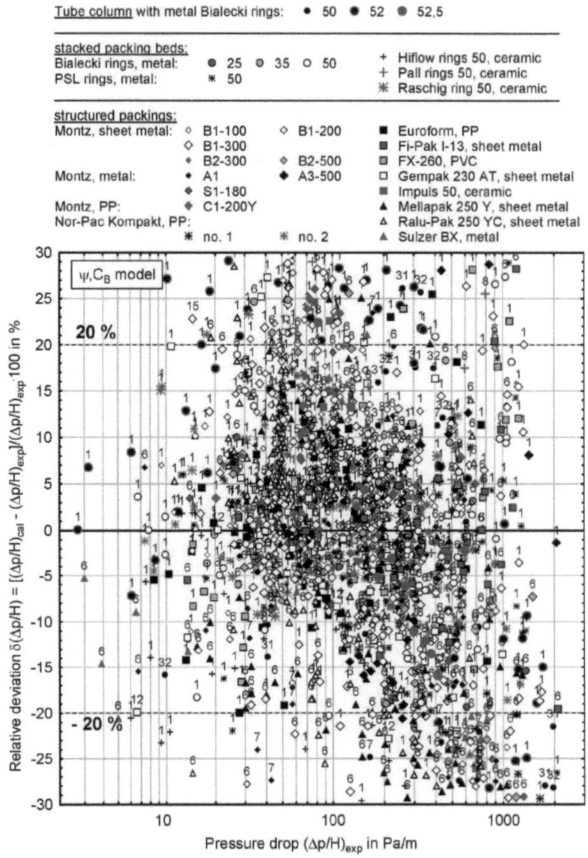

Figure 4-26. Relative deviation $\delta(\Delta p/H)$ of experimental data for determination of pressure drop $(\Delta p/H)_{exp}$ up to flooding point acc. to Eq. (4-49) and (4-54), valid for tube columns, stacked and structured packings. No. of system see Table 2-2. Test conditions see Table 4-4

Equation (4-57) for calculating the pressure drop $\Delta p/H$ for $Re_L \geq 2$ is given as:

$$\frac{\Delta p}{H} = \psi \cdot \frac{1-\varepsilon}{\varepsilon^3} \cdot \frac{F_V^2}{d_p \cdot K} \cdot \left[1 - C_{B,0} \cdot \frac{Fr_L^{1/3}}{\varepsilon}\right]^{-5} \quad \left[\mathrm{Pa m^{-1}}\right], \tag{4-57}$$

which, for C_B acc. to Eq. (4-43), leads to Eq. (4-49):

$$\frac{\Delta p}{H} = \psi \cdot \frac{1-\varepsilon}{\varepsilon^3} \cdot \frac{F_V^2}{d_P \cdot K} \cdot \left[1 - \frac{C_B}{\varepsilon} \cdot a^{1/3} \cdot u_L^{2/3}\right]^{-5} \quad \left[\mathrm{Pa m^{-1}}\right] \tag{4 - 58}$$

where $K = 1$ for tube columns with regularly filled packing elements and structured packings, and $K \leq 1$, acc. to Eq. (3-6), for randomly filled and stacked packing elements.

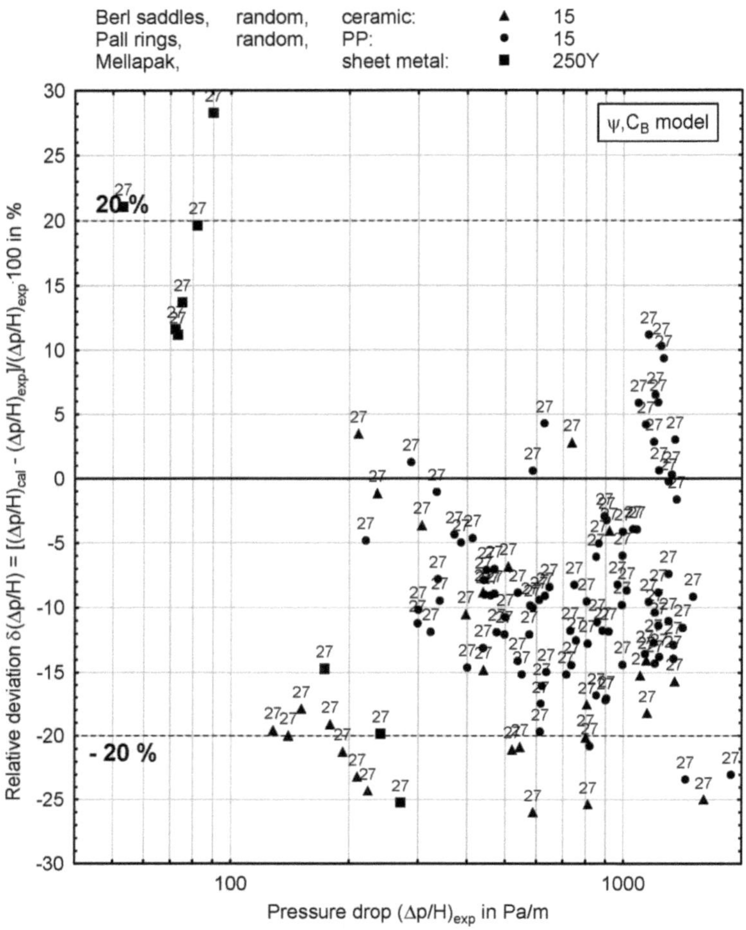

Figure 4-27. Relative deviation $\delta(\Delta p/H)$ of experimental data for determination of pressure drop $(\Delta p/H)_{exp}$ up to flooding point acc. to Eq. (4-49) and (4-54), valid for random 15 mm Pall rings made of PP, 15 mm ceramic Berl saddles and Mellapak 250Y made of metal sheet. Literature: experimental data by Krehenwinkel [56]. No. of system see Table 2-2. Test conditions see 4-4

The pressure drop $\Delta p/H$ above the loading line, $F_V/F_{V,Fl} > 0.65$, is calculated using the same Eq. (4-49). The only difference lies in the calculation of the variable C_B, which, for $0.65 < F_V/F_{V,Fl} < 1$, is as follows, acc. to Eq. (4-59):

$$C_{B,S} = C_{B,Fl} - \left(C_{B,Fl} - C_B\right) \cdot \left[1 - \left(\frac{\left(F_V/F_{V,Fl}\right) - 0.65}{0.35}\right)^{6/5}\right]^{5/6} \qquad \left[s^{2/3} m^{-1/3}\right] \tag{4-59}$$

In this case, the C_B value must be substituted by the $C_{B,S}$ value in Eq. (4-49). In order to determine the pressure drop $(\Delta p/H)_{Fl}$ at the flooding point, the following value is used to replace the C_B value, acc. to Eq. (4-53):

$$C_{B,Fl} = 0.407 \cdot \lambda_0^{-0.16} \quad \left[s^{2/3} m^{-1/3}\right], \tag{4-60}$$

see Fig. 4-16.

In order to apply the equation, presented above, for determining the pressure drop $\Delta p/H$ of irrigated packings, it is sufficient to ascertain the packing-specific geometric data a and ε as well as the operating conditions u_L and F_V. The gas velocity at the flooding point is determined using Eq. (2-69) as well as the resistance coefficient ψ, acc. to Eq. (3-14), based on the empirically derived packing constants K_1 and K_2, which are listed in Tables 3-1 and 3-2 and/or 6-1a–c, or according to the new, universally valid model acc. to Eq. (3-27).

As described above, it is possible to determine the pressure drop $\Delta p/H$ of irrigated random and structured packings relatively accurately for practical applications for the entire parameter range of the varied systems, packing elements and constructive variables d_S and H, see Eq. (4-63). The mean errors $\bar{\delta}\left(\Delta p/H\right)$ relating to the individual packing systems can be found in Tables 4-4a–e.

2. In the case of laminar liquid flow, $Re_L < 2$ and $C_C = 0.674$, the pressure drop $\Delta p/H$, acc. to Eq. (4-50):

$$\frac{\Delta p}{H} = \psi \cdot \frac{1-\varepsilon}{\varepsilon^3} \cdot \frac{F_V^2}{d_P \cdot K} \cdot \left[1 - \frac{C_C}{\varepsilon} \cdot a^{1/3} \cdot (u_L \cdot \nu_L)^{1/3}\right]^{-5} \quad \left[\mathrm{Pa\,m}^{-1}\right]$$

with K = 1 for structured packings, is dependent on the operating variables u_L, F_V, the geometric data a and ε of the respective packing element and on the viscosity of the liquid η_L. In the case of larger-diameter, perforated packing elements, the surface tension σ_L is not expected to have any influence on the pressure drop, as long as the number We_L/Fr_L is higher than 100, which means h_L is not dependent on We_L/Fr_L, acc. to Mersmann and Deixler [44].

3. The following calculation examples illustrate the application of the models, presented above, to determine the pressure drop $\Delta p/H$ throughout the entire flow range up to flooding point.

Numerical Example 4.1

Based on the operating data for the Białecki rings, mentioned in task description 2.2, we must determine the pressure drop $\Delta p/H$ per 1 m packing height.

Solution

The pressure drop $\Delta p/H$ below the loading line, $F_V/F_{V,Fl} < 0.65$, is calculated using Eq. (4-58). The following variables are required:

$$u_V = 1\,\mathrm{ms}^{-1} \quad u_L = 0.0111\,\mathrm{ms}^{-1}$$

Acc. to Eq. (3-3):

$$d_P = 6 \cdot (1 - 0.942)/238 = 1.462 \cdot 10^{-3}\mathrm{m}$$

Acc. to Eq. (3-6):

$$K = \left[1 + \frac{2}{3} \cdot \frac{1}{(1 - 0.942)} \cdot \frac{1.462 \cdot 10^{-3}}{0.15}\right]^{-1} = 0.9$$

Acc. to Eq. (3-10):

$$\mathrm{Re}_V = \frac{1.0 \cdot 0.001462}{(1 - 0.942) \cdot 15.2 \cdot 10^{-6}} \cdot 0.9 = 1492.5$$

Acc. to Eq. (3-16) and Table (2-5a), [17]:

$$\psi_1 = 10.17 \cdot \mathrm{Re}_V^{-0.17} = 2.935$$

or acc. to Eq. (3-27) and $\varphi_P = 0.208$:

$$\psi_2 = \left(\frac{725.6}{Re_V} + 3.203\right) \cdot (1 - \varphi_P) = \left(\frac{725.6}{1492.5} + 3.203\right) \cdot (1 - 0.208) = 2.922$$

Acc. to Eq. (4-58) we obtain:
for $\psi_1 = 2.935$ acc. to Eq. (3-16)

$$l\left(\frac{\Delta p}{H}\right)_1 = 2.935 \cdot \frac{1 - 0.942}{0.942^3} \cdot \frac{1.0814^2}{0.001462 \cdot 0.9} \cdot \left[1 - \frac{0.4 \cdot 238^{1/3}}{0.942} \cdot 0.0111^{2/3}\right]^{-5} =$$
$$= 365.2\ \mathrm{Pam}^{-1}$$

for $\psi_2 = 2.922$ acc. to Eq. (3-27) and (3-29)

$$l\left(\frac{\Delta p}{H}\right)_2 = 2.922 \cdot \frac{1 - 0.942}{0.942^3} \cdot \frac{1.0814^2}{0.001462 \cdot 0.9} \cdot \left[1 - \frac{0.4 \cdot 238^{1/3}}{0.942} \cdot 0.0111^{2/3}\right]^{-5} =$$
$$= 363.5\ \mathrm{Pam}^{-1}$$

Relative deviation from the experimental value, see Fig. 2-2,

$$\left[\frac{\Delta p}{H}\right]_{exp} = 343.4 \text{ Pam}^{-1}$$

$$\delta\left(\Delta p/H\right)_1 = \left[(365.2 - 343.4)/343.4\right] \cdot 100\% = +6.34\ \%$$
$$\delta\left(\Delta p/H\right)_2 = \left[(363.5 - 343.4)/343.4\right] \cdot 100\% = +5.85\ \%$$

Numerical Example 4.2

We must determine the pressure drop of the irrigated, random 25 mm Białecki packing at the flooding point, if the specific liquid load is $u_L = 0.0111\ \text{ms}^{-1}$. The physical properties are applicable to the air/water system at 1 bar and 293 K. The gas velocity $u_{V,Fl}$ at the flooding point is given as $u_{V,Fl} = 1.776\ \text{ms}^{-1}$, acc. to numerical Example 2.2.

Solution

A. Calculation of the pressure drop in the loading range, $0.65 < F_V/F_{V,Fl} < 1$, using Eq. (4-49), where the parameter $C_{B,S}$ is determined by means of Eq. (4-53). The following applies for $u_V = 0.8 \cdot u_{V,Fl}$:

$$u_V = 0.8 \cdot 1.776 = 1.421\ ms^{-1} \quad \Rightarrow F_V = 1.537\ \sqrt{Pa}$$

$$u_L = 0.0111\ \text{ms}^{-1}, \quad \lambda_0 = u_L/u_{V,Fl} = 0.0111/1.776 = 6.25 \cdot 10^{-3}$$

Acc. to Eq. (3-10):

$$\text{Re}_V = \frac{1.424 \cdot 0.001462}{(1 - 0.942) \cdot 15.2 \cdot 10^{-6}} \cdot 0.9 = 2118 \quad > \quad 2100$$

Acc. to Eq. (4-53):

$$C_{B,Fl} = 0.407 \cdot \left(6.25 \cdot 10^{-3}\right)^{-0.16} = 0.917$$

Acc. to Eq. (4-59):

$$C_{B,S} = C_{B,Fl} - \left(C_{B,Fl} - 0.4\right) \cdot \left[1 - \left(\frac{\left(F_V/F_{V,Fl}\right) - 0.65}{0.35}\right)^{6/5}\right]^{5/6}$$

$$C_{B,S} = 0.917 - (0.917 - 0.4) \cdot \left[1 - \left(\frac{0.8 - 0.65}{0.35}\right)^{6/5}\right]^{5/6} = 0.5613$$

Acc. to Eq. (3-14) with $K_1 = 4.13$ and $K_2 = -0.0522$, [17]:

$$\psi_1 = 4.13 \cdot 2118^{-0.0522} = 2.769$$

or acc. to Eq. (3-27) and $\varphi_P = 0.208$:

$$\psi_2 = \left(\frac{725.6}{Re_V} + 3.203\right) \cdot (1 - \varphi_P) = \left(\frac{725.6}{2118} + 3.203\right) \cdot (1 - 0.208) = 2.808$$

And acc. to Eq. (4-58) we obtain:
for $\psi_1 = 2.769$ acc. to Eq. (3-14):

$$\left(\frac{\Delta p}{H}\right)_1 = 2.769 \cdot \frac{1-0.942}{0.942^3} \cdot \frac{1.537^2}{0.001462 \cdot 0.9} \cdot \left[1 - \frac{0.561}{0.942} \cdot 238^{1/3} \cdot 0.0111^{2/3}\right]^{-5} =$$
$$= 951.4 \text{ Pam}^{-1}$$

and for $\psi_2 = 2.808$ acc. to Eq. (3-27):

$$\left(\frac{\Delta p}{H}\right)_2 = 2.808 \cdot \frac{1-0.942}{0.942^3} \cdot \frac{1.537^2}{0.001462 \cdot 0.9} \cdot \left[1 - \frac{0.561}{0.942} \cdot 238^{1/3} \cdot 0.0111^{2/3}\right]^{-5} =$$
$$= 964.8 \text{ Pam}^{-1}$$

$[\Delta p/H]_{exp} \cong 860$ Pam^{-1} based on Fig. 2-2a

Relative error:

$$\delta\left(\Delta p/H\right)_1 = \frac{951.4 - 860}{860} \cdot 100\ \% = +10.6\ \% \quad \text{resp.}$$
$$\delta\left(\Delta p/H\right)_2 = \frac{964.8 - 860}{860} \cdot 100\ \% = +12.2\ \%$$

B. Calculation of the pressure drop $\Delta p/H$ at the flooding point, using Eq. (4-49), for $u_{V,Fl} = 1.776$ ms^{-1}, $u_L = 0.0111$ ms^{-1} and $F_{V,Fl} = 1.921$ Pa$^{1/2}$(based on numerical Example 2.2). The parameter $C_{B,Fl}$ is determined using Eq. (4-53).

$$C_{B,Fl} = 0.407 \cdot \lambda_0^{-0.16} = 0.407 \cdot \left(6.25 \cdot 10^{-3}\right)^{-0.16} = 0.917$$

Acc. to Eq. (3-10):

$$Re_V = \frac{1.776 \cdot 0.001462}{(1 - 0.942) \cdot 15.2 \cdot 10^{-6}} \cdot 0.9 = 2650.7 \quad > \quad 2100,$$

$\Rightarrow$ acc. to Eq. (3-14) for $K_1 = 4.13$ and $K_2 = -0.0522$, [17]:

$$\psi_1 = 4.13 \cdot 2650.7^{-0.0522} = 2.737$$

or acc. to Eq. (3-27) and $\varphi_P = 0.208$:

$$\psi_2 = \left(\frac{725.6}{2650.7} + 3.203\right) \cdot (1 - 0.208) = 2.754$$

Acc. to Eq. (4-54):

for $\psi_1 = 2.737$ acc. to Eq. (3-14):

$$\left(\frac{\Delta p}{H}\right)_1 = 2.737 \cdot \frac{1 - 0.942}{0.942^3} \cdot \frac{1.921^2}{0.001462 \cdot 0.9} \cdot \left[1 - \frac{0.917}{0.942} \cdot 238^{1/3} \cdot 0.0111^{2/3}\right]^{-5} = 3173.2\ \text{Pam}^{-1}$$

for $\psi_2 = 2.754$ acc. to Eq. (3-27):

$$\left(\frac{\Delta p}{H}\right)_2 = 2.754 \cdot \frac{1 - 0.942}{0.942^3} \cdot \frac{1.921^2}{0.001462 \cdot 0.9} \cdot \left[1 - \frac{0.917}{0.942} \cdot 238^{1/3} \cdot 0.0111^{2/3}\right]^{-5} = 3193.0\ \text{Pam}^{-1}$$

$$\left[(\Delta p)/H\right]_{exp} \cong 2800\ \text{Pam}^{-1} \text{ based on Fig. 2-2a}$$

Relative errors:

$$\delta\left(\Delta p/H\right)_1 = \frac{2800 - 3173.2}{2800} \cdot 100\ \% = -13.3\ \%$$

$$\delta\left(\Delta p/H\right)_2 = \frac{2800 - 3193.0}{2800} \cdot 100\ \% = -14.0\ \%$$

Numerical Example 4.3

Based on the data given in numerical Example 3.2, the aim is to determine the pressure drop of the packing Mellapak 250Y in a column with $d_S = 1.0$ m for the test system air/water at ambient conditions. The pressure drop will be determined for a liquid load of $V_L = 15.7\ m^3\ h^{-1}$ and for the gas capacity factors $F_V = 1.6\ \sqrt{Pa}$ and $F_V = 2.3\ \sqrt{Pa}$.

The experimentally derived values are as follows, acc. to Kuźniar and Niżański [25, Chap. 3]:

$$\left(\frac{\Delta p}{H}\right)_{exp} = 100\ Pam^{-1} \quad \text{for } F_V = 1.6\ \sqrt{\text{Pa}}$$

$$\left(\frac{\Delta p}{H}\right)_{exp} = 190\ Pam^{-1} \quad \text{for } F_V = 2.3\ \sqrt{\text{Pa}}$$

Solution

Based on numerical Example 3.2, the pressure drop of the dry packing is given as:

$$\left(\frac{\Delta p_0}{H}\right)_{3.2} = 63.1 \text{ Pam}^{-1} \quad \text{for } F_V = 1.6\,\sqrt{\text{Pa}}$$

$$\left(\frac{\Delta p_0}{H}\right)_{3.2} = 126 \text{ Pam}^{-1} \quad \text{for } F_V = 2.3\,\sqrt{\text{Pa}}$$

For a liquid flow rate of 17.5 m^3h^{-1}, the specific liquid load for $d_S = 1.0$ m is $u_L = 0.0062$ $m^3m^{-2}s^{-1}$. Based on Eq. (4-49) and $C_B = 0.4$, the dimensionless pressure drop $\Delta p/\Delta p_0$ is given as:

$$\frac{\Delta p}{\Delta p_0} = \left(1 - 0.4 \cdot \frac{a^{1/3} \cdot u_L^{2/3}}{\varepsilon}\right)^{-5} =$$
$$= \left(1 - 0.4 \cdot \frac{250^{1/3} \cdot 0.0062^{2/3}}{0.975}\right)^{-5} = 1.577$$

Equations (3-1) and (4-49) for $F_V = 1.6\,\sqrt{\text{Pa}}$ lead to:

$$\frac{\Delta p}{H} = \frac{\Delta p_0}{H} \cdot \frac{\Delta p}{\Delta p_0} = 63.1 \cdot 1.577 = 99.5 \text{ Pam}^{-1}$$

and for $F_V = 2.3\,\sqrt{\text{Pa}}$:

$$\frac{\Delta p}{H} = \frac{\Delta p_0}{H} \cdot \frac{\Delta p}{\Delta p_0} = 126 \cdot 1.577 = 200.3 \text{ Pam}^{-1}$$

This results in the following deviations from the experimental data:

$$\delta\left(\frac{\Delta p}{H}\right)_{F_V=1.6} = \frac{100-99.5}{100} \cdot 100\ \% = +0.5\ \%$$

$$\delta\left(\frac{\Delta p}{H}\right)_{F_V=2.3} = \frac{190-200.3}{190} \cdot 100\ \% = -5.42\ \%$$

Note

An accurate determination of the pressure drop of irrigated packings acc. to the above model can only be guaranteed if the correlations for predicting the resistance coefficient ψ for the respective column size are known, see Sect. 3.2.3.

Annex Chapter 4

Tables Chapter 4

Table 4-3. List of relative errors $\bar{\delta}\,(h_L)$ and number of test points for investigating the total liquid hold-up for various random and structured packings

Series no.	Packing	Type	Material	Number of test points with $\bar{\delta}_i(h_L) < 20\%$			Number of test points TP	$\bar{\delta}(h_L)$ in % $h_{L,calc}$		
				Lit.1	Lit.2	Lit.3	System no.	Lit.1	Lit.2	Lit.3
1	**Pall**	25	metal	16	16	6	19/(1)	29.50	27.13	13.78
2	**rings**	38		1	1	8	21/(1)	6.90	7.10	24.22
3		35		–	2	6	6/(1)	14.40	15.42	46.22
4		35		–	1	4	4/(2)	12.50	4.20	43.60
5		50		2	1	22	24/(1)	10.10	7.90	32.30
6		58		4	7	4	16/(1)	16.30	19.80	14.50
7	**Pall**	25	PP	(1)	1	4	5/(1)	7.41	14.11	32.01
8	**rings**	25		2	10	21	37/(1)	10.40	17.10	27.40
9		35		–	0	0	7/(2)	12.60	9.02	38.00
10		35		2	0	0	5/(4)	17.90	3.40	8.20
11		35		–	1	6	7/(1)	18.30	9.02	38.04
12		35		2	2	4	5/(3)	16.70	16.30	57.95
13		35		1	0	1	4/(2)	13.90	13.10	24.70
14		50		3	3	15	15/(1)	12.40	12.90	32.40
15		50		5	3	8	16/(1)	14.00	11.50	25.10

Table 4-3. (Continued)

Series no.	Packing	Type	Material	Number of test points with $\bar{\delta}_i(h_L) < 20\%$			Number of test points TP	$\bar{\delta}(h_L)$ in % $h_{L,calc}$		
				Lit.1	Lit.2	Lit.3	System no.	Lit.1	Lit.2	Lit.3
16	**Białecki**	25	metal	–	2	6	6/(1)	2.27	15.43	28.22
17	**rings**	35		1	0	3	7/(1)	9.30	12.28	18.43
18		35		2	2	4	6/(3)	26.40	18.10	58.62
19		35		–	–	3	4/(2)	9.00	9.60	57.40
20		50		5	7	16	25/(1)	15.80	13.70	33.00
21	**Białecki**	25	metal	3	3	9	9/(1)	18.48	10.13	51.28
22	**rings**	35		1	1	5	5/(3)	14.00	12.20	40.00
23	**stacked**	35		–	1	6	7/(1)	8.43	8.50	33.03
24		50		6	6	17	24/(1)	14.20	12.30	31.00
25	**Hiflow rings**	15	PP	13	7	8	19/(1)	21.70	17.50	17.90
26	**Hiflow**	28	metal	1	2	5	10/(1)	9.34	8.04	22.41
27	**rings**	28		1	1	1	4/(1)	13.72	11.50	11.60
28		28		3	1	2	9/(1)	14.30	10.65	12.64
29		58		4	1	52	53/(1)	10.60	6.71	39.70
30		58		6	5	17	27/(1)	13.59	17.97	27.90
31	**Hiflow**	28	PP	5	1	5	18/(1)	14.70	9.61	13.80
32	**rings**	50	PP	2	0	21	29/(1)	9.80	6.90	29.80
33	**Super**	50	PVDF	0	2	3	6/(1)	6.42	18.27	20.25
34		50	PP	3	9	3	11/(1)	12.48	24.23	14.44
35		50	PP	3	6	3	11/(1)	12.48	20.88	15.05
36	**Hiflow**	90	PP	2	6	10	20/(1)	11.50	20.96	19.40
37	**rings**	90		3	0	9	9/(1)	17.80	8.80	50.50

Table 4-3. (Continued)

Series no.	Packing	Type	Material	Number of test points with $\bar{\delta}_i(h_L) < 20\%$			Number of test points TP	$\bar{\delta}(h_L)$ in % $h_{L,calc}$		
				Lit.1	Lit.2	Lit.3	System no.	Lit.1	Lit.2	Lit.3
38	**Hiflow rings**	20	ceramic	5	1	2	6/(1)	34.00	11.65	14.64
39		38		2	4	1	19/(1)	14.71	14.00	9.10
40		38		2	1	0	9/(1)	16.00	9.58	5.30
41		38		2	1	2	10/(1)	14.80	12.00	11.10
42		38		2	1	2	11/(11)	13.70	11.00	10.40
43		50		3	4	4	7/(1)	19.10	19.20	20.80
44		75		3	1	4	5/(1)	20.00	14.30	45.30
45	**Nor-Pac**	15	PP	4	5	6	13/(1)	14.30	19.60	26.70
46		25	PVDF	1	0	4	11/(1)	10.24	7.15	19.69
47		38	PP	–	2	5	5/(1)	9.63	17.53	39.36
48		38	PP	1	2	3	6/(4)	10.30	19.92	25.75
49		50	PP	10	18	8	24/(1)	16.50	24.45	14.80
50		50	PVDF	5	5	6	11/(1)	18.72	22.93	24.42
51		50	PP	2	0	5	5/(4)	17.40	4.95	49.90
52	**Nor-Pac**	22x27	PVDF	4	3	0	7/(1)	16.64	16.29	7.02
53	**VSP rings**		PP	4	3	6	7/(1)	19.64	16.29	7.02
54			metal	6	9	13	30/(1)	12.40	16.10	21.20
55			PP	2	0	7	9/(4)	9.24	5.44	34.56
56				6						
57	**Top-Pac**	Size 2	metal	0	2	6	9/(1)	6.94	16.97	26.35
58	**Envipac**	size 1A d=0.031 m	PP	9	9	2	18/(1)	19.85	18.10	13.10
59		size 1 d=0.0315 m	PP	2	7	1	8/(1)	17.29	25.76	10.72

Table 4-3. (Continued)

Series no.	Packing	Type	Material	Number of test points with $\bar{\delta}_i(h_L) < 20\%$			Number of test points TP	$\bar{\delta}(h_L)$ in % $h_{L,calc}$		
				Lit.1	Lit.2	Lit.3	System no.	Lit.1	Lit.2	Lit.3
60		size 2	PP	3	0	11	13/(1)	14.38	7.66	37.66
61		d=0.058 m		4	2	12	13/(1)	17.10	9.60	46.40
62		size 3	PP	0	3	8	11/(1)	7.77	13.48	30.64
63		d=0.080 m		2	0	9	9/(1)	14.80	8.20	46.10
64	**Dtnpac**	size 1	PP	–	0	8	11/(1)	9.26	5.87	28.73
65		d=0.045 m		3	2	11	14/(1)	14.65	10.41	27.27
		size 2								
		d=0.070 m								
66	**Intalox**	25	ceramic	3	8	3	18/(1)	14.70	21.00	11.60
67	**saddle**	35		2	2	9	16/(1)	8.30	12.30	20.70
68		35	PP	2	3	12	14/(1)	8.00	16.00	31.30
69		35		2	4	9	12/(4)	10.50	18.20	28.60
70		#40	metal	3	5	1	7/(1)	16.00	21.49	11.87
71	**Białecki rings**	12	metal	3	6	8	15/(1)	13.20	18.10	24.50
72	**Ralu Ring**	50	PP	5	9	9	23/(1)	12.01	17.30	18.20
73	**Raschig**	25	ceramic	7	2	0	9/(1)	20.78	18.65	4.77
74	**rings**	25		3	0	2	5/(1)	25.30	8.37	13.21
75		25	glass	3	4	4	5/(5)	17.13	15.59	49.35
76	**Glitsch**	1A	PP	5	9	3	14/(1)	19.41	19.77	11.38
	CMR rings	2A		1	1	7	7/(1)	15.50	10.50	46.00

Table 4-3. (Continued)

Series no.	Packing	Type	Material	Number of test points with $\bar{\delta}_i(h_L) < 20\%$			Number of test points TP	$\bar{\delta}(h_L)$ in % $h_{L,calc}$		
				Lit.1	Lit.2	Lit.3	System no.	Lit.1	Lit.2	Lit.3
77	**Hackette**	size 1	PP	1	0	8	9/(1)	11.90	5.80	37.00
78		d=0.045 m		1	1	3	12/(1)	13.00	9.40	16.90
79	**Gempak**	202A	sheet metal	3	0	12	21/(1)	10	6.80	25.20
80	**Mellapak**	250Y	sheet metal	3	4	21	21/(1)	9.30	12.60	33.90
81	**Montz packing**	B1-300	sheet metal	0	2	1	7/(1)	12.10	13.30	12.10
	Total (%)	-	-	231 (22.7)	258 (25.3)	577 (56.6)	1019 (100)	13.90	14.10	25.80

Relative measuring error for one test series:

$$\bar{\delta}\,(h_L) = \sum \frac{\delta\,(h_L)}{n_1};$$

with n_1 = number of test points in series
Numerical values for mean relative error $\bar{\delta}\,(h_L)$ apply throughout the entire operating range up to flooding point.

Systems:
1 – air/water, 1 bar, 293 K
2 – air/ethylene glycol, 1 bar, 293 K
3 – air/aqueous solution of ethylene glycol
4 – air/4% NaOH solution
5 – air/silicone oil

Table 4-4a. Determination of pressure drop in irrigated packed column in counter-current flow acc. to Eq. (4-48) or (4-50) and (4-54) throughout the entire operating range up to flooding point. List of test points and relative mean errors $\bar{\delta}\left(\Delta p/H\right)$ for tested random packing, valid for system: air/water and other aqueous systems acc. to Table 2-2, for approx. 1 bar, 293/313 K

Packing	Material	d [mm]	Number of test points TP	$\bar{\delta}_1\left(\Delta p/H\right)$ [%]	$\bar{\delta}_2\left(\Delta p/H\right)$ [%]	$\bar{\delta}\left(\Delta p/H\right)$ [%]	d_S [m]	H [m]	$Re_L \le$	Literature
Pall rings	metal plastic ceramic	15–58	873	6.82	10.47	8.14	0.3–1.0	0.9–3.7	102.5	[A, 5, 13, 26, 28, 33, 34, 35, 37, 40, 49, 55, 59]
Białecki rings	metal plastic	25–53.5	719	8.45	11.33	9.585	0.15–0.50	0.7–1.5	315	[A, 17, 18, 41, 60, 66]
I-13 rings	metal	25	77	9.63	16.55	12.28	0.3	0.72–0.73	101	[A, 17]
PSL rings	metal	50	42	6.76	2.94	6.76	0.3	1.45	100	[A]
VSP rings	metal plastic	32–58	345	10.94	7.95	12.51	0.3–0.6	1.2–2.0	150	[A, 40]
Top-Pak	metal	45, 80	93	8.03	12.15	10.11	0.3–0.45	1.4–2.0	150	[A, 37]
Hackette	plastic	45	107	7.86	29.27	12.33	0.3–0.45	1.4–2	92	[A]
Hiflow rings	metal plastic	17–90	1224	7.32	14.04	9.53	0.3–1.0	0.9–3.7	160	[A, 27, 35, 55]
Hiflow rings	ceramic	20–75	427	7.27	15.66	10.4	0.3–0.45	1.0–2.0	240	[A, 38]
Envipac	plastic	32–80	457	11.38	7.86	11.84	0.3–1.8	0.9–4.5	200	[A, 49]
Dtnpac	plastic	45, 70	212	7.75	10.39	9.43	0.3–0.6	1.08–2	100	[A, 59]
Tellerette	plastic	45, 70	84	9.34	21.99	13.50	0.15–0.45	1.3–2	135	[A]
Glitsch rings	metal plastic	25–69	536	7.02	11.56	9.23	0.3–0.45	1.1–2.03	111	[A]
Ralu Flow rings	plastic	50–58	278	7.23	7.09	7.66	0.3–0.6	1.44–2.12	112	[A]
Raschig rings	ceramic	25–50	232	12.97	12.23	11.82	0.3–0.6	0.74–1.5	230	[A, 17, 41, 42, 43, 61, 62]
Intalox Saddle	metal plastic ceramic	25–50	455	6.62	14.71	9.54	0.3–0.45	0.84–2	120	[A, 38, 61]
Nor-Pac	plastic	17–50	819	8.57	14.05	10.80	0.3–1.4	0.9–2.0	225	[A, 24, 50, 51, 52]
Mc-Pac	metal	30, 65	321	10.5	18.2	13.13	0.32–1.6	1.5–2.5	162	[A, 63]
R-/SR-Pac	ceramic	30–65	332	7.57	16.06	11.48	0.32	1.3–1.36	125	[A, 61, 62]
Total	–	–	**7572**	**8.10**	**13.5**	**10.5**	-		-	-

Table 4-4b. Determination of pressure drop in irrigated packing column in counter-current flow acc. to Eq. (4-48) or (4-50) and (4-54) throughout the entire operating range up to flooding point. List of test points and relative mean errors $\bar{\delta}\left(\Delta p/H\right)$ for tested random packing, valid for system: rectification systems acc. to Table 2-2

Packing	Material	d [mm]	Number of test points TP	System no. acc. to table 2-2	p_T [mbar]	$\bar{\delta}\left(\Delta p/H\right)$ [%]	d_S [m]	H [m]	$Re_L \leq$	Literature
Pall rings	metal	15–50	296	2, 3, 5, 6, 7, 8, 11, 12	13–1016	15.2	0.15–0.8	1.0–3.95	68	[A, 5], Fi/Mc
Pall rings	plastic ceramic	25–50	124	6, 7, 8, 11, 15	30.4–1000	16.55	0.22–0.5	1.0–1.5	66	[A, 44], Fi/Mc
Białecki rings	metal	12–50	77	5, 6, 8	66.7–1000	12.25	0.16–0.5	1.0–1.7	77	[A], Fi/Mc
Hiflow	metal plastic ceramic	20–58	139	6, 15	30.4–521	13.04	0.22–0.5	1.43–1.51	80	[A], Fi/Mc
Intalox saddle	ceramic	25, 38	17	6	64–67	19.93	0.22	1.47	26	[A]
VSP rings	metal	32, 50	28	6	66.7–133	19.19	0.22	1.48–1.52	43	[A]
Top-Pak-VFF	metal	45	8	6	66.7	10.76	0.22	1.46	32	[A]
Interpak-VFF	metal	15, 20	17	6	66.7	9.08	0.22	1.5	15	[A]
Nor-Pac	plastic	17–58	98	6, 7, 8, 15	34–1000	13.48	0.22–0.5	0.91–1.5	104	[A], Fi
Glitsch	metal	12, 40	51	6, 12	66.7	8.57	0.22	1.47–1.54	24	[A]
PSL rings	metal	50	13	6	66.7	9.96	0.22	1.54	42	[A]
Mc-Pac	metal	65	23	6	240–520	14.69	0.6	1.5	141	[A]
R-Pac	ceramic	55	15	6	240–520	15.55	0.6	1.5	80.2	[A]
SR-Pac	ceramic	65	11	6	240–480	11.58	0.6	1.5	67	[A]
Raschig rings	ceramic	8–50	88	6, 8, 11, 15	66.7–1000	11.90	0.1–0.6	0.95–1.5	84	[A, 5], Fi, Mc/Lu
Raschig Ring	metal	15–50	45	2, 5, 15	133–1000	17.52	0.5	1.5–2.0	53	[5]
Total	–	–	**1050**	–	–	**12.9**	–	–	–	–

Table 4-4c. Determination of pressure drop in irrigated packed column in counter-current flow acc. to Eq. (4-48) or (4-50) and (4-54) throughout the entire operating range up to flooding point. List of test points and relative mean errors $\bar{\delta}\left(\Delta p/H\right)$ for tested structured packing, valid for system: air/water, 1 bar, 273 K

Packing	Type/Material	Number of test points TP	$\bar{\delta}_1\left(\Delta p/H\right)$ [%]	$\bar{\delta}_2\left(\Delta p/H\right)$ [%]	$\bar{\delta}\left(\Delta p/H\right)$ [%]	d_S [m]	H [m]	$Re_L \leq$	Literature
Montz packing	B1–100 sheet metal	36	8.98	3.29	8.5	0.3	1.4	85	[A]
	B1–200 sheet metal	72	14.35	15.19	14.63	0.3	1.56	55	[A]
	B1-300 sheet metal	73	12.9	19.0	15.05	0.22	1.42	40	[A]
	B1-300 sheet metal	50	11.47	36.48	12.47	0.3	1.5	40	[A]
	B1-300 sheet metal	45	7.77	57.01	25.3	0.45	1.7	40	[A]
	A3-500 metal gauze	35	12.42	36.6	25.5	0.45	2.0	25	[A]
	C2-200 plastic	56	8.15	4.83	7.79	0.3	1.42	55	[A]
Ralu-Pak	250 YC sheet metal	84	7.21	34.3	10.22	0.22–0.45	1.8	45	[A]
Gempak	230 AT sheet metal	51	7.54	15.8	8.5	0.30	1.47	50	[A]
Mellapak	250 Y sheet metal	65	11.65	30.85	17.8	0.22	1.25	50	[A]
Euroform	plastic PP	54	7.25	14.0	9.0	0.3	1.30	200	[A]
FX-260	plastic PVC	36	9.5	16.5	12.2	0.45	2.0	50	[A]
Nor-Pac Kompakt	size 1 plastic	48	7.17	30.16	13.4	0.3	1.4	50	[A]
	size 2 plastic	58	5.65	6.35	6.15	0.3	1.4	100	[A]
50 mm PSL rings stacked	metal	54	7.07	12.41	9.25	0.273	1.43	100	[A]
25 mm Białecki rings stacked	metal	66	14.43	–	28.0	0.15	1.4	100	[A]
50 mm Białecki rings stacked	metal	61 73	11.43 13.43	11.23 17.40	12.42 14.08	0.3	1.25 1.4	200	[A, 60, 65]

Table 4-4c. (continued)

Packing	Type/Material	Number of test points TP	$\bar{\delta}_1 (\Delta p/H)$ [%]	$\bar{\delta}_2 (\Delta p/H)$ [%]	$\bar{\delta} (\Delta p/H)$ [%]	d_S [m]	H [m]	$Re_L \leq$	Literature
50 mm Hiflow rings stacked	ceramic	45	10.86	39.5	17.23	0.45	1.0	110	[A]
50 mm Raschig rings stacked	ceramic	41	6.32	12.18	7.46	0.3	1.0	50	[A, 65]
50 mm Pall rings stacked	ceramic	32	30.0	–	–	0.4	0.9	110	[A, 28]
50 mm Białecki rings tube column, 7 tubes	metal	66	18.0	30.7	20.74	0.052×7	1.25	110	[A]
Total	–	**1201**	**11.2**	**22.2**	**14.1**	–	–	–	–

Table 4-4d. Determination of pressure drop in irrigated packed column in counter-current flow acc. to Eq. (4-48) or (4-50) and (4-54) throughout the entire operating range up to flooding point. List of test points and relative mean errors $\bar{\delta}\left(\Delta p/H\right)$ for tested structured packing, valid for system: rectification systems acc. to Table 2-2

Packing	Material	d [mm]	Number of test points TP	System no. acc. to Table 2-2	p_T [mbar]	$\bar{\delta}\left(\Delta p/H\right)$ [%]	d_S [m]	H [m]	$Re_L \leq$	Literature
Białecki rings stacked	metal	25–50	60	6	37.9–267	13.23	0.15–0.22	1.5–1.7	84	[A]
Pall rings stacked	ceramic	50	14	8	1000	12.33	0.5	0.9	68	[A, 28]
Montz packing B1-300Y B2-300Y B2-500Y B1-200Y	sheet metal with seal ring; sheet metal unperforated; sheet metal without seal ring; metal gauze with seal ring	–	206	6	33–266.7	16.36	0.22	1.32–1.52	20	[A]
Mellapak and Sulzer gauze packings BX	expanded metal with seal ring; sheet metal with seal ring; metal gauze with seal ring	–	73	6	33–133	16.25	0.22–0.5	1.48	43	[A, 30, 31]
Gempak 220 AT	sheet metal with seal ring	–	21	6, 12	64–66	20.98	0.22	1.44	32	Re/Chr/Mc
Ralu-Pak 250 Y	sheet metal with seal ring	–	36	6	33–133	19.81	0.22	1.44	24	[A]
50 mm Białecki rings tube column, 27 tubes	metal	52	8	8	1000	16.01	27x0.052	0.95	76	Fi/Lu
Fi-Pak I-13F	sheet metal	–	33	6	66.7–133	12.14	0.15–0.22	1.0–1.5	40	[A]
50 mm Białecki rings tube column, 1 tube	metal	50	17 25 7	32 31 20	1000 1000 1000	11.32 18.69 19.70	0.05	1.0	70	[A]
Total	–	–	**500**	–	–	**16.07**	–	–	–	–

Table 4-4e. Determination of pressure drop in irrigated packed column in counter-current flow acc. to Eq. (4-48) or (4-50) and (4-54) throughout the entire operating range up to flooding point. List of test points and relative mean errors $\bar{\delta}\left(\Delta p/H\right)$ for tested packings in pressure range up to 30 bar

Packing	Number of test points	System no. acc. to Table 2-2	p_T [bar]	$\bar{\delta}_1\left(\Delta p/H\right)$ [%]	$\bar{\delta}_2\left(\Delta p/H\right)$ [%]	$\bar{\delta}\left(\Delta p/H\right)$ [%]	d_S [m]	H [m]	$Re_L \leq$	Literature
15 mm Pall rings	112	27	5-30	8.90	22.7	18.3	0.155	0.8	20	[56]
Mellapak 250 Y	11	27	8.9-15	18.75	38.4	22.45	0.155	0.82	40	[56]
15 mm Berl saddles	29	27	20	14.3	17.4	16.7	0.155	1.75	20	[56]
Total	**152**	–	–	**14.0**	**26.2**	**19.2**		–	–	–

$\bar{\delta}_1\left(\Delta p/H\right)$= valid below loading line $F_V/F_{V,Fl} \leq 0.65$
$\bar{\delta}_2\left(\Delta p/H\right)$= valid below flooding point and above loading line $0.65 \leq F_V/F_{V,Fl} \leq 1.0$
$\bar{\delta}\left(\Delta p/H\right)$= valid throughout the entire operating range up to flooding point $0 \leq F_V/F_{V,Fl} \leq 1$
Total number of points: Σ TP = 10475

References Chapter 4

1. Reichelt W. Strömung in Füllkörperapparaten bei Gegenstrom einer flüssigen und einer gasförmigen Phase. Verlag Chemie, Weinheim (1974)
2. Mersmann A. Thermische Verfahrenstechnik. Springer, Berlin/Heidelberg (1980)
3. Brauer H. Grundlagen der Einphasen- und Mehrphasenströmungen. Verlag Sauerländer, Aarau und Frankfurt/M. (1971)
4. Brauer H, Mewes D. Stoffaustausch einschließlich chemischer Reaktionen. Verlag Sauerländer, Aarau und Frankfurt/M. (1971)
5. Billet R. Industrielle Destillation. Verlag Chemie, Weinheim (1973)
6. Schmidt R. The lower capacity limits of packed columns. I. Chem. E. Symposium Series No. 56, EFCE Publications. Series No. 3, Vol. 2, p. 3.1/1–3.1/13
7. Schmidt R. Zweiphasenstrom und Stoffaustausch in Schüttschichten. VDI-Verlag, Düsseldorf (1972), VDI-Forschungsheft 550
8. Bolles WL, Fair JR. Performance and design of packed distillation columns. 3rd Int. Symp. on Distillation, London, April (1979). EFCE Publications Series No. 3, Vol. 2, p. 3.3/35–89. and Improved mass transfer model enhances packed-column design. Chem. Eng. (1982) No. 12, p. 109/115
9. Eckert JS. Selecting the proper distillation column packing. Chem. Eng. Progr., Vol. 66 (1970) No. 3, p. 39
10. Schumacher R. Gasbelastung und Druckverlust von berieselten Füllkörperschüttungen. „vt"-verfahrenstechnik, Vol. 10 (1976) No. 1 1, p. 727/732
11. Weiß S, Schmidt E, Hoppe K. Obere Belastungsgrenze und Druckverlust bei der Destillation in Füllkörperkolonnen. Chem. Techn., Leipzig, Vol. 27 (1975) No. 7, p. 394/396
12. Bemer GG, Kalis GAJ. A new method to predict hold-up and pressure drop in packed columns. Trans. I. Chem. Eng., Vol. 56 (1978), p. 200/205
13. Billet R, Maćkowiak J. How to use absorption date for design and scale-up of packed columns. Presentation Helsinki, 2.-4. June (1982), EFCE Working Party on Distillation, Absorption and Extraction. Fette, Seifen, Anstrichmittel, Vol. 86 (1984) No. 9, p. 349/358
14. Billet R, Maćkowiak J. Allgemeines Verfahren zur Auslegung von Füllkörperkolonnen für die Rektifikation. Presentation at the annual meeting of chemical engineers (VDI), Basel 30.09.1982
15. Billet R, Maćkowiak J. Neues Verfahren zur Auslegung von Füllkörperkolonnen für die Rektifikation. „nvt"-verfahrenstechnik, Vol. 17 (1983) No. 4, p. 203/211
16. Teutsch T. Druckverluste in Füllkörperschüttungen bei hohen Berieselungsdichten. Chem.-Ing.-Tech., Vol. 36 (1964) No. 5, p. 496/503
17. Maćkowiak J. Dissertation TU-Wrocław/Poland (1975)
18. Maćkowiak J, Ziołkowski Z. Neue Füllkörperelemente für die Destillation und Absorption. Presentation at ACHEMA 1976, June (1976), Frankfurt/Main
19. Buchanan JE. Pressure gradient and liquid hold-up in irrigated packed towers. I & EC Fundamentals, Vol. 8 (1969), p. 502/511
20. Leva M. Tower packings and packed tower design. Acron, Ohio/USA, 2nd edition (1953)
21. Mersmann A. Zur Berechnung des Flutpunktes in Füllkörperschüttungen. Chem.-Ing.-Tech., Vol. 37 (1965) No. 3, p. 218/226
22. Gelbe H. Der Flüssigkeitsinhalt und die Rektifizierwirkung beim Vakuumbetrieb in Füllkörperschüttungen. Fort. Ber. VDI, 3, No. 23, March (1968)
23. Billet R, Maćkowiak J. Neuartige Füllkörper aus Kunststoffen für thermische Stofftrennverfahren. Chemie Technik, Vol. 9 (1980) No. 5, p. 219/226
24. Billet R, Maćkowiak J. Wirksamkeit von Kunststofffüllkörpern bei der Absorption, Desorption und Vakuumrektifikation. „vt"-verfahrenstechnik, Vol. 15 (1982) No. 2, p. 67/74
25. Molzahn M, Wolf D. Destillation, Absorption, Extraktion. Gibt es noch Forschungsaufgaben?. Chem.-Ing.-Tech., Vol. 53 (1981) No. 10, p. 768/780
26. Eckert JS, Walter LF. What affects packed bed distillation Hydrocarbon Processing and Petroleum Refiner. Vol. 43 (1964) No. 2, p. 107/114
27. Internal data of the company Rauschert KG, Steinwiesen
28. Billet R, Maćkowiak J, Ługowski S, Filip S. Development and performance of Impulse Packing for gas/liquid systems. Presentation at ACHEMA 82, 6th June (1982) Frankfurt/M.. and Fette, Seifen, Anstrichmittel, Vol. 85 (1983) No. 10, p. 383/391

29. Sherwood TK, Pigford R, Wilke ChR. Mass Transfer. McGraw-Hill, New York/Düsseldorf (1975)
30. Meier W, Hunkler R, Stöcker WD. Performance of a new regular tower packing 'Mellapak'. 3rd Int. Symp. on Distillation, London, April (1979). German version: Chem.-Ing.-Techn., Vol. 51 (1979) No. 2, p. 119/122
31. Brochure by the company Sulzer, No. 22, 13.06.20.V 85-50, (1985). Trennkolonnen für Destillation und Absorption
32. Technical information of the company VFF – Vereinigte Füllkörperfabriken. Ransbach + Baumbach/Westerwald
33. Technical information of the company Raschig, Ludwigshafen/Rhein
34. Billet R. Stand, Entwicklung und Aussichten der Destillation und Rektifikation im Vergleich zu anderen Trennmethoden. Chemie-Technik, Vol. 3 (1974), p. 355/361
35. Billet R, Maćkowiak J. Hiflow-Ring ein Hochleistungsfüllkörper für Gas-Flüssig-Systeme. Teil 1: Ausführung in Kunststoff. Chemie-Technik, Vol. 11 (1984) No. 12, p. 37/46
36. Brochure "Białecki rings" by the foreign trade office CHEMAK. Warszawa, ul. Wspólna 62, Poland
37. Billet R, Maćkowiak J. Hiflow-Ring ein Hochleistungsfüllkörper für Gas-Flüssig-Systeme. Teil 2. Ausführung in Metall. Chemie-Technik, Vol. 14 (1985) No. 4, p. 91/99
38. Billet R, Maćkowiak J. Hiflow-Ring ein Hochleistungsfüllkörper für Gas-Flüssig-Systeme. Teil 3. Ausführung in Keramik. Chemie-Technik, Vol. 14 (1985) No. 5, p. 195/206
39. Billet R, Maćkowiak J. Application of modern packings in thermal separation processes. Presentation at ACHEMA 1985, 10.06.1985, Frankfurt/Main and. Chem. Eng. Technol. Vol. 11 (1988), p. 213/227
40. Billet R, Maćkowiak J, Chromik R. Hochleistungsfüllkörper für Gas-/Flüssig-Systeme. Der VSP-Ring aus Metall. Chemie-Technik 16 (1987) No. 5, p. 79/87
41. Maćkowiak J, Ziołkowski Z. Hydraulik der regellosen Białeckiringschüttung. Inż. Chem. (orig. Polish), Vol. 6 (1975) No. 1, p. 151/186
42. Blaß E, Kurtz R. Der Einfluss grenzflächenenergetischer Größen auf den Zweiphasen-Gegenstrom durch Raschigring-Füllkörpersäulen Teil 1: Flüssigkeitsinhalt. „vt"-verfahrenstechnik, Vol. 10 (1976) No. 11, p. 721/724
43. Blaß E, Kurtz R. Der Einfluss grenzflächenenergetischer Größen auf den Zweiphasen-Gegenstrom durch Raschigring-Füllkörpersäulen. Teil 2: Druckverlust am Flutpunkt. „nvt"-verfahrenstechnik, Vol. 11 (1977) No. 1, p. 44/48
44. Mersmann A, Deixler A. Packungskolonnen. Chem.-Ing.-Techn., Vol. 56 (1986) No. 1, p. 19/31
45. Kast W. Gesetzmäßigkeiten des Druckverlustes in Füllkörperkolonnen. Chem.-Ing.-Techn., Vol. 36 (1964) No. 5, p. 464/468
46. Kolev N. Wirkungsweise von Füllkörperschüttungen. Chem.-Ing.-Techn., Vol. 48 (1976) No. 12, p. 1105/1111
47. Beck R. Ein neues Verfahren zur Berechnung von Füllkörpersäulen. VFF-GmbH & Co., Baumbach/Westerwald (1969)
48. Kleinhückenkotten H. Untersuchung zur Auslegung von Füllkörperkolonnen mit geschütteter Füllung. "vt"-verfahrenstechnik, Vol. 9 (1975) No. 8, p. 275/279
49. Billet R, Maćkowiak J. Untersuchungen zur Hydraulik und zum gasseitigen Stoffübergang in Kolonnen mit Kunststoff-Füllkörpern. Chemie-Technik, Vol. 17 (1988) No. 5, p. 149/155
50. Billet R, Kozioł A, Maćkowiak J, Suder S. Maßstabübertragung in Füllkörperkolonnen bei der Absorption (engl.). Presentation at the 4th annual Italian-Yugoslavian-Austrian Chem. Eng. Conference. Triest, 24–26 September (1984) and. Fette, Seifen, Anstrichmittel, Vol. 87 (1985) No. 5, p. 201/205
51. Billet R, Maćkowiak J, Kozioł A, Suder S. Untersuchungen der Wirksamkeit von Füllkörpern aus Kunststoff in Kolonnen großer Durchmesser. Chem.-Ing.-Techn., Vol. 58 (1986) No. 11, p. 897/900
52. Maćkowiak J. Fluiddynamik von Füllkörperkolonnen. CAV – Chemie Anlagen + Verfahren (1991) No. 6, p. 66–74
53. Maćkowiak J. Bestimmung des Druckverlustes berieselter Füllkörperschüttungen und Packungen. Staub-Reinhaltung der Luft 50 (1990) p. 455–463
54. Maćkowiak J. Pressure drop in Irrigated Packed Columns. Chem. Eng. Process 29 (1991) p. 93–105
56. Krehenwinkel H. Experimentelle Untersuchungen der Fluiddynamik und der Stoffübertragung in Füllkörperkolonnen bei Drücken bis 100 bar. Dissertation, TU Berlin, December (1986)
57. Süess A, Spiegel L. Hold-up of Mellapak structured packings. Sulzer Bros. AG (Switzerland), Ltd.. Separation Columns

58. Stein A. Der dynamische Flüssigkeitsanteil in Packungskolonnen. VDI-Fortschritt-Berichte Reihe J, No. 702. VDI-Verlag GmbH, Düsseldorf 2001
59. Maćkowiak J. Einsatz von modernen Füllkörpern zur Reduzierung von Schadstoffen aus Abluft und Abwasser. Staub-Reinhaltung von Luft 50 (1990) No. 5, p. 221–227
60. Maćkowiak J. Podstawy projektowania kolumn wypełnionych usypanymi i układanymi metalowymi pierścieniami Białeckiego dla układów gas/ciecz. Inż. i Ap. Chem. (1990) No. 5, p. 3-8
61. Maćkowiak J, Ługowski Z. Geringere Apparatevolumina und Betriebskosten mit neuen Füllkörpern. vt-Verfahrenstechnik 29 (1995) No.6, p. 19–22
62. Maćkowiak J, Szust J. Hydraulika i wymiana masy w kolumnach wypełnionych ceramicznymi pierścieniami R-Pac i SR-Pac w układach gaz/ciecz. Inż. Chem. i Procesowa 18 (1997) No. 4, p. 675–691
63. Maćkowiak J. Mc-Pac-ein neuer metallischer Füllkörper für Gas-Flüssigkeitssysteme. Chem. Ing. Tech. 73 (2001) No. 1-2, p. 74–79
64. Billet R, Schultes M. Modelling of pressure drop in packed columns. Inż. Chem. i Proc. Vol. 1 (1990) No.1, p. 17–30
65. Maćkowiak J, Suder S. Hydraulika i wymiana masy w kolumnie wypełnionej układanymi pierścieniami Białeckiego (orig. Polish). Inż. Chem. i. Procesowa Vol. 8 (1977) No. 3, p. 651–664
66. Możeński C, Kucharski E. Hydraulika wypełnień pod zwiększonym ciśnieniem (orig. Polish). Inż. Chem. i Procesowa, Vol. 3 (1986) No. 3, p. 373/384

Pressure Drop of Irrigated Random and Structured Packings Based on the Law of Resistance for Two-Phase Flow

5.1 Introduction

Chapter 4 presented a model for determining the pressure drop of irrigated random and structured packings. This model is applicable, if the law of resistance, i.e. function $\psi = f(Re_V)$, for single-phase flow through the packing, is known. There are cases, in which experimental pressure drop data is only available for irrigated packings, which means the model presented in Chap. 4 cannot be applied unrestrictedly. In the case of tube columns and packings with a flow channel angle other than $\alpha = 30°$ or $45°$, for which no C_B values for determining the pressure drop, acc. to Eq. (4-45), are available, the calculation of the pressure drop can only be carried out with insufficient accuracy. Here, it is advisable to apply the method developed by Maćkowiak [1, 2], which is based on the law of resistance $\psi_{VL} = f(Re_L)$ for gas/liquid two-phase flow.

5.2 Deriving the Model for Determining the Pressure Drop of Irrigated Random and Structured Packings

In packed columns with two-phase flow, the formation of films and runlets occurs on the surface of the packing elements and within the packing. A formation of droplets is also possible, especially in the case of random packings with larger-diameter packings elements. The droplets fall through the packing, causing a reduction of the void spaces within the packed column. As a result, the effective void fraction ε_e of the packed column in two-phase flow is reduced by the liquid hold-up h_L, compared to a packing with single-phase gas flow, $\varepsilon_e = \varepsilon - h_L$. Analogously to the correlation in single-phase flow, the initial Eq. (3-8) is used to calculate the pressure drop $\Delta p/H$ per 1 m packing height in two-phase flow. The following applies:

J. Maćkowiak, *Fluid Dynamics of Packed Columns*,
Chemische Technik/Verfahrenstechnik, DOI 10.1007/b98397_5,

$$\frac{\Delta p}{H} = \psi_{VL} \cdot \frac{1-(\varepsilon - h_L)}{(\varepsilon - h_L)^3} \cdot \frac{F_V{}^2}{d_P \cdot K} \quad \left[\mathrm{Pam}^{-1}\right] \quad \text{for } u_V >> u_L \tag{5-1}$$

A simple converstion of Eq. (5-1) now leads to the following Eq. (5-2):

$$\frac{\Delta p}{H} = \psi_{VL} \cdot \frac{1-\varepsilon}{\varepsilon^3} \cdot \frac{F_{V2}}{d_P \cdot K} \cdot \left(1 + \frac{h_L}{1-\varepsilon}\right) \cdot \left(1 - \frac{h_L}{\varepsilon}\right)^{-3} \quad \left[Pam^{-1}\right] \tag{5-2}$$

In order to determine the pressure drop acc. to Eq. (5-2), the only data required is the packing-specific data, such as specific geometric packing surface area a, void fraction ε, resistance coefficient ψ_{VL} for two-phase flow and liquid hold-up h_L. The method based on Eq. (5-2) therefore offers the possibility of determining the presure drop throughout the entire operating range up to the flooding point. More information on determining the pressure drop within the operating range of packed columns can be found in Sect. 4.2.

5.3
Law of Resistance $\psi_{VL} = f(Re_L)$ for Packed Columns with Two-Phase Flow – Deriving the Model

Contrary to expectations, the resistance coefficient ψ_{VL} in counter-current two-phase flow under certain operating conditions has been found to be merely a function of the Reynolds number Re_L, acc. to Eq. (4-18) [1, 2]:

$$Re_L = \frac{u_L}{a \cdot \nu_L} \tag{4-18}$$

The resistance coefficient ψ_{VL} is dependent on the type and also the size of the packing elements [1, 2] and is given as:

$$\psi_{VL} = \mu \cdot f_{VL}(Re_L) \tag{5-3}$$

with

$$f_{VL}\,(Re_L) = A \cdot Re_L^B \tag{5-4}$$

The parameter μ in Eq. (5-3) is a packing-specific form factor. A and B are packing parameters, which were determined for a given reference packing element [1, 2].

The resistance coefficient ψ_{VL} in two-phase flow is likely to be the highest in the case of non-perforated, ceramic Raschig rings, whereas for lattice packing elements, the resistance coefficient ψ_{VL} is expected to be smaller, depending on the size of the perforations in the packing surface. For the purpose of evaluating the experimental data, it was useful to determine the numerical value of the form factor μ for ceramic Raschig rings:

$$(\mu)_{Raschig-Ring} \equiv 1 \tag{5-5}$$

Acc. to this definition, the μ values for modern packing elements are expected to be below one, i.e. $\mu < 1$.

The constant A and the exponent B as well as the numerical values of the packing-specific form factors μ were determined on the basis of approx. 10,000 experimental data items from the available database, using the minimisation procedure. The numerical values of the form factors μ are listed in Tables 6-1a–c.

Based on the evaluation of the experimental data for the packed columns tested, the following equation was found for calculating the resistance coefficient ψ_{VL} [1, 2]:

$$\psi_{VL} = \mu \cdot 5.4 \cdot \mathrm{Re}_L^{-0.14} \tag{5-6}$$

Equation (5-6) is applicable for a liquid flow range of $0.3 < \mathrm{Re}_L < 12.3$.

For higher Reynolds numbers $\mathrm{Re}_L > 12.3$, an approximately constant resistance coefficient ψ_{VL} was found by experiment and given as:

$$\psi_{VL} \cong \mu \cdot 3.8 \tag{5-7}$$

The following approximate Eq. (5-8) is valid for the flow range of $0 \leq \mathrm{Re}_L < 0.3$ and based on $\psi = f(\mathrm{Re}_V)$, acc. to (3-14):

$$\psi_{VL} = \mu \cdot (\psi + (6.4 - \psi)) \cdot \frac{\mathrm{Re}_L}{0.3} \tag{5-8}$$

5.4 Deriving the Equation for the Calculation of the Pressure Drop of Irrigated Random Packings

In the case of larger-diameter packing elements, e.g. in absorption with a mostly mass transfer resistance in one phase, the liquid flow is mostly turbulent, with $\mathrm{Re}_L > 12.3$. Here, Eqs. (5-2) and (5-6) lead to the following equations for calculating the pressure drop of irrigated packed columns with any type of column internal:

$$\frac{\Delta p}{H} = \mu \cdot 5.4 \cdot \mathrm{Re}_L^{-0.14} \cdot \left(\frac{1-\varepsilon}{\varepsilon^3}\right) \cdot \frac{F_V{}^2}{d_P \cdot K} \cdot \left(1 + \frac{h_L}{1-\varepsilon}\right) \cdot \left(1 - \frac{h_L}{\varepsilon}\right)^{-3} \quad \left[\mathrm{Pam}^{-1}\right] \tag{5-9}$$

This leads to Eq. (5-10) for $\mathrm{Re}_L > 12.3$:

$$\frac{\Delta p}{H} \cong 3.8 \cdot \mu \cdot \left(\frac{1-\varepsilon}{\varepsilon^3}\right) \cdot \frac{F_V{}^2}{d_P \cdot K} \cdot \left(1 + \frac{h_L}{1-\varepsilon}\right) \cdot \left(1 - \frac{h_L}{\varepsilon}\right)^{-3} \quad \left[\mathrm{Pam}^{-1}\right] \tag{5-10}$$

For randomly filled and stacked packing elements, $K \neq 1$, and for structured packings and tube columns, $K = 1$.

5.5
Comparing Calculated and Experimental Values Throughout the Entire Operating Range of Packed Columns

Tables 5-1a–c show the results of the evaluation of approx. 10,000 experimental data items for numerous types of packed columns. The mean relative error in these tables is given as $\bar{\delta}_1 \left(\Delta p/H\right)$. It is used to determine the pressure drop of irrigated packings in the operating range below the loading line, i.e. for $F_V \leq 0.65\ F_{V,Fl}$. The numerical values of $\bar{\delta}_2 \left(\Delta p/H\right)$ are applicable for the operating range between the loading line and the flooding point, whilst $\bar{\delta} \left(\Delta p/H\right)$ is valid for the entire operating range up to and including the flooding point.

The physical properties of the tested systems as well as the operating parameters and geometric data were in the following ranges:

$$\left.\begin{array}{lcl} 6.3 \leq & \mathbf{Re_L} & \leq 200 \text{ (and/or } Re_{L,max} \text{ acc. to Tables 5-1a – 5-1c} \\ 54 \leq & \mathbf{a} & \leq 550\,\text{m}^2\,\text{m}^{-3} \\ 0.008 \leq & \mathbf{d} & \leq 0.09\,\text{m} \\ 0.025 \leq & \mathbf{d_S} & \leq 1.8\,\text{m} \\ 660 \leq & \boldsymbol{\rho_L} & \leq 1260\,\text{kgm}^{-3} \\ 0.03 \leq & \boldsymbol{\rho_V} & \leq 40\,\text{kgm}^{-3} \\ 14 \leq & \boldsymbol{\sigma_L} & \leq 74.6\,\text{mNm}^{-1} \\ 0.2 \leq & \boldsymbol{\eta_L \cdot 10^3} & \leq 8\,\text{kgm}^{-1}\text{s}^{-1} \\ 6.5 \leq & \boldsymbol{\eta_V \cdot 10^6} & \leq 18.2\,\text{kgm}^{-1}\text{s}^{-1} \\ 0.63 \leq & \boldsymbol{\varepsilon} & \leq 0.987\,\text{m}^3\,\text{m}^{-3} \\ 2 \leq & \boldsymbol{\Delta p/H} & \leq 4000\,\text{Pam}^{-1} \end{array}\right\} \tag{5-11}$$

Figures 5-1, 5-2, 5-3, 5-4 and 5-5 show the comparison between the calculated pressure drop data, acc. to Eq. (5-9), and the experimental data, on the examples of metal Pall rings.

Figures 5-6, 5-7, 5-8, 5-9, 5-10, 5-11, 5-12, 5-13, 5-14 and 5-15 show the spread of the experimental pressure drop data for various random packings: Pall, Białecki and Hiflow rings, Mc-Pac and Envipac, VSP rings, Hackette, Top-Pak, Ralu-Flow, Ralu rings, Intalox saddles, Raschig rings, SR-Pak, R-Pak, Nor-Pac at al. as well as for structured packings.

5.6
Evaluation of Results

The concurrence between the calculated and experimental data was found to be very good, particularly in the case of randomly filled packing elements. As could be expected, the experimental data below the loading line was verified more accurately by calculation,

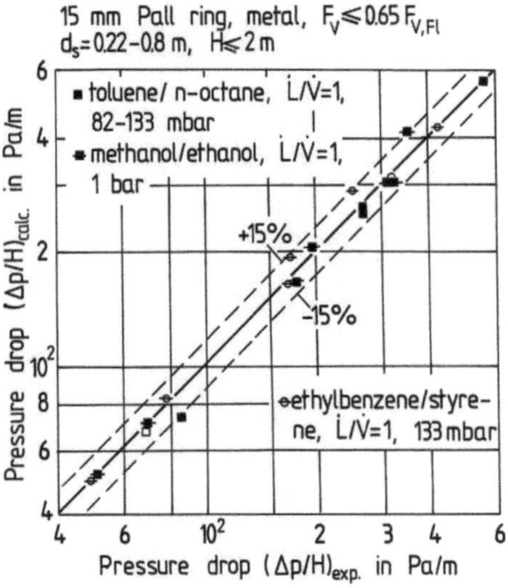

Figure 5-1. Comparison of experimentally determined pressure drop Δp/H with calculated values acc. to Eqs. (5-9) and (5-10) for 15 mm metal Pall rings in the range below the loading line $F_V/F_{V,Fl} \leq 0.65$, valid for various rectification systems

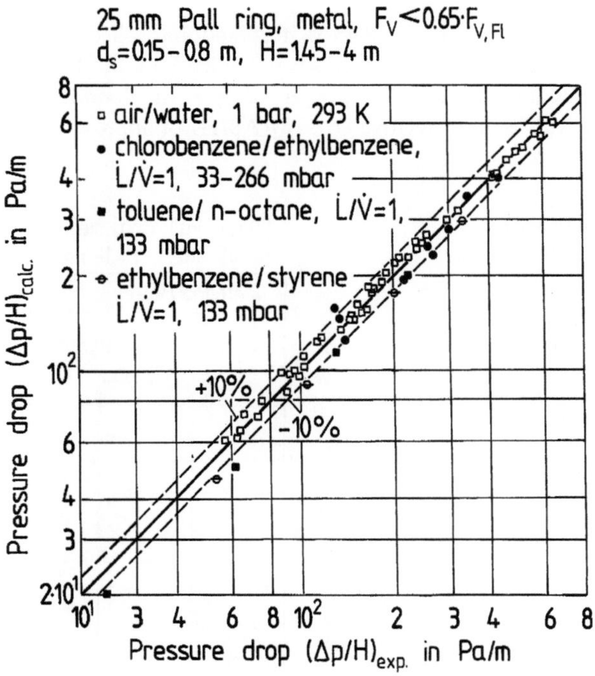

Figure 5-2. Comparison of experimentally determined pressure drop Δp/H with calculated values acc. to Eqs. (5-9) and (5-10) for 25 mm metal Pall rings in the range below the loading line $F_V/F_{V,Fl} \leq 0.65$, valid for various rectification systems

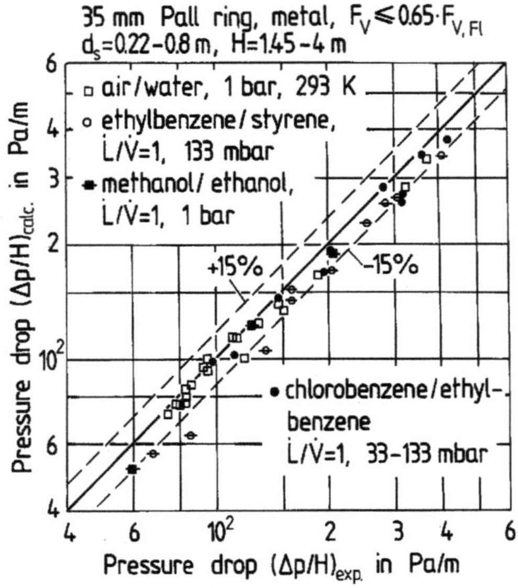

Figure 5-3. Comparison of experimentally determined pressure drop $\Delta p/H$ with calculated values acc. to Eqs. (5-9) and (5-10) for 35 mm metal Pall rings in the range below the loading line $F_V/F_{V,Fl} \leq 0.65$, valid for various rectification systems

with a mean error $\bar{\delta}_1(\Delta p/H)$ of 8.45 % for randomly filled packing elements and 11.33 % for structured packings, as shown in Tables 5-1a, b. The experimental values above the loading line, including the pressure drop at the flooding point, were verified by experiment with a mean relative error $\bar{\delta}_2(\Delta p/H)$ of approx. ± 12 % for randomly

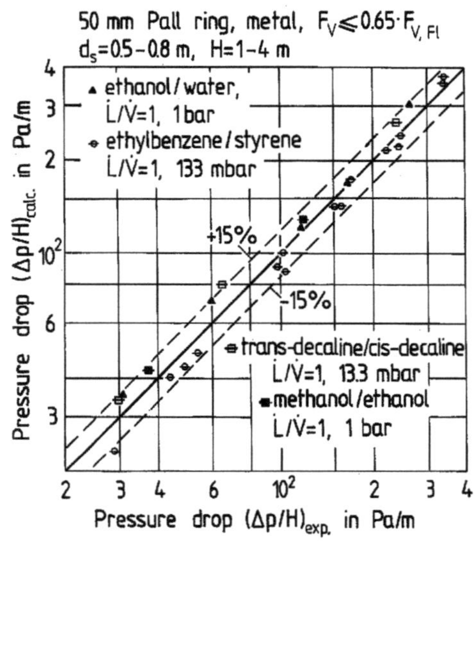

Figure 5-4. Comparison of experimentally determined pressure drop $\Delta p/H$ with calculated values acc. to Eqs. (5-9) and (5-10) for 50 mm metal Pall rings in the range below the loading line $F_V/F_{V,Fl} \leq 0.65$, valid for various rectification systems

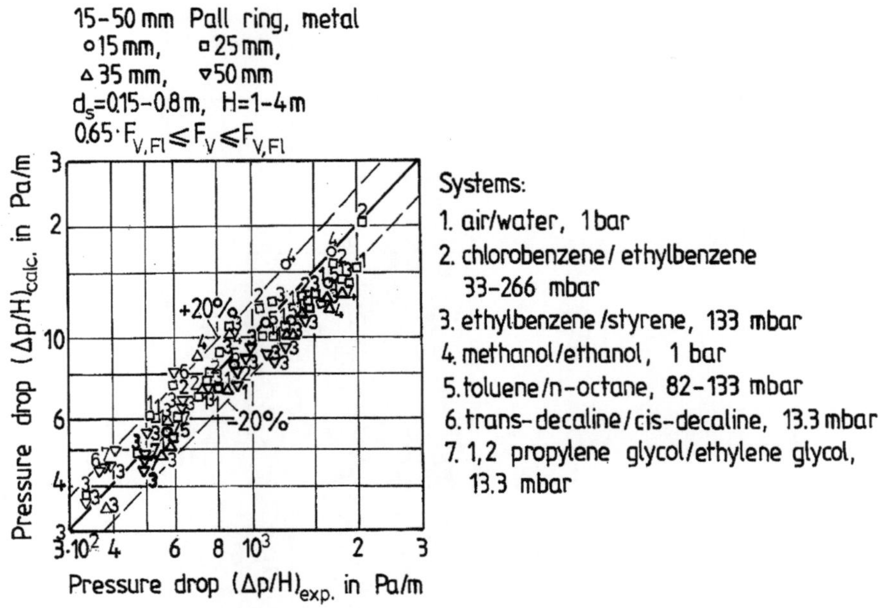

Figure 5-5. Comparison of experimentally determined pressure drop Δp/H with calculated values acc. to Eqs. (5-9) and (5-10) for 15–50 mm metal Pall rings in the range above the loading line $0.65 < F_V/F_{V,Fl} \leq 1$, valid for various rectification systems

filled packing elements, see Table 5-1a. It must be noted, however, that the equation for calculating the pressure drop (Δp/H) within the load range takes into account the uncertainties in connection with the determination of the gas velocity at the flooding point $u_{V,Fl}$ and the liquid hold-up $h_{L,S}$and/or $h_{L,Fl}$. The relative error in the determination of the pressure drop of is higher for irrigated structured packings than it is for random packings, see Table 5-1b. This is basically due to the use of seal rings in the test columns.

Which of the models, presented in Chaps. 4 and 5, is selected for determining the pressure drop Δp/H of packed columns is down to the user's preference. However, the decision should be based on the accuracy of the calculated pressure drop Δp/H value. A comparison of the accuracy of the pressure drop data can be found in Tables 4-4a–e and 5-1a–c.

The advantage of the model, presented above, for determining the pressure drop of irrigated packings, acc. to Eqs. (5-9) and (5-10), also lies in the fact that just one packing-specific constant μ is sufficient to allow for a direct comparison between the pressure drop of the individual packing and other types of packings. In addition, the calculation of the flooding point of irrigated packings does not necessarily require knowledge of the pressure drop, but it does require the resistance coefficient ψ in single phase flow.

The application of the model for determining the pressure drop, acc. to Eqs. (5-9) and (5-10), which is presented above, is now illustrated for various operating conditions, using a numerical example.

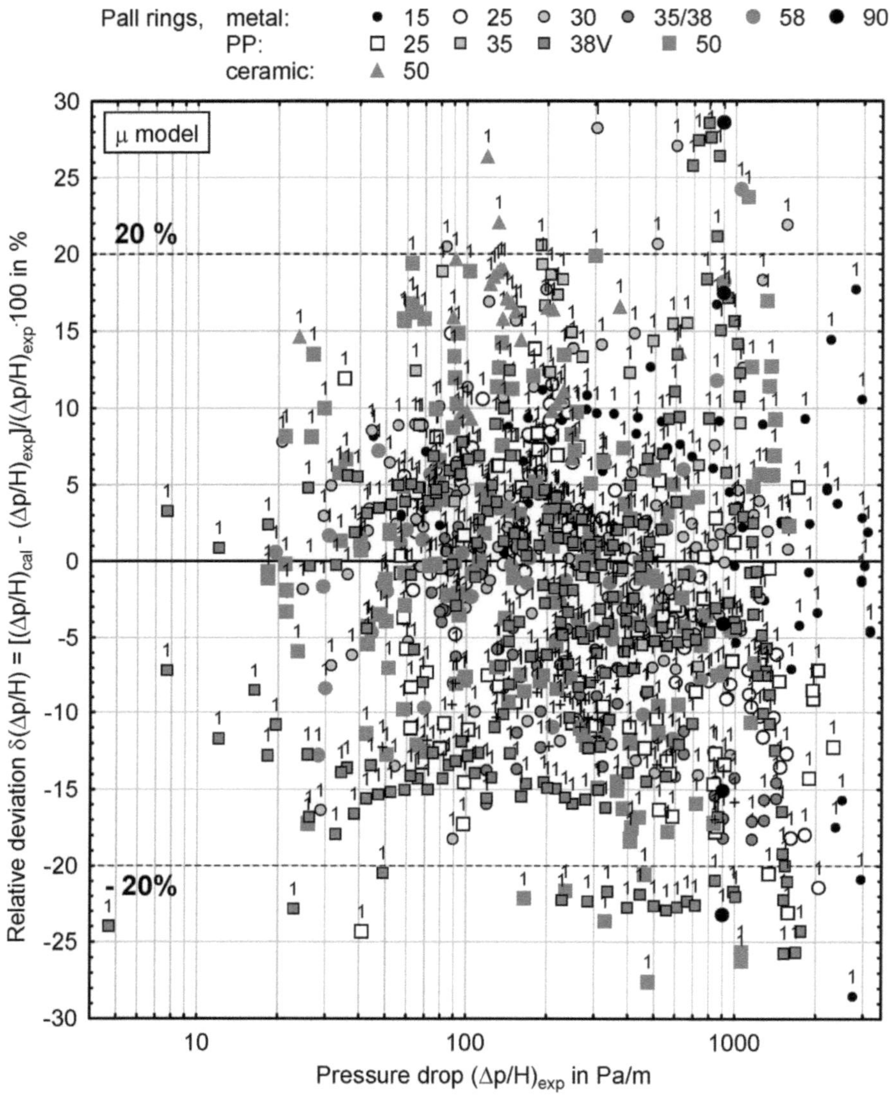

Figure 5-6. Relative deviation $\delta(\Delta p/H)$ of experimental data for determination of pressure drop $(\Delta p/H)_{exp}$ up to flooding point acc. to Eqs. (5-9) and (5-10), valid for 15–90 mm Pall rings made of metal, plastic and ceramic. No. of system see Table 2-2. Test conditions see Table 5-1a–c

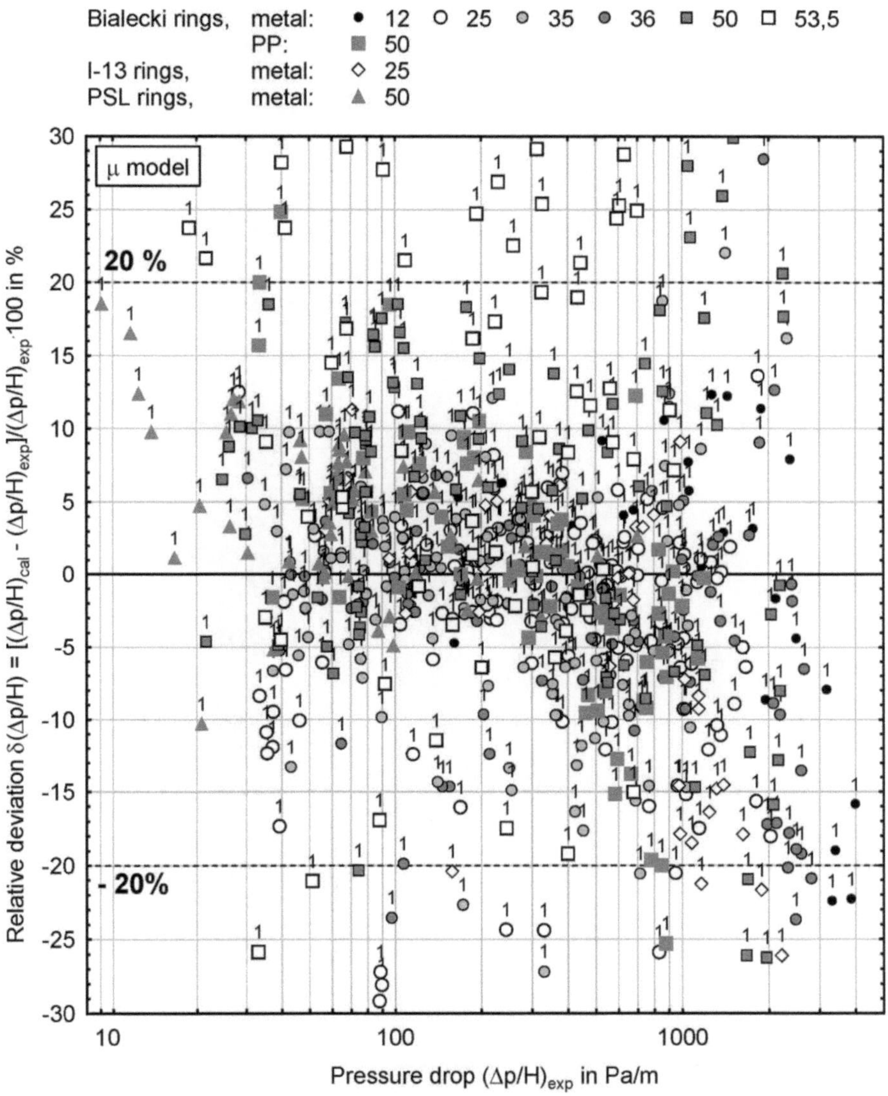

Figure 5-7. Relative deviation $\delta(\Delta p/H)$ of experimental data for determination of pressure drop $(\Delta p/H)_{exp}$ up to flooding point acc. to Eqs. (5-9) and (5-10), valid for random Białecki rings made of metal and plastic and for PSL rings and I-13 rings made of metal. No. of system see Table 2-2. Test conditions see Table 5-1a–c

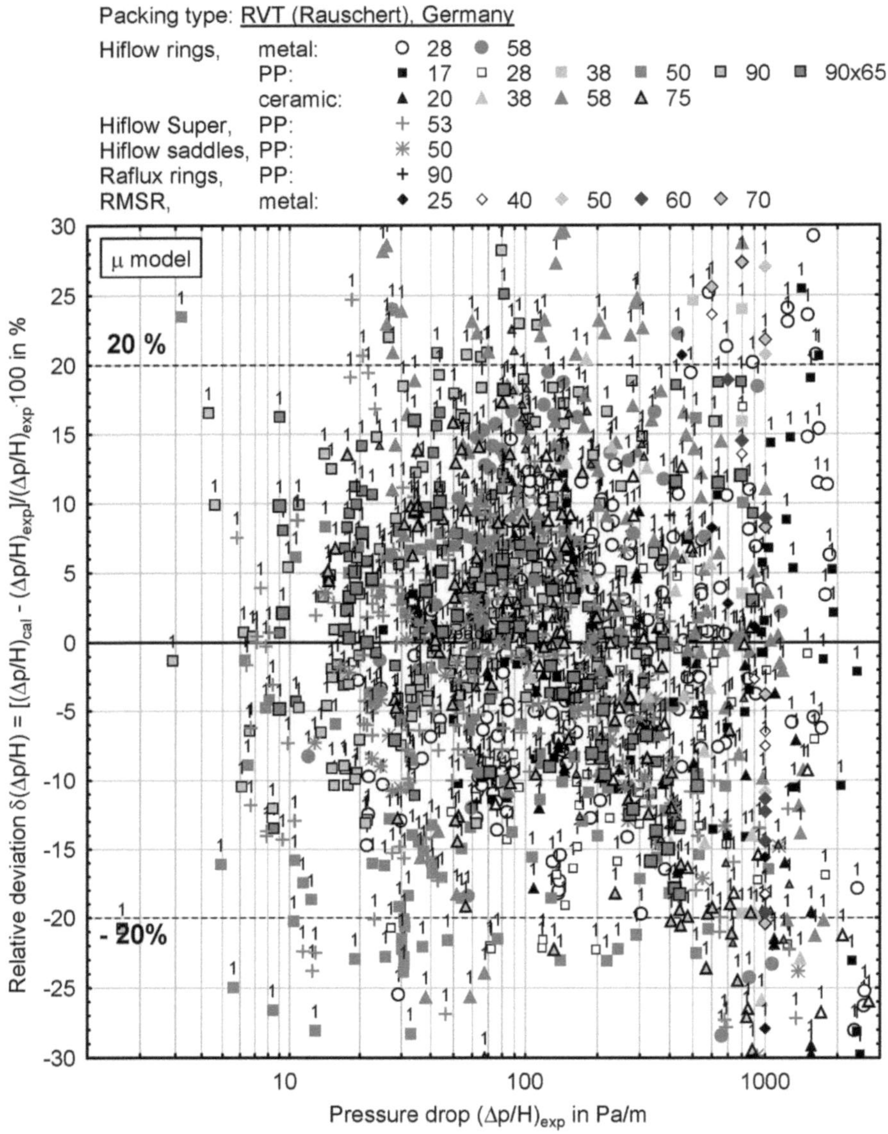

Figure 5-8. Relative deviation $\delta(\Delta p/H)$ of experimental data for determination of pressure drop $(\Delta p/H)_{exp}$ up to flooding point acc. to Eqs. (5-9) and (5-10), valid for 15–90 mm Hiflow rings and other random packings by RVT (Rauschert) made of metal, plastic and ceramic. No. of system see Table 2-2. Test conditions see Table 5-1a–c

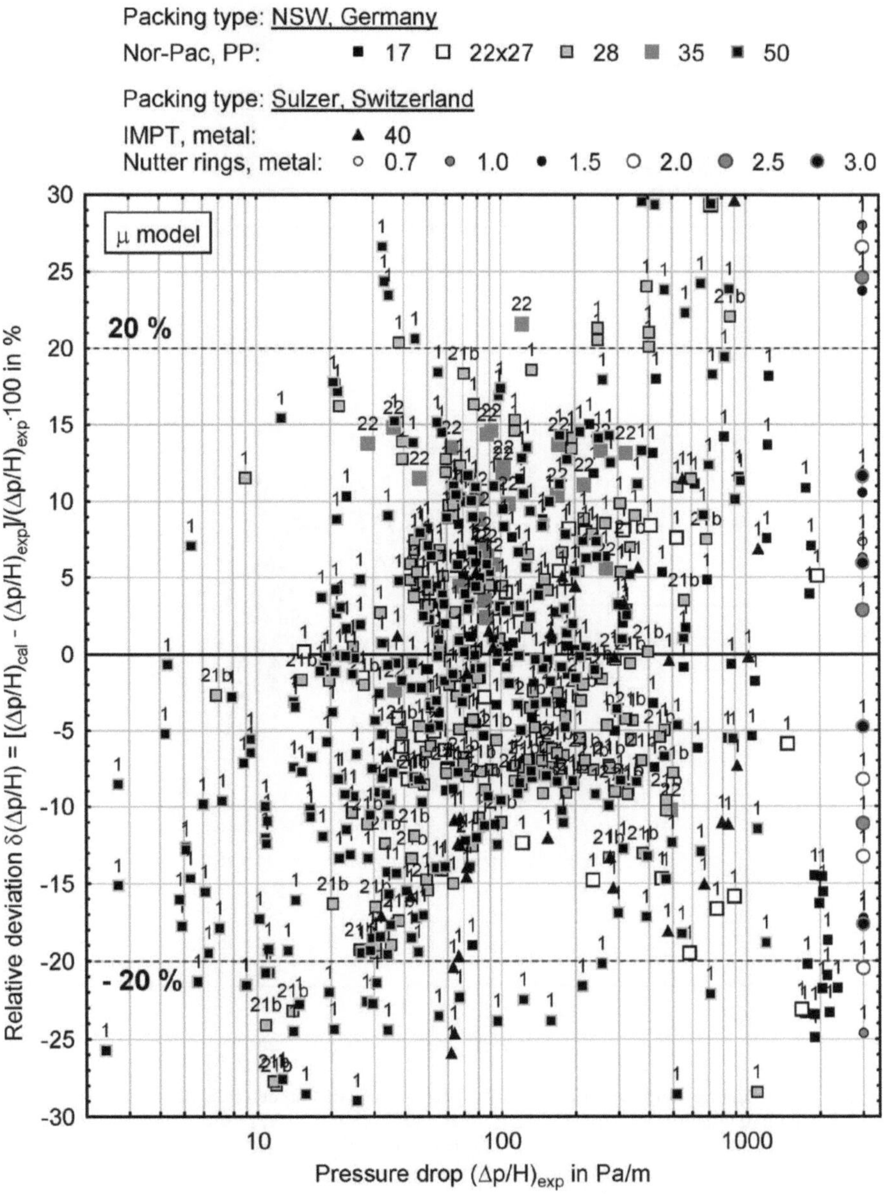

Figure 5-9. Relative deviation $\delta(\Delta p/H)$ of experimental data for determination of pressure drop $(\Delta p/H)_{exp}$ up to flooding point acc. to Eqs. (5-9) and (5-10), valid for plastic Nor-Pac packing by NSW and metal IMPT and Nutter rings by Sulzer. No. of system see Table 2-2. Test conditions see Table 5-1a–c

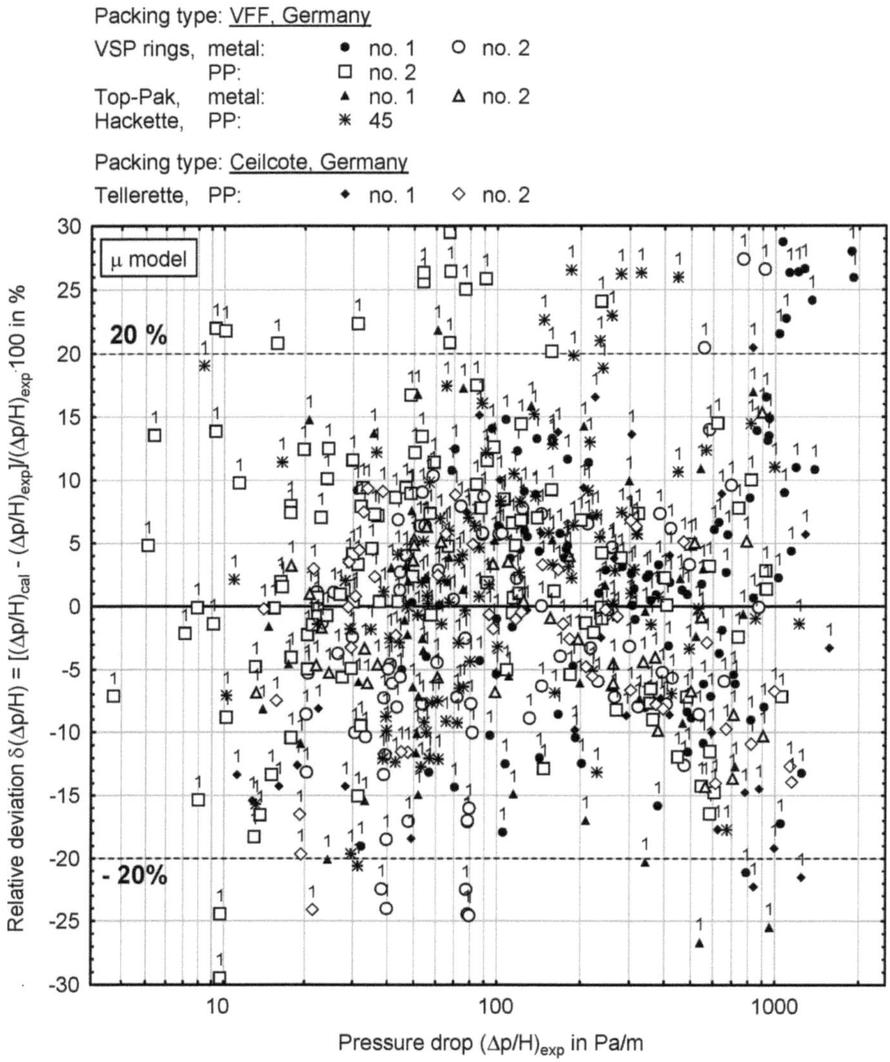

Figure 5-10. Relative deviation $\delta(\Delta p/H)$ of experimental data for determination of pressure drop $(\Delta p/H)_{exp}$ up to flooding point acc. to Eqs. (5-9) and (5-10), valid for VSP rings, Top-Pak and Hackette by VFF made of metal and plastic and for plastic Tellerette by Ceilcote. No. of system see Table 2-2. Test conditions see Table 5-1a–c

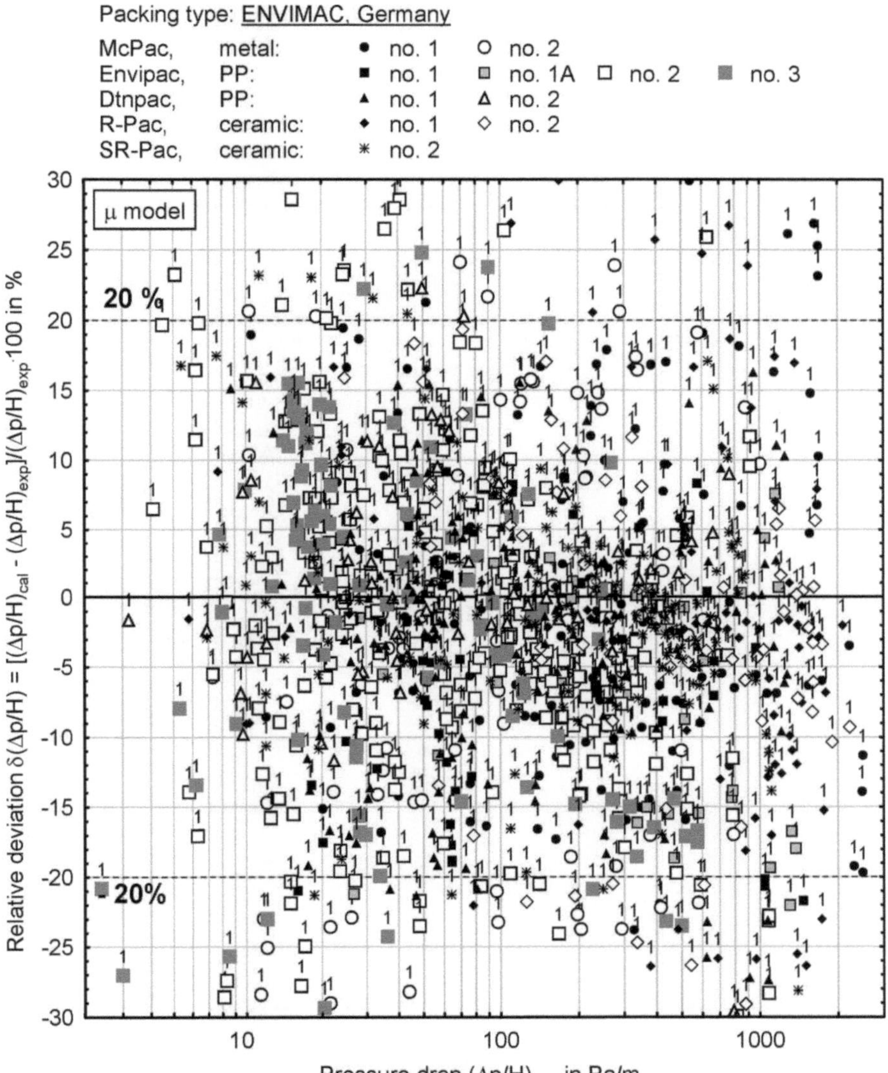

Figure 5-11. Relative deviation $\delta(\Delta p/H)$ of experimental data for determination of pressure drop $(\Delta p/H)_{exp}$ up to flooding point acc. to Eqs. (5-9) and (5-10), valid for Envipac, Dtnpac, Mc-Pac, R-Pac, SR-Pac by ENVIMAC made of metal, plastic and ceramic. No. of system see Table 2-2. Test conditions see Table 5-1a–c

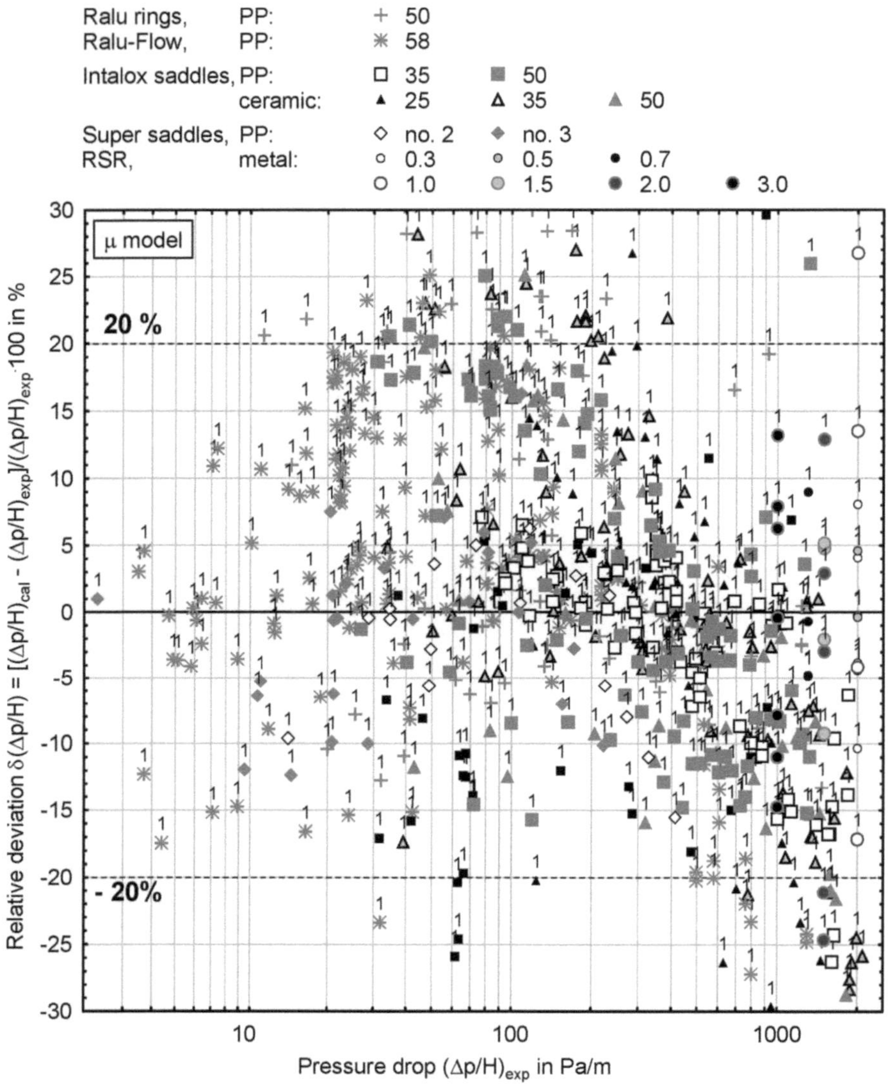

Figure 5-12. Relative deviation $\delta(\Delta p/H)$ of experimental data for determination of pressure drop $(\Delta p/H)_{exp}$ up to flooding point acc. to Eqs. (5-9) and (5-10), valid for saddle packing made of plastic and ceramic, for plastic Ralu rings and Ralu-Flow and for metal Super rings by Raschig. No. of system see Table 2-2. Test conditions see Table 5-1a–c

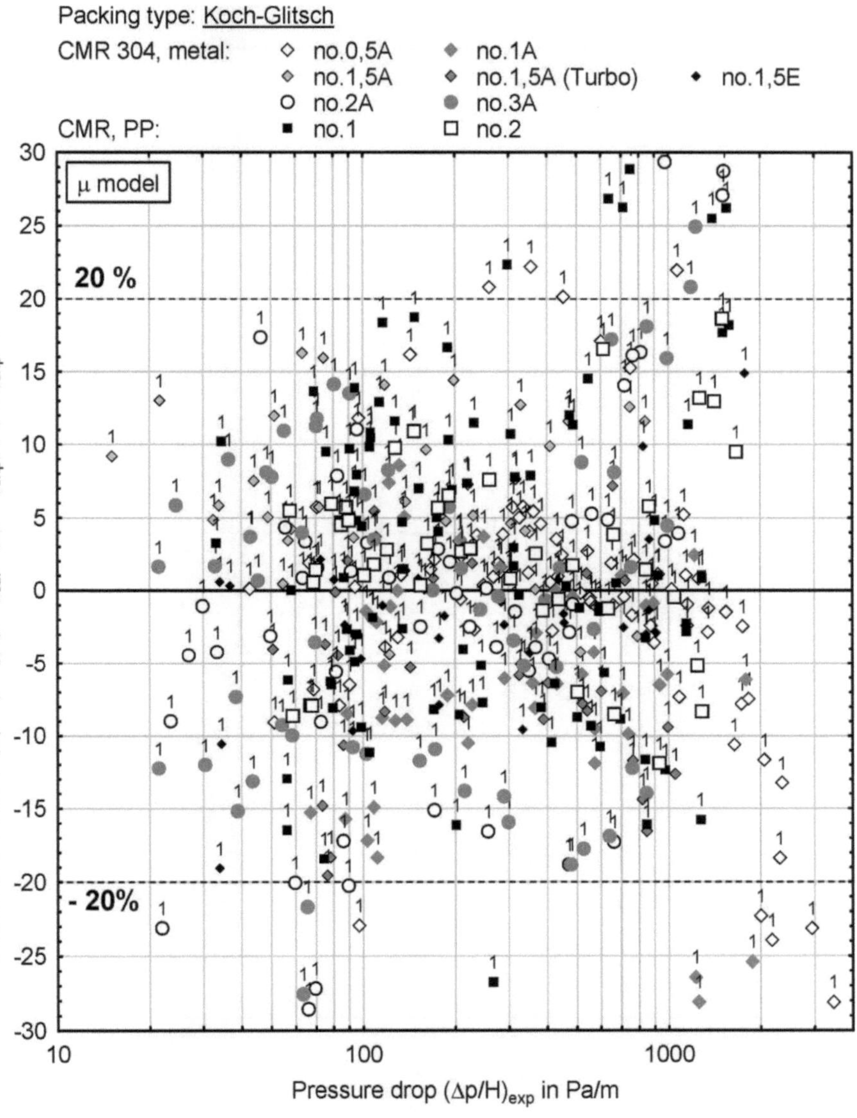

Figure 5-13. Relative deviation $\delta(\Delta p/H)$ of experimental data for determination of pressure drop $(\Delta p/H)_{exp}$ up to flooding point acc. to Eqs. (5-9) and (5-10), valid for CMR rings by Koch-Glitsch made of metal and plastic. No. of system see Table 2-2. Test conditions see Table 5-1a–c

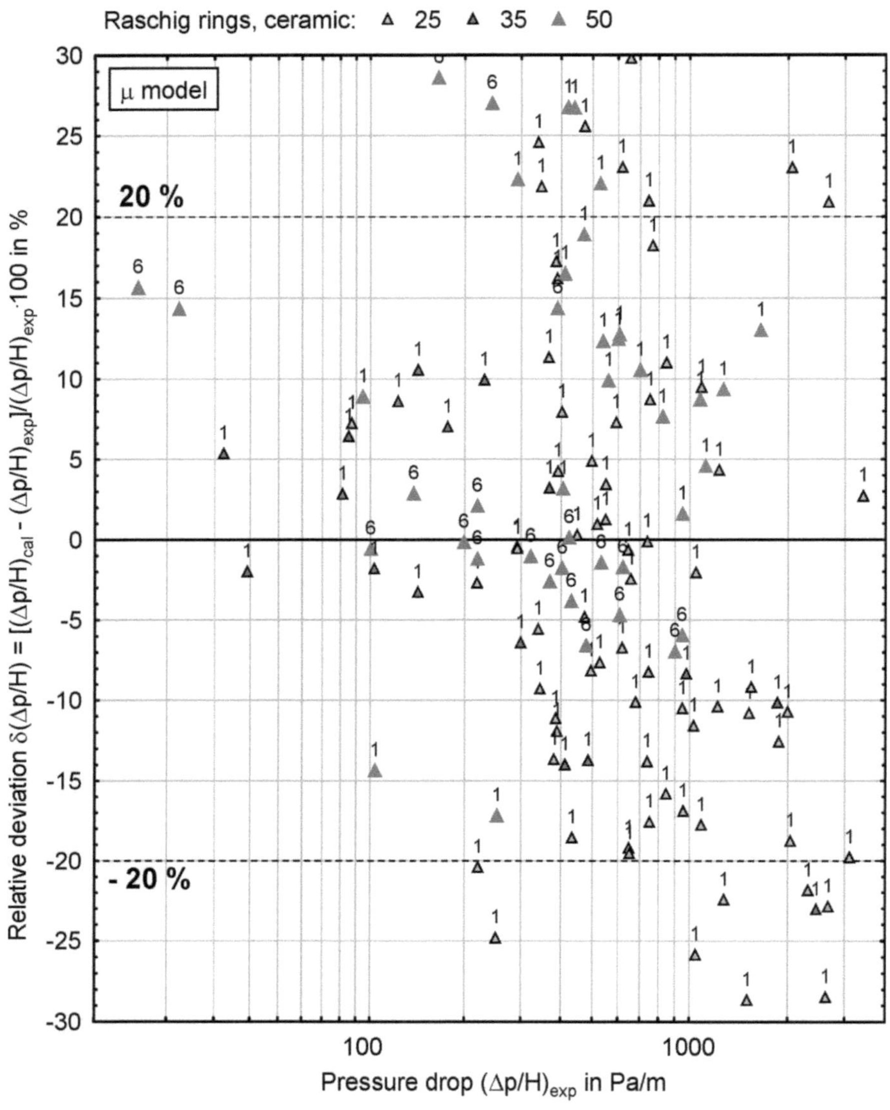

Figure 5-14. Relative deviation $\delta(\Delta p/H)$ of experimental data for determination of pressure drop $(\Delta p/H)_{exp}$ up to flooding point acc. to Eqs. (5-9) and (5-10), valid for Raschig rings made of ceramic. No. of system see Table 2-2. Test conditions see Table 5-1a–c

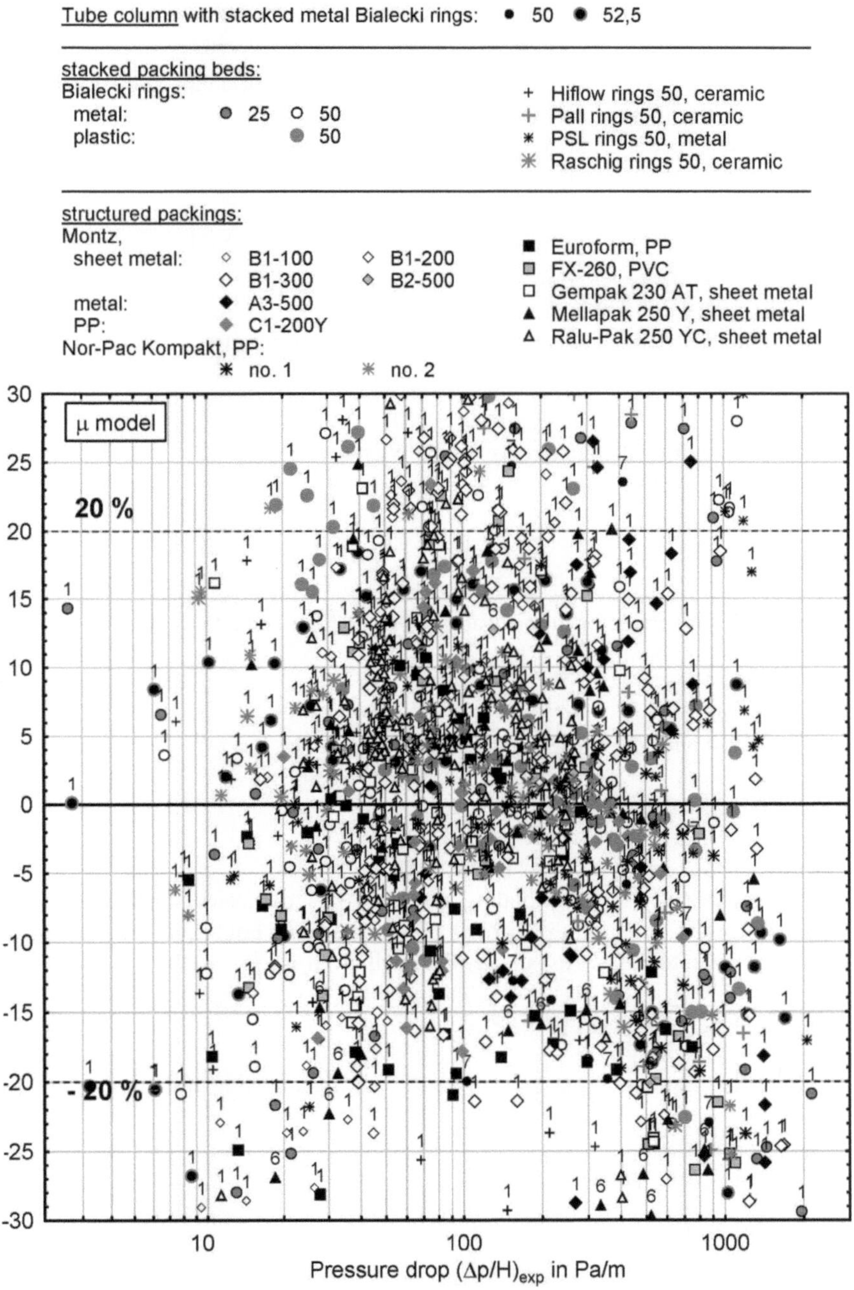

Figure 5-15. Relative deviation $\delta(\Delta p/H)$ of experimental data for determination of pressure drop $(\Delta p/H)_{exp}$ up to flooding point acc. to Eqs. (5-9) and (5-10), valid for tube columns, structured and stacked packings. No. of system see Table 2-2. Test conditions see Table 5-1a–c

Numerical Example 5.1

The aim is to determine the pressure drop of an irrigated packing, consisting of 50 mm plastic Pall rings in a column with a diameter of 1 m and a height of H = 3.5 m. The column is operated at a gas velocity of $u_V = 0.92\ ms^{-1}$ and a specific liquid load of $u_L = 5.56\ 10^{-3}\ ms^{-1}$. The test system is air/water at 1 bar and 293 K. What is the pressure drop, if the gas velocity is increased up to $2.3\ ms^{-1}$?

The technical data of the 50 mm Pall rings made of PP is as follows:

- geometric packing surface area $a = 112.0\ m^2m^{-3}$
- void fraction $\varepsilon = 0.929\ m^3m^{-3}$
- experimentally derived pressure drop values, acc. to Bornhütter, [3] references Chap. 5, Fig. 3-9b:

$$\begin{aligned} (\Delta p/H)_{exp} &= \ \ 90\ \text{Pam}^{-1} \quad \text{for } u_V = 0.92\ \text{ms}^{-1} \\ (\Delta p/H)_{exp} &= 650\ \text{Pam}^{-1} \quad \text{for } u_V = 2.3\ \text{ms}^{-1} \end{aligned}$$

Solution

The physical properties for the air/water system at 1 bar and 293 K are:

- $\rho_L = 998.2\ kgm^{-3}$
- $g = 9.81\ ms^{-2}$
- $\sigma_L = 0.0724\ Nm^{-1}$
- $\rho_V = 1.17\ kgm^{-3}$
- $\eta_L = 1 \cdot 10^{-3}\ Pas$

Table 6-1a shows that $\psi_{Fl} \cong 2.42$ and $\mu = 0.572$.

The gas velocity at the flooding point $u_{V,Fl}$ is calculated iteratively, based on Eq. (2-69), and is given as $u_{V,Fl} = 2.756\ ms^{-1}$. The liquid hold-up at the flooding point is calculated acc. to Eqs. (2-46) and (2-43) and is given as $h^0_{L,Fl} = 9.05 \cdot 10^{-2}\ m^3m^{-3}$ and

$$h_{L,Fl} = h^0_{L,Fl} \cdot \varepsilon = 8.41 \cdot 10^{-2} m^3 m^{-3}.$$

The Reynolds number Re_L, acc. to Eq. (4-18), has the following numerical value:

$$Re_L = \frac{5.56 \cdot 10^{-3}}{1 \cdot 10^{-6} \cdot 112.0} = 49.64 > 12.3$$

The resistance coefficient ψ_{VL}, acc. to Eq. (5-7), is given as:

$$\psi_{VL} = 3.8 \cdot \mu = 3.8 \cdot 0.572 = 2.174$$

The following variables are required for calculating the pressure drop $\Delta p/H$, acc. to Eq. (5-10):

– the particle diameter d_P, acc. to Eq. (3-5):

$$d_P = 6 \cdot \frac{(1-\varepsilon)}{a} = 6 \cdot \frac{(1-0.929)}{112.0} = 3.804 \cdot 10^{-3}\ \mathrm{m}$$

– the wall factor K, acc. to Eq. (3-6):

$$K = \left(1 + \frac{2}{3} \cdot \frac{1}{(1-0.929)} \cdot \frac{3.804 \cdot 10^{-3}}{1}\right)^{-1} = 0.966$$

– the dimensionless irrigation density B_L, acc. to Eq. (4-16)

$$B_L = \left(\frac{10^{-3}}{998.2 \cdot 9.81^2}\right)^{1/3} \cdot \frac{5.56 \cdot 10^{-3}}{0.929} \cdot \frac{(1-0.929)}{0.929 \cdot 3.804 \cdot 10^{-3}} = 2.63 \cdot 10^{-4}$$

– the liquid hold-up h_L, acc. to Eq. (4-35):

$$h_L = 4.39 \cdot \left(2.63 \cdot 10^{-4}\right)^{0.575} = 3.84 \cdot 10^{-2}\ \mathrm{m^3 m^{-3}}$$

Compared to the experimental value given by Bornhütter [3] $h_{L,exp} = 3.51 \cdot 10^{-2}\ \mathrm{m^3 m^{-3}}$, the deviation amounts to:

$$\delta\,(h_L) = \frac{h_{L,\exp} - h_{L,\mathrm{calc}}}{h_{L,\exp}} \cdot 100 = \frac{3.5 - 3.84}{3.5} \cdot 100 = 9.7\ \%$$

The numerical value of the pressure drop $\Delta p/H$, acc. to Eq. (5-9), at a relative column load of

$$\frac{u_V}{u_{V,Fl}} = \frac{0.92}{2.756} = 0.334 < 0.65$$

is given as:

$$\frac{\Delta p}{H} = 3.8 \cdot 0.572 \cdot \frac{1-0.929}{0.929^3} \cdot \frac{0.92^2 \cdot 1.17}{3.804 \cdot 10^{-3} \cdot 0.966} \cdot \left(1 + \frac{0.035}{1-0.929}\right) \cdot \left(1 - \frac{0.035}{1-0.929}\right)^{-3}$$
$$= 86.9\ \mathrm{Pa m^{-1}}$$

The experimental pressure drop of a column with $d_S = 1.0$ was found to be $(\Delta p/H)_{exp} =$ approx. 90 $\mathrm{Pam^{-1}}$ [3], which leads to a relative deviation from the calculated value of:

$$\delta\,(\Delta p/H) = \frac{90 - 86.9}{90} \cdot 100 = 3.44\ \%.$$

The pressure drop above the loading line at a high dynamic gas load is calculated as:

$$\frac{u_V}{u_{V,Fl}} = \frac{F_V}{F_{V,Fl}} = \frac{2.3}{2.756} = 83.5 \cdot 10^{-2} = 83.5\ \%$$

Equation (2-46) leads to the following correlation, for $h^0_{L,Fl} = 9.05 \cdot 10^{-2}\ \mathrm{m^3m^{-3}} \Rightarrow h_{L,Fl} = 8.41 \cdot 10^{-2}\ \mathrm{m^3m^{-3}}$:

$$h_{L,S} = 0.0841 - (0.0841 - 0.0385) \cdot \sqrt{1 - \left(\frac{0.835 - 0.65}{0.35}\right)^2} = 4.54 \cdot 10^{-2}\ \mathrm{m^3m^{-3}}$$

Based on Eq. (5-10), the numerical value of the pressure drop of irrigated Pall rings is given as:

$$\frac{\Delta p}{H} = 3.8 \cdot \mu \cdot \left(\frac{1-\varepsilon}{\varepsilon^3}\right) \cdot \frac{F_V^2}{d_p \cdot K} \cdot \left(1 + \frac{h_L}{1-\varepsilon}\right) \cdot \left(1 - \frac{h_L}{\varepsilon}\right)^{-3}$$

$$= 3.8 \cdot 0.572 \cdot \frac{1-0.929}{0.929} \cdot \frac{2.3^2 \cdot 1.17}{3.804 \cdot 10^{-3} \cdot 0.966} \cdot$$

$$\left(1 + \frac{0.0454}{1-0.959}\right) \cdot \left(1 - \frac{0.0454}{0.929}\right)^{-3}$$

$$= 617.7\ \mathrm{Pam^{-1}}$$

The experimental pressure drop value, acc. to Bornhütter, was found to be approx. 650 $\mathrm{Pam^{-1}}$ [3], which results in a relative error $\delta_2(\Delta p/H)$ of:

$$\delta_2\left(\Delta p/H\right) = \frac{617.7 - 650}{650} \cdot 100 = -4.96\ \%$$

The pressure drop at the flooding point for $u_V = u_{V,Fl} = 2.756\ \mathrm{ms^{-1}}$ and for $h_{L,Fl} = 0.0841\ \mathrm{m^3m^{-3}}$ is calculated as:

$$\frac{\Delta p}{H} = 3.8 \cdot 0.572 \cdot \frac{1-0.929}{0.929} \cdot \frac{2.756^2 \cdot 1.17}{0.003804 \cdot 0.966} \cdot \left(1 + \frac{0.0841}{0.071}\right) \cdot \left(1 - \frac{0.0841}{0.929}\right)^{-3}$$

$$= 1351.8\ \mathrm{Pam^{-1}}$$

$$\left(\frac{\Delta p}{H}\right)_{exp} = 1250\ \mathrm{Pam^{-1}}$$

$$\delta\left(\Delta p/H\right) = \frac{(1250 - 1351.8)}{1250} \cdot 100 = -8.145\ \%$$

Annex Chapter 5

Tables Chapter 5

Table 5-1a. Determination of pressure drop in irrigated packed column in counter-current flow acc. to Eq. (5-9) or (5-10) (μ-model) throughout the entire operating range up to flooding point. List of test points and relative mean errors $\bar{\delta}\,(\Delta p/H)$ for tested random packing, valid for system: air/water, 1 bar, 273 K

Packing	Material	d[mm]	Number of test points TP	$\bar{\delta}_1\,(\Delta p/H)$ [%]	$\bar{\delta}_2\,(\Delta p/H)$ [%]	$\bar{\delta}\,(\Delta p/H)$ [%]	d_S[m] / H [m]	μ model	max. Re_L
Pall rings	metal	15–58	441	5.35	8.93	6.67	0.3–0.45 0.9–2.0	0.5	102.5
Pall rings	metal	58	25	0	0	0	0.3 1.4	0.3	100
Pall rings Glitsch+ collars	metal	86	737	19.21	11.23	0.3 1.4	0.26	60	60
Pall rings	PP	25, 35, 50	259	5.64	10.69	6.70	0.3–0.45 1.2–2.0	0.572	120
Pall rings	ceramic	50	62	11.55	13.95	13.28	0.3–0.45 1–1.1	0.572	100
Białecki rings	metal	36, 50	194	7.37	14.25	9.94	0.3–0.45 0.8–1.45	0.515	125
Białecki rings	metal PP	25–53.5	525	10.43	13.63	11.22	0.15–0.45 0.7–1.25	0.485	315
Białecki rings	metal (1975)	25	12	5.66	19.78	12.51	0.3 0.9	1.05	12
I-13 rings	metal	25	77	6.45	12.24	8.49	0.3 0.72–0.73	0.562	101
PSL rings	metal	50	42	5.00	5.52	5.01	0.3 1.45		100
VSP rings	metal	32, 50	203	8.23	15.45	11.20	0.3–0.45 1.45–2.0	0.27	120

Table 5-1a. (continued)

Packing	Material	d[mm]	Number of test points TP	$\bar{\delta}_1\,(\Delta p/H)$ [%]	$\bar{\delta}_2\,(\Delta p/H)$ [%]	$\bar{\delta}\,(\Delta p/H)$ [%]	d_S[m] / H [m]	μ model	max. Re_L
VSP rings	PP	50	142	16.51	2.08	16.58	0.45–0.6 1.1–2.0	0.225	180
Top-Pak	metal	size 2 80	37	4.14	7.76	5.51	0.45 2	0.27	150
Top-Pak	metal	size 1 45	56	9.63	13.38	11.12	0.3 1.4	0.37	106
Hackette	PP	size 1 45	107	7.05	11.24	8.62	0.3–0.45 1.4–2	0.27	92
Hiflow rings	metal	28	177	8.52	13.79	9.21	0.3–0.6 0.92–2.0	0.268	61
Hiflow rings	metal	58	113	7.81	11.03	8.75	0.45 2.0	0.175	110
Hiflow rings	PP	28	85	8.59	6.83	8.2	0.3 0.9–1.4	0.333	61
Hiflow rings	PP	50, 90	270	11.24	9.91	10.22	0.3–1.0 0.9–2.0	0.19	160
Hiflow saddles	PP	50	63	3.36	13.72	5.66	0.45 2.0	0.297	135
Hiflow rings	PP	50, 90 × 65	190	9.93	10.49	10.19	0.3–0.45 1.47–2.0	0.19	150
Hiflow rings	PP	17	142	4.51	7.48	6.17	0.3 0.9–1.45	0.403	45
Hiflow rings Super	PP	53	210	6.54	8.41	5.82	0.3–0.45 1.4–2.0	0.276	140
Hiflow rings	ceramic	20–50	249	12.09	17.09	11.74	0.3–0.45 1.0–1.45	0.50	140
Hiflow rings	ceramic	75	53	9.7	19.95	13.76	0.45 2.0	0.315	180
Hiflow rings	ceramic	20, 75	64	7.75	15.94	10.46	0.3–0.45 1.4–2.0	0.408 0.369	240

Table 5-1a. (continued)

Packing	Material	d[mm]	Number of test points TP	$\bar{\delta}_1\,(\Delta p/H)$ [%]	$\bar{\delta}_2\,(\Delta p/H)$ [%]	$\bar{\delta}\,(\Delta p/H)$ [%]	d_S[m] / H [m]	μ model	max. Re_L
Envipac	PP	size 1A, 32	38	4.76	10.64	7.70	0.3 1.4	0.420	71
Envipac	PP	size 1, 32	68	11.58	5.04	12.14	0.3–1.8 0.9–1.42	0.318	91
Envipac	PP	size 2, 58	367	12.2	10.0	12.4	0.3–1.8 1.1–4.5	0.20	121
Envipac	PP	size 3, 80	109	10.93	9.7	11.25	0.3–0.6 1.44–2.0	0.261	203
Dtnpac	PP	size 1 45	146	6.86	7.91	7.34	0.45–0.6 1.1–2.0	0.318	91
Dtnpac	PP	size 2 70	66	8.13	2.22	8.28	0.3–0.6 1.08–1.42	0.2427	100
Tellerette	PP	sizes 1+2 45, 70	81	9.35	8.86	9.10	0.15–0.45 1.3–2	0.30 0.27	135
Glitsch rings	metal	size 0.5	87	4.46	17.06	9.24	0.3 1.1	0.383	31
Glitsch rings	metal	sizes 1A, 1.5A, 1.5E	185	6.66	10.86	8.03	0.3–0.45 1.4–2.0	0.247	57
Glitsch rings	metal	sizes 2+3	115	8.64	13.58	10.55	0.45 2.0	0.14 0.175	111
Glitsch rings	PP	size 1	107	7.98	14.84	11.13	0.3–0.45 1.4–2.03	0.35	90
Glitsch rings	PP	size 2	42	4.92	6.58	5.71	0.45 2.03	0.5	110
Ralu rings	PP	50	50	4.56	6.6	5.13	0.45 2.0	0.293	100

Table 5-1a. (Continued)

Packing	Material	d[mm]	Number of test points TP	$\bar{\delta}_1\,(\Delta p/H)$ [%]	$\bar{\delta}_2\,(\Delta p/H)$ [%]	$\bar{\delta}\,(\Delta p/H)$ [%]	d_S[m] / H [m]	μ model	max. Re_L
Ralu-Flow	PP	Nr. 2 58	228	10.28	7.23	10.43	0.3–0.6 1.44–2.12	0.215	112
Raschig rings	ceramic	25–50	191	21.05	20.74	20.03	0.3–0.6 0.74–1.5	1	230
Intalox saddles	PP	35, 50	159	9.31	7.18	8.64	0.3–0.45 0.84–2.0	0.72 0.652	111
Intalox saddles	ceramic	25–50	193	11.65	18.60	15.02	0.3–0.45 0.87–2.0	0.705	120
Intalox-Super	PP	size 2 50	19	4.45	27.3	6.86	0.3 1.34	0.375	48
Intalox-Super	PP	size 3 80	25	5.54	0	5.54	0.6 2.0	0.255	58
Intalox	metal	40	46	10.35	28.3	18.57	0.45 2.0	0.273	85
Nor-Pac	PP	size 0.5 17	81	4.52	13.72	9.15	0.3 0.9	0.300	35
Nor-Pac	PP	sizes 1, 1½, 2; 28, 38, 50	523	7.6	12.30	8.8	0.3–1.4 1.0–2.0	0.20	225
Nor-Pac	PP	22 × 27	40	6.42	20.77	10.72	0.3 1.4	0.26	45
Mc-Pac	metal	size 1 30	180	9.75	17.47	12.51	0.32–0.35 1.5–1.65	0.210	63
Mc-Pac	metal	size 2 65	149	14.19	20.79	17.67	0.32–1.6 2.5–3.0	0.165	162
R-Pac	ceramic	size 1+2 30, 50	187	14.34	12.59	13.55	0.32 3.0	0.585 0.625	154
SR-Pac	ceramic	size 2 65	145	7.37	8.31	7.83	0.32 1.35	0.370	185
Total			**8098**	8.45	11.93	9.50			

Table 5-1b. Determination of pressure drop in irrigated packed column in counter-current flow acc. to Eqs. (5-9) and (5-10) (μ-model) throughout the entire operating range up to flooding point. List of test points and relative mean errors $\bar{\delta}\left(\Delta p/H\right)$ for tested stacked and structured packing, valid for system: air/water, 1 bar, 273 K

Packing	Material	d [mm]	Number of test points TP	$\bar{\delta}_1\left(\Delta p/H\right)$ [%]	$\bar{\delta}_2\left(\Delta p/H\right)$ [%]	$\bar{\delta}\left(\Delta p/H\right)$ [%]	d_S [m] / H [m]	μ-Model	max. Re_L
Hiflow rings stacked	ceramic	50	45	13.4	38.37	19.13	0.45 / 1.0	0.0895	110
Raschig rings stacked	ceramic	50	41	5.75	12.4	7.05	0.3 / 1.0	0.14	80
Nor-Pac-compact packing	PP	25, 50	106	5.65	23.1	10.18	0.3 / 1.4	0.450	95
25 mm Pall rings tube column	metal	25	16	16.29	30.2	20.63	0.025 / 1.0	0.025	76
25 mm Białecki rings stacked	metal (1978) $N = 6350\ m^{-3}$	25	66	14.97	22.03	16.68	0.15 / 1.40	0.185	61
50 mm Białecki rings stacked	metal (1974) $N = 6174\ m^{-3}$	53.2	61	11.46	8.85	12.33	0.3 / 1.4	0.35	100
	metal (1990) $N = 7654\ m^{-3}$	50	64	7.23	16.08	10.13	0.3 / 1.45	0.185	100
	PP (1974) $N = 8140\ m^{-3}$	50	73	12.9	14.04	13.09	0.3 / 1.0	0.40	200
50 mm PSL rings stacked	metal $N = 7882\ m^{-3}$	50	54	6.40	14.89	9.86	0.273 / 1.45	0.197	100
52 mm Białecki rings tube column	metal (7 tubes) $N = 9422\ m^{-3}$	52	66	10.69	32.39	15.29	7 × 0.052 / 1.25	0.32	110
50 mm Białecki rings tube column	metal (1 tube) $N = 10193\ m^{-3}$	50	20	16.60	16.60	16.60	0.050 / 1.0	0.60	100

Table 5-1b. (continued)

Packing	Material	d [mm]	Number of test points TP	$\bar{\delta}_1\left(\Delta p/H\right)$ [%]	$\bar{\delta}_2\left(\Delta p/H\right)$ [%]	$\bar{\delta}\left(\Delta p/H\right)$ [%]	d_S [m] / H [m]	μ-Model	max. Re_L
50 mm stacked Pall rings stacked	ceramic $N = 7550\ m^{-3}$	50	32	18.04	24.45	22.05	0.46 0.9	0.275	110
Gempak	sheet metal	230 AT Y	51	10.53	16.2	11.1	0.3 1.47	0.12	50
Ralu-Pak	sheet metal	250 Y 250 Y	63 63	7.73 9.82	28.60 30.06	9.99 12.07	0.45/2.0 0.3/1.8	0.085 0.085	50 50
Euroform	PP	110 Y	54	8.79	19.93	11.69	0.3 1.4	1.45	185
Mellapak	sheet metal	250 Y	5 63	8.62 28.90	17.18 22.65	12.04 26.90	0.15/0.85 0.22/1.26	0.152 0.12	25 80
Montz packing	PP	C1-200Y	56	8.37	3.65	7.86	0.3 1.42	0.230	60
Montz gauze packing BX	metal gauze	A3-500	35	9.92	19.24	14.98	0.45 2.0	0.077	25
Montz packing	sheet metal	B1-300 Y	73 50 45	13.06 14.14 19.50	17.18 20.72 49.0	14.52 14.50 30.20	0.22/1.42 0.3/1.5 0.3/1.6	0.12 0.12 0.105	40 40 40
		B1-200 Y	72	14.41	20.56	16.46	0.3 1.6	0.12	60
		B1-100 Y	36	12.97	1.65	12.01	0.3 1.6	0.12	90
		B2-500 Y	5	9.98	36.5	15.28	0.22 1.4	0.152	5
FX-VFF packing	PP	FX 260	36	10.13	30.46	18.03	0.45 2.0	0.0623	50
Total			**1350**	**11.33**					

Table 5-1c. Determination of pressure drop in irrigated packing column in counter-current flow acc. to Eqs. (5-9) and (5-10) (μ-model) throughout the entire operating range up to flooding point. List of test points and relative mean errors $\bar{\delta}\,(\Delta p/H)$ for tested packings, valid for pressure range up to 30 bar [4]

Packing	Material	p[bar]	Number of test points TP	$\bar{\delta}_1\,(\Delta p/H)$ [%]	$\bar{\delta}_2\,(\Delta p/H)$ [%]	$\bar{\delta}\,(\Delta p/H)$ [%]	$\frac{d_S\,[m]}{H\,[m]}$	μ-Model	max. Re_L
15 mm Pall rings	PP	5–30	112	9.75	23.9	19.03	0.155 0.8	0.567	20
Mellapak 250 Y	sheet metal	7.25–15	12	8.62	17.18	12.04	0.155 0.8	0.152	40
Total			**124**	9.19	20.5	15.54			

Total number of test points $\Sigma TP = 9572$.

References Chapter 5

1. Maćkowiak J. Bestimmung des Druckverlustes berieselter Füllkörperschüttungen und Packungen. Staub-Reinhaltung der Luft 50 (1990), p. 455–463
2. Maćkowiak J. Pressure drop in Irrigated Packed Columns. Chem. Eng. Process. (1991), p. 93–105
3. Bornhütter H. Stoffaustausch von Füllkörperschüttungen unter Berücksichtigung der Flüssigkeitsströmungsform. Dissertation, TU Munich (1991)
4. Krehenwinkel H. Experimentelle Untersuchungen der Fluiddynamik und der Stoffübertragung in Füllkörperkolonnen bei Drücken bis zu 100 bar. Dissertation TU Berlin 1986

CHAPTER 6

Fluid Dynamics of Packed Columns for Gas/Liquid Systems – Summary of Results

6

6.1 General Information

One the major objectives of this book was to consolidate the extensive research material available on the fluid dynamics of packed columns in counter-current two-phase flow and to extend the application range, compared to the first German edition of this book. This was possible by deriving some generally valid equations for individual packing-specific parameters, such as gas velocity at flooding point $u_{V,Fl}$, liquid hold-up h_L at operating and flooding point and pressure drop $\Delta p_0/H$ of dry and $\Delta p/H$ of irrigated random and structured packings.

This work has looked at calculation principles for approx. 200 classic as well as modern randomly filled and stacked lattice packing elements, tube columns and structured packings of various types and sizes. Today, these packings are used not only in rectification in the vacuum and normal pressure range, but also in many areas of high-pressure rectification and high-pressure absorption as well as in waste-air and wastewater treatment and groundwater treatment.

The design of these plants is based on the so-called *SBD model*, see Chap. 2, which can be used to determine the gas velocity at the flooding point for various types of internals, if the resistance coefficient ψ of the dry packing is known.

If the resistance coefficient ψ is known, it is also possible to calculate the pressure drop of dry and irrigated packings for different operating conditions.

The law of resistance $\psi = f(Re_V)$, see Chap. 3, is the basis for the entire fluid dynamic design of packed columns and for the calculation of the following parameters:

- gas velocity at the flooding point, see Eq. (2-69)
- pressure drop of the dry packing, see Eq. (3-29)
- pressure drop of the irrigated packing in the operating range below the flooding point, see Eq. (4-49) and (4-50)
- pressure drop of the irrigated packing at the flooding point, see Eq. (4-54)

The limit of validity of the derived correlations was verified based on 32 mixtures with different physical properties, see Table 2-2. The experiments, which were drawn on

J. Maćkowiak, *Fluid Dynamics of Packed Columns*,
Chemische Technik/Verfahrenstechnik, DOI 10.1007/b98397_6,

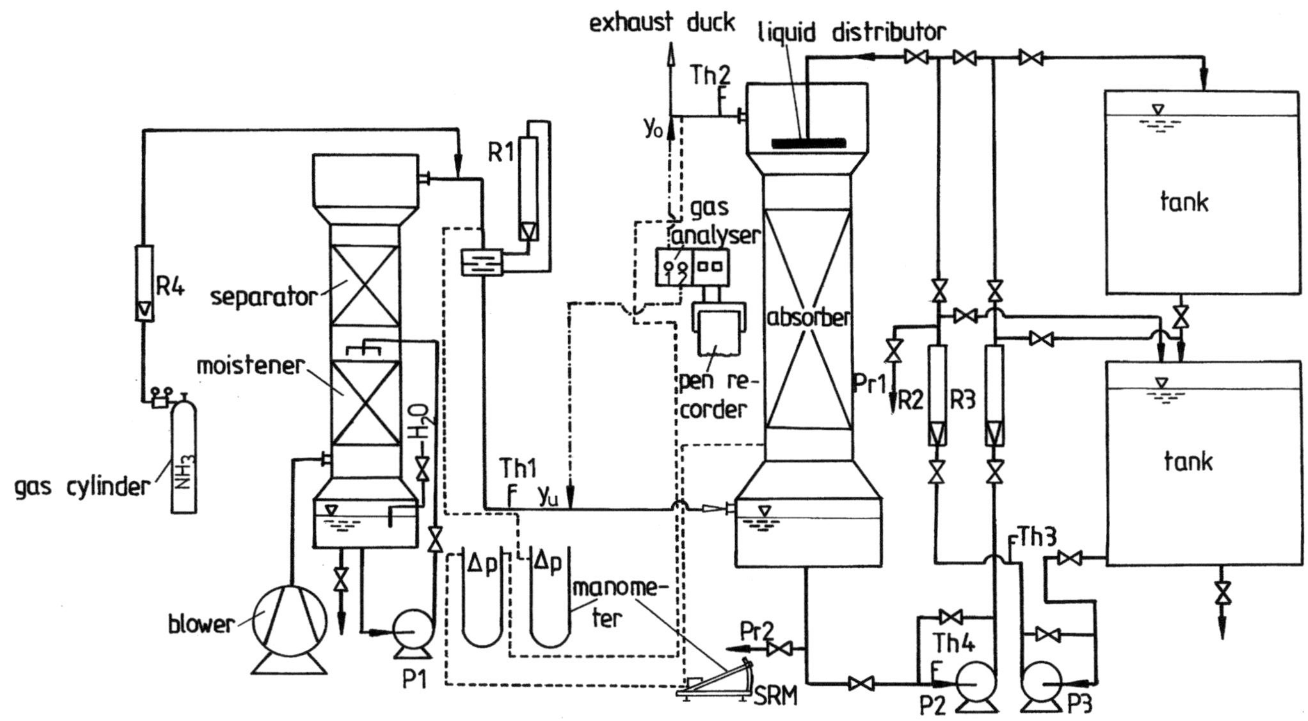

Figure 6-1. Diagram of test plant for investigation of packings with air/water system and absorption system NH_3-air/water under normal conditions. Column diameter $d_S = 150$, 225 and 300 mm, packed bed height $H = 0.5-2$ m

Figure 6-2. Picture of test plant, shown as diagram in Fig. 6-1, for investigation of packings under normal conditions. $d_S = 0.45$ m, $H = 1.45–2.0$ m, system: NH_3-air/water

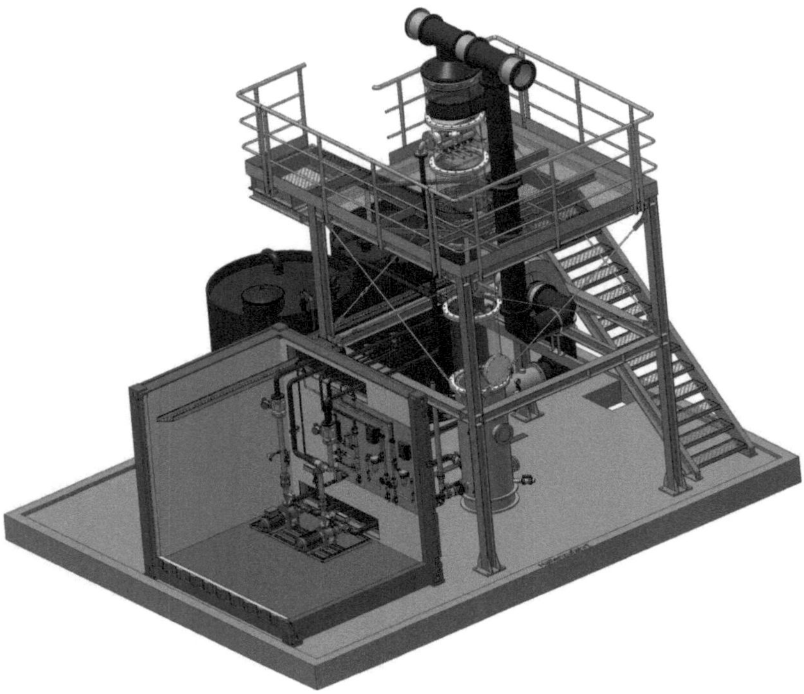

Figure 6-3. Three dimensional drawing of ENVIMAC test plant for investigation of packings with air/water system and absorption systems under normal conditions. Column diameter d_S = 500/600/800 mm, packed bed height H = 1–3.2 m

Figure 6-4. Picture of ENVIMAC test plant, shown as drawing in Fig. 6-3, for investigation of packings under normal conditions. $d_S = 0.6$ m, H = 3.2 m, system: air/water

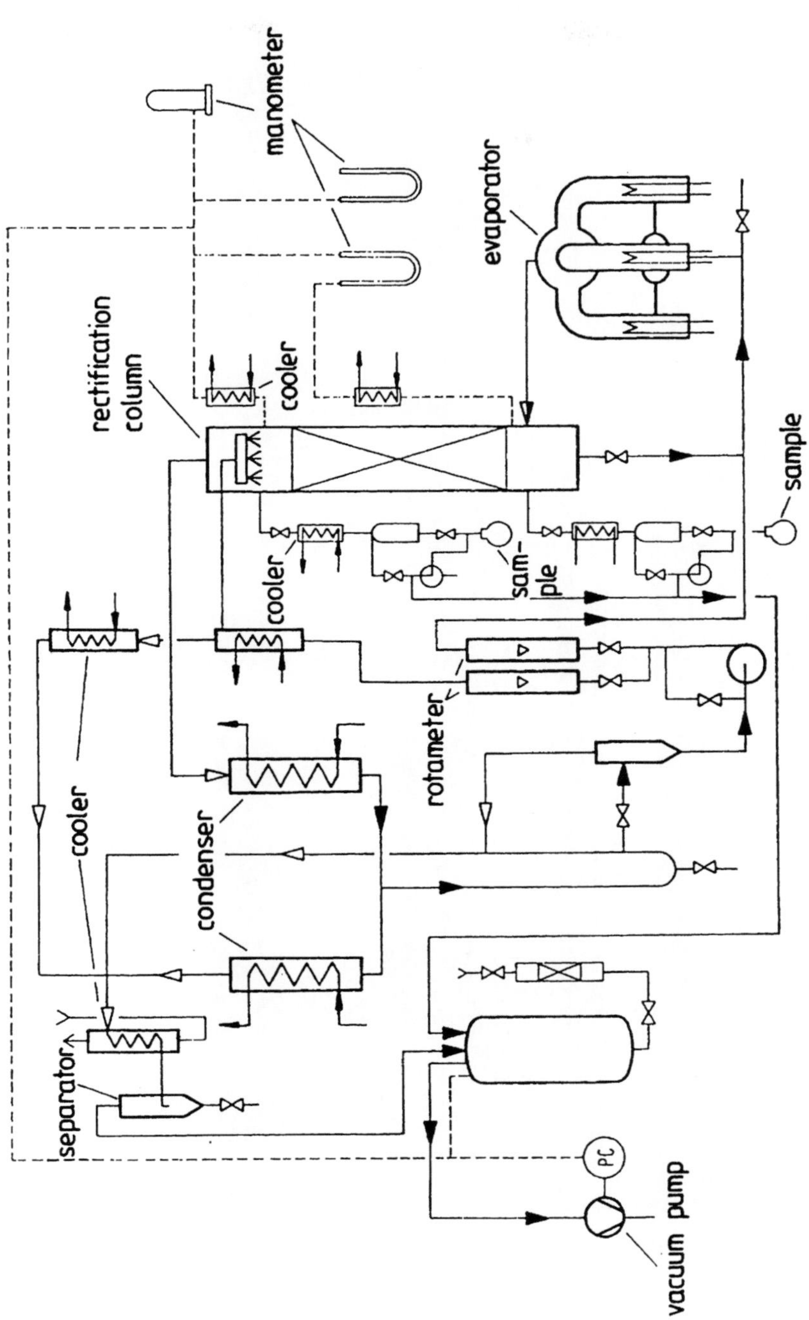

Figure 6-5. Diagram of test plant for investigation of packings used for vacuum rectification. Column diameter $d_S = 150$ and 220 mm, packed bed height $H = 1.0–1.7$ m

Figure 6-6. Picture of rectification test plant DN 200, shown as diagram in Fig. 6-5

for the numerical evaluation, were based on a variety of physical properties, operating parameters F_V and u_L, packing types, sizes and materials as well as column diameters and packing heights. Figures 6-1, 6-2, 6-3, 6-4, 6-5 and 6-6 show diagrams and pictures of the plants used for some of the experiments carried out by the author.

The evaluation of experimental data also took into account numerous publications of the 90s, which include some experimental results of studies carried out at larger plants, with $d_S = 0.6$, 1, 1.2, 1.6, 1.8 and 3 m, see Chap. 2 [6, 7, 66–69, 80] or at higher pressures, see Chap. 2 [39, 68, 69, 75].

6.2 Determining the Flooding Point

The suspended bed of droplets (SBD) model, which was developed further in this work, describes the processes at the flooding point at low, moderate and high phase flow ratios in the range of $10^{-4} < \lambda_0 < 1$ with an accuracy of $< \pm 8$–12% for 80–90% of the experimental data evaluated. At higher pressures up to 100 bar, the accuracy is $\leq \pm 15\%$, see Sect. 2.2. For the purpose of determining the gas velocity at the flooding point, a general, dimensionless flooding point correlation, see Eq. (2-73), was derived and verified by evaluating approx. 1200 experimental data points. It is applicable to any types of column internals:

- random packings
- stacked packing elements
- structured packings
- tube columns with stacked packings

The dimensionless Eq. (2-73) leads to the following dimension-dependent Eq. (2-69):

$$u_{V,Fl} = 0.8 \cdot \cos\alpha \cdot \psi_{Fl}^{-1/6} \cdot \varepsilon^{6/5} \cdot \left[\frac{d_h}{d_T}\right]^{1/4} \cdot \left[\frac{d_T \cdot \Delta\rho \cdot g}{\rho_V}\right]^{1/2} \cdot \left(1 - h_{L,Fl}^0\right)^{7/2} \cdot K_{\rho V}\ \left[\text{ms}^{-1}\right]$$

with

$$\psi_{Fl} = \psi_0 \cdot (1 - \varphi_P) = \left(\frac{725.6}{\text{Re}_V} + 3.203\right)(1 - \varphi_P)$$

acc. to Eqs. (3-26) and (3-28).

This equation was validated for randomly filled and stacked packing elements, structured packings and tube columns for resistance coefficients ψ_{Fl} of 0.15 to 8.5 for single-phase flow through the packed bed. Equation (2-69) for determining the flooding point was found to be suitable in the vacuum and normal pressure range as well as in the range of high pressures of up to 100 bar.

6.3 Liquid Hold-Up at Flooding Point

In order to predict the liquid hold-up $h^0_{L,Fl}$ at the flooding point, it is necessary to ascertain the phase flow ratio at the flooding point $\lambda_0 = V_L/V_{V,Fl}$.

Based on experimental data, it was possible in the course of this work to derive a new equation, which describes the liquid hold-up at the flooding point $h^0_{L,Fl}$ only based on the phase flow ratio at the flooding point $\lambda_0 = V_L/V_{V,Fl}$, independent of the type, material and size of the packing or the physical properties.

In the first edition of this book (1991), it was possible to validate Eq. (2-46) by means of experimental liquid hold-up data up to a phase flow ratio of $\lambda_0 \approx 0.030$. Thanks to the availability of more recent research material, mainly from the 90s, the application range has now been extended to include very high phase flow ratios. The equation is now applicable to phase flow ratios between $\lambda_0 = 10^{-4}$ and $\lambda_0 \cong 1$. This range is of particular interest for high-pressure absorption and high-pressure rectification processes. The generally valid Eq. (2-46) for determining the liquid hold-up at the flooding point is as follows:

$$h^0_{L,Fl} = \frac{\sqrt{\lambda_0^2 \cdot (m+2)^2 + 4 \cdot \lambda_0 \cdot (m+1) \cdot (1-\lambda_0)} - (m+2) \cdot \lambda_0}{2 \cdot (m+1) \cdot (1-\lambda_0)} \qquad \left[m^3 m^{-3}\right]$$

The parameter m depends to a small extend on the liquid flow and the phase flow ratio at the flooding point λ_0, see Eqs. (2-43) and (2-44).

For $Re_L \geq 2$, Eq. (2-46) leads to Eq. (2-47) for $\lambda_0 < 0.030$ and $m = -0.8$. This applies to random and structured packings of any type and size.

The parameter $m = -0.88$ is valid for small Reynolds numbers of $0.5 < Re_L < 2$, Eq. (2-49).

The gas velocity at the flooding point for any type of packing can be calculated, acc. to Eq. (2-69), if the physical properties of the system to be separated and the phase flow ratios λ_0 at the flooding point as well as the resistance coefficient ψ of the packing in single-phase gas flow are known. In the turbulent flow range of the gas phase, $Re_V > 2100$, at which packed columns are normally operated, the parameter of the phase flow ratio is approximately constant. The numerical values ψ_m for the packings tested are listed in Tables 6-1a–c. The experimental effort required for designing packed columns is therefore reduced to the experimental determination of the *single parameter* ψ_m. The required tests can be carried out using air at ambient conditions. Based on the evaluation of the experimental data contained in the database, it was possible to verify the transferability of the law of resistance $\psi = f(Re_V)$ to any type of system and operating pressure.

6.4 Pressure Drop and Liquid Hold-Up

The flow of gas as well as liquid in packed columns containing large random and structured packings is usually turbulent at the operating point below the flooding point. In this range, the viscosity forces acting on the liquid, which occurs in the form of films, runlets and droplets, can be neglected, which simplifies the correlations used for describing the pressure drop of irrigated packed beds $\Delta p/H$ as well as the liquid hold-up h_L.

The liquid hold-up h_L, which occurs in the form of films, runlets and droplets, reduces the void space available for the gas phase, i.e. the free, effective void fraction ε_{eff} of the packing. According to the general law of resistance, Eq. (3-2), which applies to channel flow, this results in an increase in the pressure drop of the gas in two-phase flow. If the gas capacity factor F_V remains constant, the pressure drop $\Delta p/H$ increases with the liquid load u_L, see Fig. 2-2a, which is due to an increase in the liquid hold-up of the packing. The pressure drop of the packing, however, decreases, as the size of the packing increases.

These correlations are described by the equations for calculating the pressure drop $\Delta p/H$, developed in this work. It has proved useful to describe the pressure drop $\Delta p/H$ of irrigated packings in relation to the following individual ranges:

(a) below the loading line, $F_V/F_{V,Fl} \leq 0.65$
(b) between the loading line and the flooding point, $0.65 < F_V/F_{V,Fl} < 1$
(c) at the flooding point $F_V/F_{V,Fl} \equiv 1 \pm 0.05$

The range below the loading line (range a) is characterised by a parallel run of the pressure drop curves for different liquid loads u_L and the $\Delta p_0/H$ curve for single-phase flow, see Fig. 2-1. Below the loading line, the gas has practically no influence on the liquid flow. In the range close to or on the loading line, which is often used as a design point for packed columns, both phases are usually in turbulent flow, i.e.:

(a1) $Re_L \geq 2$ and $Re_V \geq 2100$.

The following flow conditions may also occur:

(a2) $Re_L \geq 2$ and $Re_V < 2100$
(a3) $Re_L < 2$ and $Re_V \geq 2100$
(a4) $Re_L < 2$ and $Re_V < 2100$

The ranges a1 and a2 are very significant for the design of vacuum and normal pressure distillation columns as well as for most cases of application in gas absorption with packing elements of $d \geq 0.025$–0.100 m, which are commonly used for the separation of low-viscosity mixtures.

6.4.1 Pressure Drop Below the Loading Line

The pressure drop $\Delta p/H$ of irrigated packings for cases a1 and a2 is given by Eq. (4-48) and/or Eq. (4-49), derived from the channel model:

$$\frac{\Delta p}{H} = \psi \cdot \frac{1-\varepsilon}{\varepsilon^3} \cdot \frac{F_V^2}{d_P \cdot K} \cdot \left[1 - C_B \cdot \frac{a^{1/3}}{\varepsilon} \cdot u_L^{2/3}\right]^{-5} \quad \left[\mathrm{Pa\,m^{-1}}\right],$$

with $\psi = \psi_0 \cdot (1 - \varphi_P) = \left(\frac{725.6}{Re_V} + 3.203\right)(1 - \varphi_P)$

acc. to Eqs. (3-26) and (3-28).

In the case of random or structured packings with a flow channel angle of 45° (type Y), the following applies:

$$C_B = 0.4\,s^{2/3}\,m^{-1/3},$$

whereas for stacked packing elements and type X structured packings:

$$C_B = 0.325\,s^{2/3}\,m^{-1/3}.$$

A generally valid Eq. (3-8) was derived to calculate the pressure drop of dry, non-irrigated, random and structured packings.

Here, the resistance coefficient ψ is dependent on the Reynolds number Re_V as well as on the type, material, form factor φ_P and size of the packing. The numerical values of the form factor φ_P used for calculating the resistance coefficient ψ, acc. to Eq. (3-27) and/or for Eq. (4-48), are listed in Tables 6-1a–c.

Based on the constant value of the parameter $C_B = 0.4 s^{2/3} m^{-1/3}$, it is possible to verify by calculation 85% of the experimental pressure drop values for irrigated packing elements of any type or size as well as for type Y structured packings (for the latter, the wall factor $K \equiv 1$ must be substituted into Eq. (4-48) or (4-49), see Chap. 3). The calculation is performed with a relative error of less than ± 15 %. This was the result of the evaluation of approx. 10000 experimental data items of various systems. The resistance coefficients ψ of the tested packing elements for the transition range and the turbulent flow range of the gas phase are listed in Tables 6-la–c.

In the operating range below the loading line, $F_V < 0.65 \cdot F_{V,Fl}$, it is possible to determine the pressure drop of irrigated packings $\Delta p/H$ using Eq. (4-50) for a liquid flow of $Re_L < 2$ and any given Reynolds number Re_V. It applies to cases a3 and a4.

$$\frac{\Delta p}{H} = \psi \cdot \frac{1-\varepsilon}{\varepsilon^3} \cdot \frac{F_V^2}{d_P \cdot K} \cdot \left[1 - \left[\frac{3}{g}\right]^{1/3} \cdot \frac{a^{2/3}}{\varepsilon} \cdot (\nu_L \cdot u_L)^{1/3}\right]^{-5} \quad \left[\text{Pam}^{-1}\right]$$

This correlation has been validated up to $Re_L \geq 0.1$ and for $F_V \leq 0.75 \cdot F_{V,Fl}$, see Fig. 4-14.

6.4.2 Liquid Hold-Up Below the Loading Line

Figure 2-2b shows the liquid hold-up h_L relating to the packing volume V_S as a function of the gas capacity factor F_V for the liquid loads u_L shown in Fig. 2-2a (more information on the liquid load can be found in Sect. 4.2). Up to the loading line, the liquid hold-up h_L is dependent on the liquid load u_L and the size of the packing, assuming a turbulent liquid flow, $Re_L \leq 100$, through the random or structured packing with a flow channel angle of 45°. In this case, it is possible to describe the total liquid hold-up sufficiently accurately for practical applications using the following dimensionless Eq. (4-32), $\delta(h_L) < \pm 20\%$:

$$h_L = C_P \cdot Fr_L^{1/3} \quad [\text{m}^3\,\text{m}^{-3}]$$

where

$Fr_L = (u_L^2 \cdot a)/g$ acc. to Eq. (4-31) and $C_P = 0.57$ acc. to Eq. (4-33a) for randomly filled packing elements and type Y structured packings and/or $C_P = 0.465$ for stacked packing elements and type X structured packings.

For turbulent liquid flow, the constant C_P in Eq. (4-32) is not dependent on the type, size or material of the packing. It was determined by experiments carried out in the course of this work. Equation (4-32) is easy to use and, based on a given liquid load and We/Fr_L numbers ranging between 100 and 1300, only requires a knowledge of the geometric packing surface area a.

In the laminar range, $Re_L < 2$, however, the viscosity forces must not be neglected. Based on the modified model by Bemer and Kalis [12], Eq. (4-20),

$$h_L = \frac{3}{4} \cdot \left[\frac{3}{g}\right]^{1/3} \cdot a^{2/3} \cdot (\nu_L \cdot u_L)^{1/3} \quad \left[m^3 m^{-3}\right]$$

it is possible to verify by calculation the experimental liquid hold-up data for this flow range, with a satisfactory accuracy of $\delta(h_L) < \pm 20\%$. The factor 3/4 in Eq. (4-20) was also determined based on experiments carried out in the course of this work.

In the model chosen by Mersmann and Deixler [44], $h_L = f(B_L)$, with the We_L/Fr_L parameter, the experimental data is also described with a satisfactory accuracy of approx. ±20%, using Eq. (4-34):

$$h_L = 2.2 \cdot B_L^{1/2} \quad \left[m^3 \cdot m^{-3}\right]$$

An evaluation of the experimental results of this work has shown that this equation is applicable to a We_L/Fr_L number range of around 28.8 to 1300. Based on the results on the liquid hold-up, it can be concluded that by modifying the models known from literature [12, 19, 44], the total liquid hold-up below the loading line can be determined with satisfactory accuracy for practical applications.

6.4.3 Pressure Drop and Liquid Hold-Up in the Range Between the Loading Line and the Flooding Point

When describing the pressure drop and the liquid hold-up for counter-current two-phase flow, it is recommended to use the gas capacity factor F_V relating to the maximum value at the flooding point $F_{V,Fl}$. The liquid hold-up h_L as well as the parameter C_B, Fig. 4-15, are dependent on the relative gas capacity $F_V/F_{V,Fl}$. The liquid hold-up can be determined using Eq. (4-39), and the parameter C_B based on Eq. (4-55). The calculation of the pressure drop above 65% of the flooding point, based on Eqs. (4-48) and (2-72), can be performed with an accuracy of approx. ±20%, Fig. 4-17.

The advantage of determining the pressure drop of irrigated random and structured packings acc. to Eq. (4-48) lies in the fact that this method is applicable to a wide range of packing parameters and physical properties as well as a wide operating range. In addition, it does not require a knowledge of the liquid hold-up, see Sect. 4.3.

6.4.4 Pressure Drop at Flooding Point

The pressure drop $(\Delta p/H)_{Fl}$ at the flooding point can be determined using Eq. (4-54):

$$\left(\frac{\Delta p}{H}\right)_{Fl} = \psi_{Fl} \cdot \frac{1-\varepsilon}{\varepsilon^3} \cdot \frac{F_V^2}{d_P \cdot K} \cdot \left[1 - 0.407 \cdot \lambda_0^{-0.16} \cdot \frac{a^{1/3} \cdot u_L^{2/3}}{\varepsilon}\right]^{-5} \quad \left[\text{Pam}^{-1}\right]$$

which is valid for $Re_L \geq 2$ and for phase flow ratios at the flooding point of $\lambda_0 < 0.020$.

Equation (4-54) can be used to verify 85% of the experimental values with an accuracy of $< \pm 20\,\%$ and 95% of the values with an accuracy of $< \pm 30\%$. When designing new plants, it is very important to ascertain the pressure drop at the flooding point for regulation purposes and stress calculations. For practical applications, a quick estimation of the pressure drop at the flooding point can also be performed using Mersmann's diagram, see Figs. 2-5a–d. The plotting of the experimental values for the random and structured packings tested led to the following pressure drop values $\Delta p/(\rho \cdot g \cdot H)$:

(a) For 15–20 mm randomly filled packing elements with a specific surface area of $a \approx 300\ m^2m^{-3}$:

$$\left(\frac{\Delta p}{\rho_L \cdot g \cdot H}\right)_{Fl} \approx 0.30 \quad \text{see Fig. 2-5a.}$$

(b) For randomly filled 1″ packing elements with $a \approx 220\ m^2m^{-3}$:

$$\left(\frac{\Delta p}{\rho_L \cdot g \cdot H}\right)_{Fl} \approx 0.25 \quad \text{see Fig. 2-5a.}$$

(c) For randomly filled 1.5″ packing elements with $a \approx 140\ m^2m^{-3}$:

$$\left(\frac{\Delta p}{\rho_L \cdot g \cdot H}\right)_{Fl} \approx 0.15 \quad \text{see Fig. 2-5b.}$$

(d) For classic, randomly filled 2″ packing elements with $a = 95-120\ m^2m^{-3}$ as well as for structured packings with $a = 100-300\ m^2m^{-3}$:

$$\left(\frac{\Delta p}{\rho_L \cdot g \cdot H}\right)_{Fl} \approx 0.10 \quad \text{see Fig. 2-5c,d.}$$

(e) For 2″-3″ lattice packing elements with $a = 50-110\ m^2\ m^{-3}$:

$$\left(\frac{\Delta p}{\rho_L \cdot g \cdot H}\right)_{Fl} \approx 0.05 \quad \text{see Fig. 2-5d.}$$

6.5 Pressure Drop Calculation Acc. to the ψ_{VL} Model Presented in Chapter 5

Chapter 5 presented a model, derived from Eq. (5-2), for determining the pressure drop of any types of irrigated packings, based on the resistance coefficient for two-phase flow ψ_{VL}:

$$\frac{\Delta p}{H} = \psi_{VL} \cdot \frac{1-\varepsilon}{\varepsilon_3} \cdot \frac{F_V^2}{dp \cdot K} \cdot \left(1 + \frac{h_L}{1-\varepsilon}\right) \cdot \left(1 + \frac{h_L}{\varepsilon}\right)^{-3} \quad \left[\mathrm{Pam}^{-1}\right]$$

The equation is valid throughout the entire operating range up to the flooding point. The resistance coefficient for two phase flow ψ_{VL}, acc. to Eqs. (5-6), (5-7) and (5-8), is a function of the Reynolds number of the liquid and the form factor μ of packings, $\psi_{VL} = f(\mu, Re_L)$.

The form factors μ for the individual packings can be found in Tables 6-1a–c. They provide a direct comparison between the predicted pressure drops for the individual types of packings without requiring any calculations or experiments.

The application of the model, based on Eq. (5-9) and/or (5-10), is recommended in particular for operating ranges above the loading line and at the flooding point. In these cases, the correlations presented in Chap. 5 should not be used, as they are not within the validity range.

For the purpose of comparing the results of the pressure drop calculation, acc. to the models presented above, with those shown in Chap. 4, Eq. (4-48), as well as Eq. (5-2), Tables 5-1a-c contain a list of the relative errors for the individual flow ranges and types of packings. They indicate which of the correlations for determining the pressure drop are preferable.

6.6 Notes on Tables Containing Technical Data of Random and Structured Packings as Well as Model Parameters $\psi_{Fl}/\psi_{Fl,m}$ for Determining Flooding Point and Pressure Drop

The numerical examples at the end of each chapter are given as a quantitative illustration of the derived correlations. They correspond to the calculations used as a basis for the hydraulic design of plants.

Tables 6-1a–c contain the numerical values of the model parameters as well as the technical data for approx. 200 various types of randomly filled and stacked packing elements and structured packings. These can be used as a basis for the fluid dynamic design of packed columns according to the model presented in this book.

The individual Tables 6-1a–c also show the technical data of the individual packing elements a_0, ε_0, N_0, G_0, according to the manufacturer's specifications, the numerical values of the parameters K_1, K_2 and K_3, K_4 for determining the resistance coefficient ψ as well as arithmetically determined ψ_m values for the turbulent flow range for the Reynolds numbers $Re_V \in (2100–10000)$. The form factor data $\varphi_{P,exp}$ for all tested packings is listed in Table 6-1a–c. With regards to the structured packings, the parameters K_1 to K_4, which are dependent on the column diameter, are specified in order to determine the resistance coefficient. If the influence of the column diameter d_S on $\Delta p/H$ is not taken into account, the differences may be up to 65% when determining the pressure drop and up to 10% for the flooding point, as explained in Chap. 3. There are separate numerical values listed as reference values for K_1 to K_4, relating to structured packings with larger column diameters from $d_S \geq 1$ m.

The most important literature references used for developing the model of the fluid dynamic design of packed columns are listed in the following. They refer to individual random and structured packings, see – when not other mentioned – Chap. 4:

- metal Pall rings, [5–8, 13–15, 26, 40, 55, 56, 66] and new data [A]
- plastic Pall rings, [13–15, 23, 24, 55, 56, Chap. 2: 96] and new data [A]
- Pall rings made of ceramic etc. [28, 34] and new data [A]
- Białecki rings made of metal and plastic etc. [17, 18, 36, 41, 60, 65, 66, 81-Chap. 2] and new data [A]
- stacked Białecki rings made of metal etc. [18, 65] and new data [A]
- metal Hiflow rings etc. [37, 55] and new data [A]
- Hiflow rings made of plastic etc. [35, 55] and new data [A]
- Hiflow rings made of ceramic etc. [38] and new data [A]
- NSW (Nor-Pac) rings [23, 24, 50, 51] and new data [A]
- ceramic Intalox saddles, new data [A]
- saddle packings, new data [A]
- Raschig rings made of ceramic etc. [7, 17, 56, 61, 62] and new data [A]
- VSP rings [40, 57] and Top Pak, as well as VFF Pall rings, metal [A]
- Glitsch rings CMR [A]
- Plastic ENVIPAC and DTNPAC [49, 54, 59, A]
- VSP ring and Intalox saddle, size 3, plastic [A]
- Mellapak Sulzer packings types 250Y, 250X, 500Y made of sheet metal [56, 57], sheet-metal Montz packing B1-100 to B2-500Y,X and PP, Ralu-Pak and other structured packings [30, 31, 39] as well as new data [A]
- metal Raschig rings [5]
- Mc-Pac, R-Pac, SR-Pac [Chap. 2: 79, 80, 82, 83; Chap. 3: 34, 35]
- Ralu-Super rings [Chap. 2: 86]
- metal Nutter rings [Chap. 2: 95]
- Raschig Super rings [Chap. 2: 86, 94]
- RVT metal saddle rings [Chap. 2: 93]

6.7 Validity Range of Correlations

The experimental validation of the correlations developed to calculate the flooding point $u_{V,Fl}$ and the pressure drop of dry $\Delta p_0/H$ and irrigated packings $\Delta p/H$ was achieved by means of a systematic analysis of the random and structured packings, shown in Figs. 1-1a and 1-1b.

More information on the parameters used for the experiments can be found at the end of Chaps. 2, 3, 4 and 5. Below is a summary of the most important parameters for the variables required for the design process, $u_{V,Fl}$, $\Delta p_0/H$, $\Delta p/H$, h_L, $h^0_{L,Fl}$:

40	$\leq$	Re_V	$\leq$	35000
0.15	$\leq$	Re_L	$\leq$	200
5	$\leq$	d_S/d	$\leq$	56
0.0133	$\leq$	p_T	$\leq$	100 bar
0.025	$\leq$	d_S	$\leq$	3.0 m
0.4	$\leq$	H	$\leq$	7 m
390	$\leq$	ρ_L	$\leq$	1800 kgm^{-3}
0.03	$\leq$	ρ_V	$\leq$	130 kgm^{-3}
14	$\leq$	σ_L	$\leq$	80 mNm^{-1}
0.3	$\leq$	$\eta_L \cdot 10^3$	$\leq$	91 $kgm^{-1}s^{-1}$
6	$\leq$	$\eta_V \cdot 10^6$	$\leq$	18.2 $kgm^{-1}s^{-1}$
0	$\leq$	$u_L \cdot 10^3$	$\leq$	70 ms^{-1}
0	$<$	$\Delta p/H$	$\leq$	$4 \cdot 10^4$ Pam^{-1}

Number of experimental data items:

- Flooding points: more than 1200
- Hold-up data: more than 1100
- Pressure drop data: more than 10.500
- Number of packings: more than 200
- Number of systems: 32

6.8 FDPAK Programme for Fluid Dynamic Design of Columns with Modern Random and Structured Packings

6.8.1 Programme Information

In 1990, the first DOS programme was developed to supplement this work on the fluid dynamics of columns containing random and structured packings, which has since been improved and is now available as a WINDOWS version with graphic representation of the calculation results.

This FDPAK programme provides chemical engineers with a fast calculation the hydraulic parameters of any types of column internals (more than 200) and allows the user to assess the individual types of packings in terms of their maximum capacity, pressure drop and liquid hold-up. The experience of the past 20 years has shown that the programme is also suitable for teaching purposes at colleges and universities.

Thanks to the programme, it is possible to determine the gas velocity at the flooding point, pressure drop, percentage of distance from flooding point as well as liquid hold-up at operating and flooding point, based on a given column diameter (upgrading of existing plants) as well as physical and operational variables.

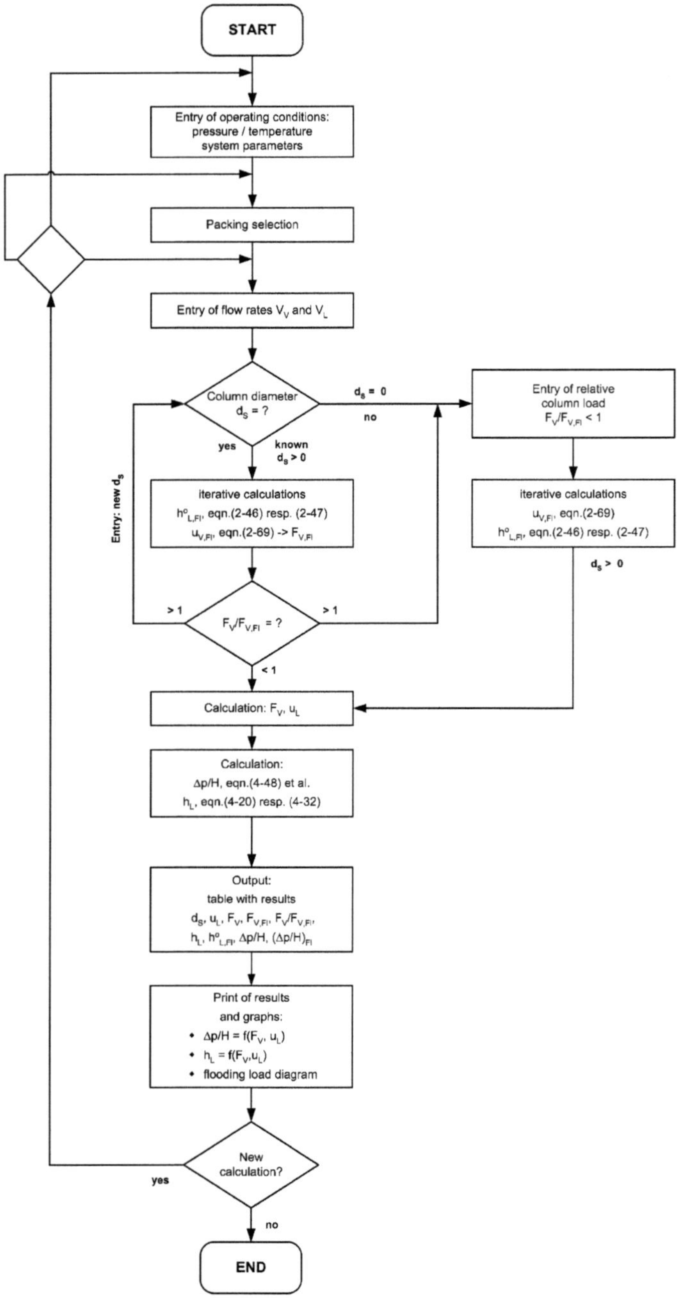

Figure 6-7. Block diagram of program FDPAK for hydraulic dimensioning of packed columns

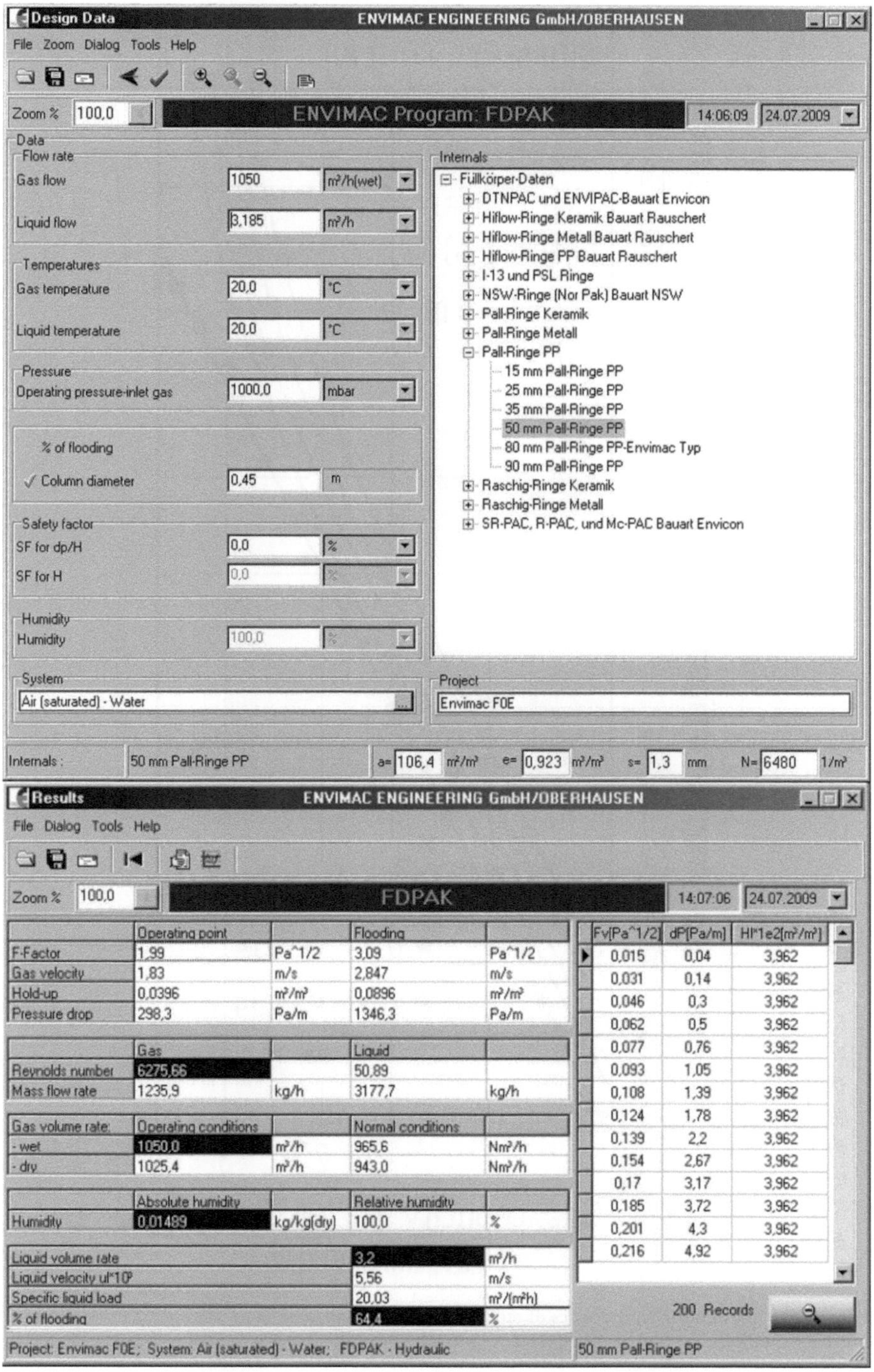

Figure 6-8. FDPAK for Windows – individual work templates

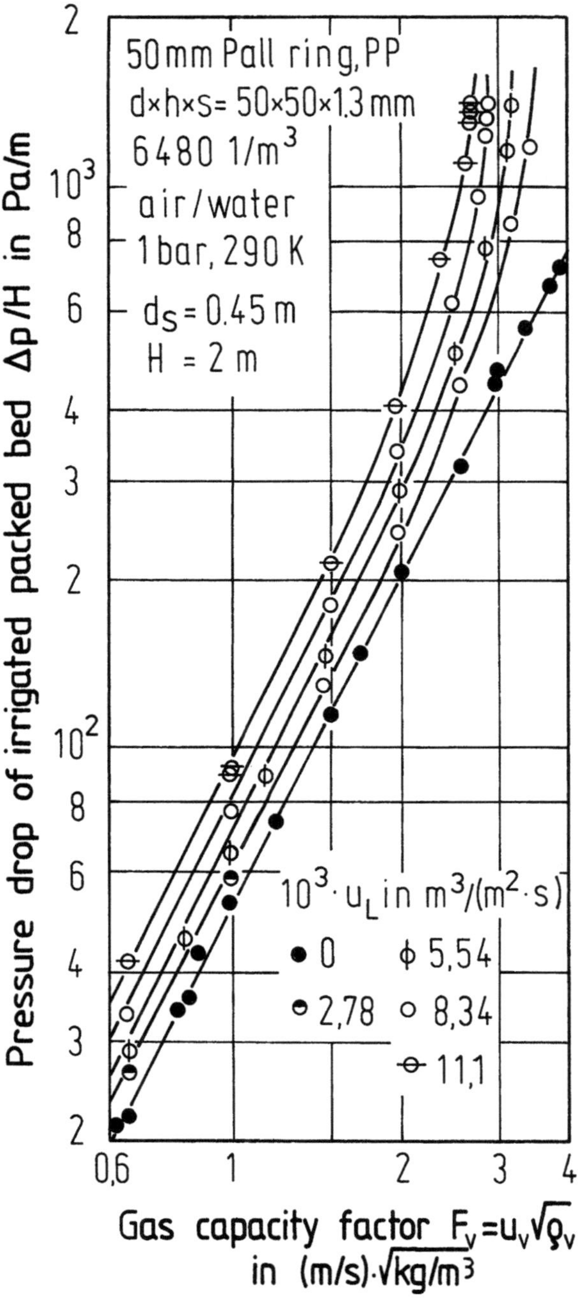

Figure 6-9. Hydraulic characteristics of random 50 mm Pall rings, valid for the system air/water under normal conditions. Pressure drop $\Delta p/H$ as a function of the gas capacity factor F_V, u_L with the specific liquid load u_L as a parameter

24.07.2009 ENVIMAC

ENVIMAC Engineering GmbH; Im Erlengrund 27; 46149 Oberhausen; Tel: +49-208/941 044-0; Fax: +49 208/941 044-100

FDPAK

System: Air (saturated) - Water - FDPAK - Hydraulic

Design Data

Gas flow	1050 m³/h(wet)
Liquid flow	3,185 m³/h
Operating pressure-inlet gas	1000,0 mbar
Gas temperature	20,0 °C
Liquid temperature	20,0 °C
Properties of the system	
Liquid density	997,7 kg/m³
Gas density	1,177 kg/m³
Liquid viscosity	1,025 mPas
Gas viscosity	0,0179 mPas
Surface tension	0,0728 N/m
Internals:	50 mm Pall-Ringe PP
	a=106,4m²/m³; e=0,923m³/m³; s=1,3mm; N=6480 1/m³
Column diameter	0,45 m

Result

Hydraulics:	Operating point	Flooding
F-Factor	1,99 Pa^1/2	3,09 Pa^1/2
Gas velocity	1,83 m/s	2,847 m/s
Hold-up	0,0396 m³/m³	0,0896 m³/m³
Pressure drop	298,3 Pa/m	1346,3 Pa/m
	Gas	Liquid
Reynolds number	6275,66	50,89
Mass flow rate	1235,9 kg/h	3177,7 kg/h
Gas volume rate:	Operating conditions	Normal conditions
- wet	1050,0 m³/h	965,6 Nm³/h
- dry	1025,4 m³/h	943,0 Nm³/h
	Absolute humidity	Relative humidity
Humidity	0,01489 kg/kg(dry)	100,0 %
Liquid volume rate	3,2 m³/h	
Liquid velocity ul*10³	5,56 m/s	
Specific liquid load	20,03 m³/(m²h)	
% of flooding	64,4	
Column diameter	0,45 m	

Project: Envimac F0E

Figure 6-10. Printout of dimensioning results for 50 mm Pall rings column, PP, using FDPAK for Windows – results report

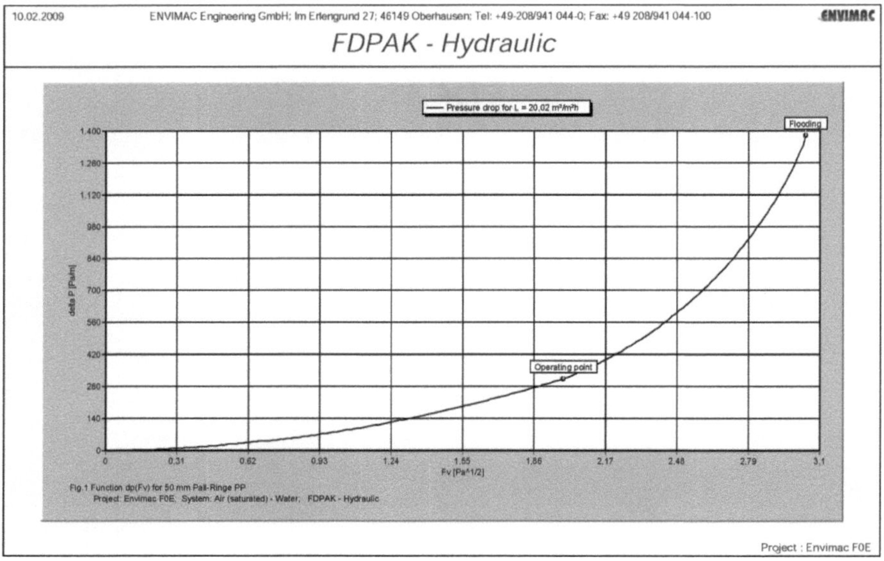

Figure 6-11. Printout of dimensioning results for 50 mm Pall rings column, PP, using FDPAK for Windows – pressure drop diagram $\Delta p/H = f(F_V, u_L = \text{const.})$

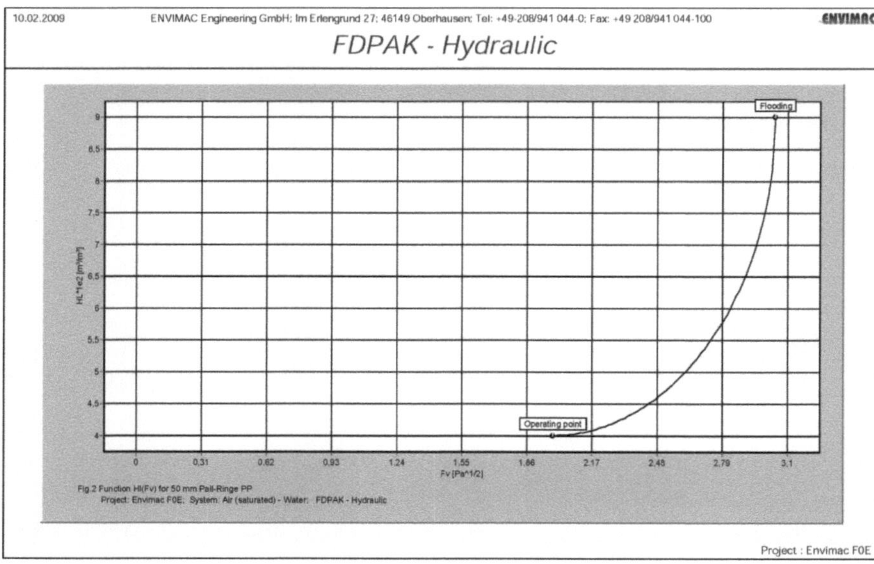

Figure 6-12. Printout of dimensioning results for 50 mm Pall rings column, PP, using FDPAK for Windows – hold-up diagram $h_L = f(F_V, u_L = \text{const.})$

The column diameter d_S is calculated iteratively, based on a given relative column load $F_V/F_{V,Fl}$ (as a percentage of the gas velocity at the flooding point).

The flow chart in Fig. 6-7 shows the procedure of the fluid dynamic design of packed columns.

Once the programme has been started, a type of packing is selected from the list, for which the volumetric flow rates of the gas $\dot{V}_V$ and the liquid $\dot{V}_L$ are input. The programme then calculates the gas velocity at the flooding point $u_{V,Fl}$ and/or the gas capacity factor at the flooding point $F_{V,Fl}$, acc. to the SBD model, Eq. (2-69), the pressure drop $\Delta\rho/H$, acc. to Eq. (4-48) etc., for the selected operating range as well as the liquid hold-up below the flooding point h_L and at the flooding point $h_{L,Fl}$. The programme can be used for the design of a known packed column with a given column diameter d_S. This is useful, e.g., for the upgrading for columns. By selecting the relative percentage of the column load $F_V/F_{V,Fl} < 100$, the column diameter d_S is determined iteratively. This function is used for new designs.

Finally, it is possible to select a new operating point or a new type of packing in order to perform a new calculation. The units of the input values can be selected prior to the calculations.

The calculation results are then displayed on the screen in the form of table, see Fig. 6-8, and diagrams, which can be printed out, see Figs. 6-10, 6-11 and 6-12. Thanks to its simple structure, the programme does not require a special user manual.

Table 6-2 shows the example of results of fluid dynamic calculations for column with 50 mm Pall rings made of plastic and comparison of the results with the experimental data shown in Fig. 6-9. The printout of the design, produced by the FDPAK programme, is shown in Figs. 6-10, 6-11 and 6-12.

Table 6-2. Comparison of experimentally determined data and calculation using the FDPAK programme valid for 50 mm Pall rings, PP and system: air/water, 1 bar, 293 K, $d_S = 0.45$ m, $V_V = 1050$ $m^3\,h^{-1}$, $V_L = 3.185$ $m^3\,h^{-1}$

Parameter	Unit	Calculation	Experiment	δ (i) [%]
F_V	$\sqrt{Pa}$	1.99	1.99	–
$F_{V,Fl}$	$\sqrt{Pa}$	3.09	3.13	−1.28
$h^0_{L,Fl}\cdot 10^2$	m^3m^{-3}	8.96	9.67	−7.34
$h_L\cdot 10^2$	m^3m^{-3}	3.96	3.60	+10.28
$\Delta\rho/H$	Pam^{-1}	298	290	+2.76
$(\Delta\rho/H)_{Fl}$	Pam^{-1}	1346	1350	−0.30
$u_L\cdot 10^3$	ms^{-1}	5.56	5.56	–

6.8.2 Conclusions

Thanks to the FDPAK programme, it is easy to follow the steps for calculating the hydraulic parameters $F_{V,Fl}$, $u_{V,Fl}$, $\Delta p/H$, h_L, $h^0_{L,Fl}$, based on the numerical examples given in this work. In addition, the FDPAK programme is an excellent tool for the upgrading of existing plants or the design of new plants. It is available in German, English and Polish.

Annex Chapter 6

Tables Chapter 6

Table 6-1a. List of required technical data of investigated packings for dimensioning of packing columns. Constants K_i for determination of resistance coefficient ψ_{Fl} acc. to Eq. (3-14) or form factors φ_P for determination of resistance coefficient of dry packing ψ acc. to Eq. (3-26), gas velocity at flooding point $u_{V,Fl}$ acc. to Eq. (2-69) and pressure drop $\Delta p/H$ acc. to Eqs. (4-48) and (4-54). List of form factors μ for determination of pressure drop $\Delta p/H$ acc. to ψ_{VL} – model; Eq. (5-9, 5-10)

Packing	Material	Size			$N \cdot 10^{-3}$ $[1\ m^{-3}]$	$a\ [m^2 m^{-3}]$	$\varepsilon\ [m^3 m^{-3}]$	$G\ [kg/m^3]$	ψ_{Fl} (a) $Re_V < 2100$		ψ_{Fl} (b) $Re_V \geq 2100$		$\psi_{Fl,m}$ (c)	$C_{Fl,0}$ [–]	μ-factor [–]	φ_P-factor [–]
		$d \cdot 10^3$ [m]	$h \cdot 10^3$ [m]	$s \cdot 10^3$ [m]					K_1	K_2	K_3	K_4				
Pall rings	metal	15	15	0.4	224.0	360.0	0.936	512	10	−0.18	3.23	−0.0343	2.42	0.566	0.500	0.280
by Raschig,	(V2A,V4A)	25	25	0.5	51.5	215.0	0.942	464	10	−0.18	3.23	−0.0343	2.42	0.566	0.500	0.280
VFF, rvt		25	25	0.4	51	215.0	0.956	352	10	−0.18	3.23	−0.0343	2.42	0.566	0.500	0.280
		35	35	0.6	19.0	145.0	0.948	416	10	−0.18	3.23	−0.0343	2.42	0.566	0.500	0.280
		38	38	0.6	16.0	151.8	0.953	376	10	−0.18	3.23	−0.0343	2.42	0.566	0.500	0.280
		50	50	0.8	6.1	110.0	0.952	384	10	−0.18	3.23	−0.0343	2.42	0.566	0.500	0.280
		80	80	1.2	1.6	78.0	0.960	320	10	−0.18	3.23	−0.0343	2.42	0.566	0.500	0.280
		58	58	0.6	6.0	105.0	0.970	240	8	−0.18	2.85	−0.0453	1.95	0.566	0.500	0.417
Pall rings	metal	30 oK	30	0.5	32.5	168.9	0.957	344	10	−0.17	4.13	−0.0522	2.62	0.566	0.260	0.244
by Glitsch	30 P	30 mK	30	0.3	28.5	180.0	0.975	200	8	−0.18	2.85	−0.0453	1.95	0.566	0.270	0.390
Pall rings	plastic	15	15	1.0	215.00	350	0.880	108	10	−0.18	3.23	−0.0343	2.42	0.566	0.572	0.309
by Raschig,	(PP)	25	25	1.3	51.00	220	0.890	99	10	−0.18	3.23	−0.0343	2.42	0.566	0.572	0.309
VFF, rvt		35	35	1.3	18.00	160	0.905	85.5	10	−0.18	3.23	−0.0343	2.42	0.566	0.572	0.309
		50	49	1.3	6.70	110	0.920	72	10	−0.18	3.23	−0.0343	2.42	0.566	0.572	0.309
Pall rings	plastic	37	37	1.3	15.50	148.0	0.912	81.3	10	−0.18	3.23	−0.0343	2.42	0.566	0.572	0.309
by ENVIMAC	(PP)	50	50	1.5	6.70	110.0	0.922	72.0	10	−0.18	3.23	−0.0343	2.42	0.566	0.572	0.309
		80	80	1.6	1.50	66.0	0.957	45.0	10	−0.18	3.23	−0.0343	2.42	0.566	0.572	0.309
Pall rings	plastic	90	90	2.0	1.24	90	0.94	62.5	5.585	−0.021	1.76	−0.0510	1.09	0.566	0.509	0.664
by Raschig	(PP)															

Table 6-1a. (continued)

Packing	Material	Size $d \cdot 10^3$ [m]	$h \cdot 10^3$ [m]	$s \cdot 10^3$ [m]	$N \cdot 10^{-3}$ [$1\ m^{-3}$]	a [$m^2 m^{-3}$]	ε [$m^3 m^{-3}$]	G [kg/m^3]	ψ_{Fl} (a) $Re_V < 2100$ K_1	K_2	ψ_{Fl} (b) $Re_V \geq 2100$ K_3	K_4	$\psi_{Fl,m}$ (c)	$C_{Fl,0}$ [–]	μ-factor [–]	φ_P-factor [–]
Raflux type by rvt	plastic (PP)	90	90	0.8	1.24	86	0.96	38	6.576	−0.21	2.07	−0.0510	1.23	0.566	0.397	0.604
Pall rings by Raschig	ceramic	25	25	3.0	39.9	220	0.73	670	10.00	−0.18	3.23	−0.0343	2.42	0.566	0.572	0.280
		50	50	5.0	6.4	120	0.77	623	12.68	−0.23	3.81	−0.0812	1.93	0.566	0.572	0.430
Białecki rings	metal (prod. 1974)	12	12	0.3	420	388.2	0.932	544	11.50	−0.18	3.77	−0.0343	2.82	0.566	1.050	0.158
		27	25	0.5	52.0	225.0	0.945	440	10.17	−0.17	4.13	−0.0522	2.66	0.566	0.485	0.208
		38	35	0.6	19.0	155.0	0.950	400	10.17	−0.17	4.13	−0.0522	2.66	0.566	0.485	0.208
		53.5	50	0.8	6.5	110.0	0.965	280	10.17	−0.17	4.13	−0.0522	2.66	0.566	0.485	0.208
	metal (prod. 1986)	50	50	0.6	6.7	112.4	0.967	264	11.50	−0.18	3.77	−0.0343	2.82	0.566	0.515	0.158
		36	35	0.6	19.0	164.0	0.956	352	11.50	−0.18	3.77	−0.0343	2.82	0.566	0.515	0.158
		25	25	0.6	52.0	223.0	0.954	368	11.50	−0.18	3.77	−0.0343	2.82	0.566	0.515	0.158
	ceramic	100	100	10	–	69.5	0.766	514.8	2.381	−0.04	2.38	−0.038	1.73	0.566	–	0.493
Białecki rings	plastic (PP)	49	49	1.6	6.4	115	0.93	63	10.17	−0.17	4.13	−0.0522	2.66	0.566	0.485	0.028
PSL rings	metal (Al)	50	50	0.4	6.5	115.2	0.9754	66.42	10	−0.18	3.23	−0.0343	2.42	0.566	0.336	0.320
	(V2A, V4A)	50	50	0.4	6.5	115.2	0.9754	200	10	−0.18	3.23	−0.0343	2.42	0.566	0.366	0.320
Glitsch rings CMR-304	metal (0.5A)	16	5	–	560.8	356.8	0.955	360	6.53	−0.148	4.330	−0.0920	2.00	0.566	0.383	0.415
	(1.0A)	25	8	–	160.0	234.7	0.971	232	5.40	−0.140	3.241	−0.0733	1.75	0.566	0.247	0.475
	(1.5A turbo)	34	11	0.4	60.8	176.5	0.974	208	5.40	−0.140	3.241	−0.0733	1.75	0.566	0.247	0.475
	(1.5A)	34	11	0.4	60.8	170.0	0.976	192	5.40	−0.140	3.241	−0.0733	1.75	0.566	0.247	0.475
	(2.0A)	43	14	0.4	33.0	150.7	0.989	88	8.12	−0.180	2.903	−0.0445	1.98	0.566	0.140	0.437
	(2.5A)	51	17	0.4	10.0	123.0	0.989	96	8.12	−0.180	2.903	−0.0445	1.98	0.566	0.175	0.437
	(3.0A)	66	22	0.5	–	101.2	0.988	96	8.11	−0.180	2.903	−0.0445	1.98	0.566	0.175	0.437

Table 6-1a. (continued)

Packing	Material	Size d·10^3 [m]	h·10^3 [m]	s·10^3 [m]	N·10^{-3} [1 m^{-3}]	a [m^2m^{-3}]	ε [m^3m^{-3}]	G [kg/m^3]	ψ_{Fl} (a) $Re_V < 2100$ K_1	K_2	ψ_{Fl} (b) $Re_V \geq 2100$ K_3	K_4	$\psi_{Fl,m}$ (c)	$C_{Fl,0}$ [–]	μ-factor [–]	φ_P-factor [–]
Glitsch	plastic (size 1A)	44	16.5	1.0	25.5	196.2	0.935	58.5	5.0	−0.14	3.00	−0.0733	1.58	0.566	0.350	0.496
rings	(PP) (size 2A)	69	24.6	0.9	6.3	130.2	0.941	53.1	10.0	−0.18	3.23	−0.0343	2.42	0.566	0.500	0.345
CMR-304																
Top-Pak	metal (V4A)	45	38	0.5	6.85	105	0.975	118	8.0	−0.18	2.85	−0.045	1.95	0.566	0.370	0.474
type VFF	size 1 (Al)	45	38	1.1	6.85	105	0.956	190	8.0	−0.18	2.85	−0.045	1.95	0.566	0.370	0.424
	size 2 (V4A)	80	63	0.7	2.80	75	0.980	160	9.1	−0.18	3.37	−0.045	2.21	0.566	0.270	0.424
VSP rings	metal (size 1)	32	25	0.3	33.5	200	0.972	224	9.1	−0.18	3.37	−0.050	2.20	0.566	0.270	0.38
type VFF	(size 2)	50	40	0.4	7.8	104	0.980	160	9.1	−0.18	3.37	−0.050	2.20	0.566	0.270	0.38
Interpak	metal	10	10	0.3	2000.0	620	0.920	656	12.78	−0.18	5.90	−0.079	3.00	0.566	–	0.107
type VFF		15	15	0.3	565.0	350	0.954	362	12.78	−0.18	5.90	−0.079	3.00	0.566	–	0.107
		20	20	0.4	202.5	225	0.962	301	12.78	−0.18	5.90	−0.079	3.00	0.566	–	0.107
		30	30	0.5	–	160	0.965	275	12.78	−0.18	5.90	−0.079	3.00	0.566	–	0.107
Raschig	ceramic	8	8	1.5	500.0	550	0.650	770	34.6	−0.307	6.05	−0.069	3.05	0.566	1.000	0.000
rings		15	15	2.7	200.0	292	0.667	732.6	34.6	−0.307	6.05	−0.069	3.05	0.566	1.000	0.000
		25	25	4.0	46.0	177	0.693	675.4	34.6	−0.307	6.05	−0.069	3.05	0.566	1.000	0.000
		35	35	4.0	18.5	140	0.710	638	34.6	−0.307	6.05	−0.069	3.05	0.566	1.000	0.000
		50	50	5.0	6.4	98	0.730	594	34.6	−0.307	6.05	−0.069	3.05	0.566	1.000	0.000
		53	50	6.0	6.0	100.5	0.772	636	34.6	−0.307	6.05	−0.069	3.05	0.566	1.000	0.000
		100	100	10.0	0.844	52.5	0.761	525.8	–	–	3.14	−0.0343	2.30	0.566	1.000	0.000

Table 6-1a. (continued)

Packing	Material	Size $d \cdot 10^3$ [m]	$h \cdot 10^3$ [m]	$s \cdot 10^3$ [m]	$N \cdot 10^{-3}$ [$1\ m^{-3}$]	a [$m^2 m^{-3}$]	ε [$m^3 m^{-3}$]	G [kg/m^3]	ψ_{Fl} (a) $Re_V < 2100$ K_1	K_2	ψ_{Fl} (b) $Re_V \geq 2100$ K_3	K_4	$\psi_{Fl,m}$ (c)	$C_{Fl,0}$ [–]	μ-factor [–]	φ_P-factor [–]
Raschig rings	glass	8	8	0.7	500	500	0.726	685	17.508	−0.209	5.264	−0.0525	3.38	0.566	–	0.000
		10	10	1.0	300	450	0.740	650	17.508	−0.209	5.264	−0.0525	3.38	0.566	–	0.000
		15	15	1.5	225	375	0.771	572.5	17.508	−0.209	5.264	−0.0525	3.38	0.566	–	0.000
		20	20	2.0	97.9	243.6	0.793	517.5	17.508	−0.209	5.264	−0.0525	3.38	0.566	–	0.000
		25	25	2.0	54.1	203	0.829	427.5	17.508	−0.209	5.264	−0.0525	3.38	0.566	–	0.000
		40	40	4.0	20	160	0.870	325	17.508	−0.209	5.264	−0.0525	3.38	0.566	–	0.000
		50	50	4.0	6.5	120	0.856	360	17.508	−0.209	5.264	−0.0525	3.38	0.566	–	0.000
Raschig rings	metal (V2A,V4A)	15	15	0.5	240.0	350	0.92	640	36.245	−0.19	12.0	−0.0455	8.18	0.566	2.000	0.000
		25	25	0.8	51.5	220	0.92	640	36.245	−0.19	12.0	−0.0455	8.18	0.566	2.000	0.000
		35	35	1.0	19.0	150	0.93	560	36.245	−0.19	12.0	−0.0455	8.18	0.566	2.000	0.000
		50	50	1.0	6.5	110	0.95	400	36.245	−0.19	12.0	−0.0455	8.18	0.566	2.000	0.000
Nutter rings by Sulzer	metal (no. 0.7)	–	–	0.3	167400	226	0.978	176	11.196	−0.206	3.518	−0.051	2.31	0.693	0.867	0.326
	(no. 1.0)	–	–	0.3	67100	168	0.978	178	11.196	−0.206	3.518	−0.051	2.31	0.693	0.867	0.326
	(no. 1.5)	–	–	0.3	26800	124	0.978	181	11.196	−0.206	3.518	−0.051	2.31	0.693	0.867	0.326
	(no. 2.0)	–	–	0.3	13600	96	0.979	173	11.196	−0.206	3.518	−0.051	2.31	0.693	0.867	0.326
	(no. 2.5)	–	–	0.3	8800	83	0.982	145	11.196	−0.206	3.518	−0.051	2.31	0.693	0.867	0.326
	(no. 3.0)	–	–	0.3	4200	66	0.984	133	11.196	−0.206	3.518	−0.051	2.31	0.693	0.867	0.326
Intalox saddles by VFF	ceramic (1/2″)	12.5	–	–	387.0	327	0.648	774.4	6.664	−0.188	3.969	−0.120	1.465	0.566	0.705	0.560
	(3/4″)	20	13	0.8	136.9	252	0.698	664.4	6.664	−0.188	3.969	−0.120	1.465	0.566	0.705	0.560
	(1.0″)	25	22	3.8	68.0	197	0.704	651.2	15.5	−0.255	3.800	−0.071	2.10	0.566	0.705	0.340
	(1.5″)	38	37	6.1	20.0	133	0.743	565.4	15.5	−0.255	3.800	−0.071	2.10	0.566	0.705	0.340
	(2.0″)	50	42	6.1	9.0	99	0.770	506	15.5	−0.255	3.800	−0.071	2.10	0.566	0.705	0.340
	plastic (1.0″)	25	–	–	90.0	255	0.89	99	–	–	8.11	−0.108	3.20	0.566	0.720	0.048
	(1.5″)	38	–	–	26.0	170	0.91	81	–	–	8.11	−0.108	3.20	0.566	0.720	0.114
	(2.0″)	50	42	2.5	8.5	120	0.91	81	10	−0.18	3.23	−0.034	2.45	0.566	0.652	0.280

Table 6-1a. (continued)

Packing	Material	Size $d \cdot 10^3$ [m]	$h \cdot 10^3$ [m]	$s \cdot 10^3$ [m]	$N \cdot 10^{-3}$ [$1\,m^{-3}$]	a [$m^2 m^{-3}$]	ε [$m^3 m^{-3}$]	G [kg/m^3]	ψ_{Fl} (a) $Re_V < 2100$ K_1	K_2	ψ_{Fl} (b) $Re_V \geq 2100$ K_3	K_4	$\psi_{Fl,m}$ (c)	$C_{Fl,0}$ [–]	μ-factor [–]	φ_P-factor [–]
Intalox Super saddles	plastic	50	–	–	5.80	110	0.94	54	3.210	−0.075	3.210	−0.075	1.7	0.566	0.375	0.485
		90	–	–	1.25	90	0.96	36	2.315	−0.075	2.305	−0.075	1.2	0.566	0.255	0.640
RMSR (Rauschert metal saddle rings) by rvt	metal	25	–	0.4	–	235	0.96	305	7.397	−0.206	2.325	−0.051	1.44	0.566	0.482	0.555
		40	–	0.4	–	170	0.97	241	10.328	−0.206	3.246	−0.051	2.05	0.566	0.422	0.378
		50	–	0.5	–	115	0.97	198	7.397	−0.206	2.325	−0.051	1.44	0.566	0.429	0.555
		60	–	0.5	–	90	0.98	159	7.397	−0.206	2.325	−0.051	1.44	0.566	0.311	0.555
		70	–	0.5	–	67	0.98	116	5.853	−0.206	1.893	−0.051	1.09	0.566	0.504	0.647
		70	–	0.6	–	67	0.98	138	5.853	−0.206	1.839	−0.051	1.09	0.566	0.504	0.647
IMTP #40	metal	40	27	0.6	52.4	128.7	0.984	128	8.75	−0.18	3.17	−0.050	2.00	0.566	0.273	0.395
I-13 rings TU Wrocław/PL	metal (V2A)	25	25	0.5	51.5	218.5	0.948	416	10.17	−0.17	4.13	−0.0522	2.60	0.566	0.562	0.165
Berl saddles	ceramic	10	10	–	620	386	0.595	891	6.664	−0.188	3.969	−0.12	1.465	0.566	–	0.570
		15	15	–	228	303	0.595	891	7.08	−0.188	3.442	−0.0937	1.566	0.566	–	0.541

(a) $\psi_{Fl} = K_1 \cdot Re_V^{K_2}$ *for* $Re_V < 2100$; (b) $\psi_{Fl} = K_3 \cdot Re_V^{K_4}$ *for* $Re_V \geq 2100$; (c) $Re_V \in \{2100 - 10000\}$
oK = with collar
mK = without collar
prod. = production year

V4A ≡ stainless steel 1.4571 (SS 316Ti)
V2A ≡ stainless steel 1.4301 (SS 304)

Koch-Glitsch Italia S.r.l – Bergamo (Italy)
Sulzer Chemtech AG – Winterthur (Switzerland)
Raschig GmbH – Ludwigshafen (Germany)
rvt ≡ RVT Process Equipment GmbH – Steinwiesen (Germany)
VFF ≡ Vereinigte Füllkörper-Fabriken GmbH & Co. KG – Ransbach-Baumbach (Germany)

Table 6-1b. List of technical data of packings with highly perforated wall for dimensioning of packing columns. Constants K_i for determination of resistance coefficient ψ_{Fl} acc. to Eq. (3-14) or form factors φ_P for determination of resistance coefficient of dry packing ψ acc. to Eq. (3-26), gas velocity at flooding point $u_{V,Fl}$ acc. to Eq. (2-69) and pressure drop $\Delta p/H$ acc. to Eqs. (4-48) and (4-54). List of form factors μ for determination of pressure drop $\Delta p/H$ acc. to ψ_{VL} – model; Eq. (5-9, 5-10)

Packing	Material	Size $d \cdot 10^3$ [m]	$h \cdot 10^3$ [m]	$s \cdot 10^3$ [m]	$N \cdot 10^{-3}$ [1 m^{-3}]	a [m^{2}m^{-3}]	ε [m^{3}m^{-3}]	G [kg/m^3]	ψ_{Fl} (a) $Re_V < 2100$ K_1	K_2	ψ_{Fl} (b) $Re_V \geq 2100$ K_3	K_4	$\psi_{Fl,m}$ (c)	$C_{Fl,0}$ [–]	μ-factor [–]	φ_P-factor [–]
Hiflow rings by rvt	metal	28	25	0.4	37.2	185.0	0.965	280	4.28	−0.118	2.875	−0.066	1.65	0.566	0.268	0.509
		58	50	0.5	5.1	97.3	0.976	192	4.28	−0.118	2.875	−0.066	1.65	0.566	0.175	0.63
	plastic (PP, PVDF)	17	17	1.0	184.0	275	0.910	81	4.178	−0.145	2.610	−0.0846	1.250	0.566	0.403	0.54
		28	28	1.3	45.50	190	0.920	72	4.915	−0.145	2.25	−0.0418	1.600	0.566	0.330	0.54
		38	38	1.5	–	150	0.94	58	6.048	−0.206	1.901	−0.0510	1.210	0.566	0.380	0.618
		50	50	1.5	6.40	110	0.930	59	3.065	−0.145	1.392	−0.0418	0.980	0.566	0.190	0.700
		90	90	1.6	1.25	65	0.965	31.5	3.300	−0.145	1.500	−0.0418	1.055	0.566	0.200	0.694
		90	65	1.6	1.80	78	0.956	39.6	1.175	−0.034	1.175	−0.0434	0.857	0.566	0.190	0.736
	ceramic prod. (1982, 1984)	20	20	2.5	118.7	280.0	0.763	574	4.215	−0.145	1.600	−0.065	1.600	0.566	0.408	0.587
		38	35	3.0	17.1	110.0	0.831	372	4.215	−0.145	3.047	−0.065	1.725	0.566	0.500	0.54
		50	55	4.0	5.0	86.7	0.815	407	5.619	−0.145	3.047	−0.065	1.725	0.566	0.500	0.54
		75	65	4.0	2.0	56.8	0.860	308	2.400	−0.065	2.400	−0.065	1.310	0.566	0.369	0.586
	ceramic prod. (1988)	20	22	3.2	90.0	234.5	0.821	393.8	5.166	−0.144	2.717	−0.06	1.64	0.566	0.500	0.492
		38	37	5.0	13.5	121.0	0.784	475.2	5.166	−0.144	2.717	−0.06	1.64	0.566	0.500	0.55
		75	70	10	2.0	70.0	0.760	528	1.475	−0.034	1.475	−0.0343	1.07	0.566	0.315	0.672
Hiflow super rings by rvt	plastic (PP, PVDF)	53	50	2.5	6.20	84	0.940	54	4.178	−0.145	2.610	−0.0846	1.260	0.566	0.276	0.618
Hiflow saddles by rvt	plastic (PP)	50	35	2.0	10.0	86.5	0.938	55.8	2.62	−0.069	2.62	−0.069	1.46	0.566	0.276	0.583
Ralu rings by Raschig	metal (size 2)	50	–	–	6.76	105	0.972	224	10.0	−0.180	3.23	−0.0343	2.42	0.566	–	0.600
	plastic (PP)	38	38	1.2	12.0	150	0.934	59.4	17.5	−0.321	3.40	−0.107	1.38	0.566	0.293	0.600
		55	51	2.9	5.7	110	0.940	54	17.5	−0.321	3.40	−0.107	1.38	0.566	0.293	0.600

Table 6-1b. (continued)

Packing	Material	Size $d \cdot 10^3$ [m]	Size $h \cdot 10^3$ [m]	Size $s \cdot 10^3$ [m]	$N \cdot 10^{-3}$ [$1\,m^{-3}$]	a [$m^2 m^{-3}$]	ε [$m^3 m^{-3}$]	G [kg/m^3]	ψ_{Fl} (a) $Re_V < 2100$ K_1	ψ_{Fl} (a) $Re_V < 2100$ K_2	ψ_{Fl} (b) $Re_V \geq 2100$ K_3	ψ_{Fl} (b) $Re_V \geq 2100$ K_4	$\psi_{Fl,m}$ (c)	$C_{Fl,0}$ [–]	μ-factor [–]	φ_P-factor [–]
Ralu-Flow by Raschig	plastic (size 2) (PP)	58	58	1.6	4.57	96	0.96	36	2.26	−0.112	1.40	−0.050	0.92	0.566	0.215	0.705
Raschig Super rings by Raschig	metal (no. 0.3)	18	–	–	180	315	0.960	340	9.531	−0.206	2.995	−0.051	2.24	0.566	0.41	0.430
	(no. 0.5)	20	–	–	145	250	0.975	275	9.531	−0.206	2.995	−0.051	2.24	0.566	0.35	0.430
	(no. 0.7)	25	–	–	46.5	175	0.975	185	7.143	−0.206	2.245	−0.051	1.49	0.566	0.36	0.570
	(no. 1.0)	30	–	–	32	160	0.980	165	7.143	−0.206	2.245	−0.051	1.49	0.566	0.64	0.570
	(no. 1.5)	38	–	–	13.75	115	0.975	170	5.879	−0.206	1.848	−0.051	1.18	0.566	0.43	0.650
	(no. 2.0)	50	–	–	9.5	97.6	0.985	165	7.143	−0.206	2.245	−0.051	1.49	0.566	0.40	0.570
	(no. 3.0)	75	–	–	4.3	80	0.982	150	5.879	−0.206	1.848	−0.051	1.18	0.566	0.27	0.650
Hackette by VFF	plastic (PP)	42	48	2.5	12.4	135	0.93	63	10.98	−0.27	2.22	−0.065	1.285	0.566	0.270	0.665
VSP rings by VFF	plastic (size 2) (PP)	58	55	4.1	5.0	75.9	0.952	384	1.633	−0.05	1.633	−0.05	1.0	0.566	0.225	0.694
Dtnpac by ENVIMAC	plastic (size 1)	47	18	1.5	29.0	135.3	0.920	72	5.40	−0.159	2.382	−0.052	1.54	0.566	0.318	0.550
	(PP, PVDF) (size 2)	70	25	2.0	10.0	110.0	0.937	56.7	4.792	−0.187	0.697	−0.0513	1.075	0.566	0.243	0.676
Envipac by ENVIMAC	plastic (size 1A)	32	30	1.0	42.5	156.2	0.932	61.2	2.817	−0.070	2.120	−0.0200	1.79	0.566	0.42	0.434
	(PP) (size 1)	33	27	1.5	53.0	138.9	0.936	57.6	6.050	−0.187	2.140	−0.0513	1.400	0.566	0.318	0.590
	(size 2)	60	54	2.0	6.8	81.6	0.957	38.8	4.792	−0.187	1.697	−0.0513	1.075	0.566	0.215	0.676
	(size 3)	90	80	2.0	2.5	60.0	0.958	40.0	4.792	−0.187	1.697	−0.0513	1.075	0.566	0.261	0.676
Mc-Pac by ENVIMAC	metal (size 1)	30	15	0.3	63.5	185.0	0.975	200	5.4	−0.159	2.382	−0.052	1.53	0.59	0.210	0.532
	(size 2)	65	30	0.4	7.7	90.6	0.982	144	5.4	−0.159	2.382	−0.052	1.53	0.59	0.165	0.532
R-Pac by ENVIMAC	ceramic (size 1)	30	10.5	3.0	56.0	186.0	0.771	504	5.40	−0.159	2.382	−0.052	1.53	0.566	0.585	0.540
	(size 2)	55	13	4.0	22.0	131.3	0.808	433	9.9702	−0.205	2.4453	−0.022	2.00	0.566	0.625	0.400
	(size 3)	80	30	8.0	3.0	58.5	0.814	440						0.566		0.400

Table 6-1b. (continued)

Packing	Material	Size $d \cdot 10^3$ [m]	$h \cdot 10^3$ [m]	$s \cdot 10^3$ [m]	$N \cdot 10^{-3}$ [$1\,m^{-3}$]	a [$m^2 m^{-3}$]	ε [$m^3 m^{-3}$]	G [kg/m^3]	ψ_{Fl} (a) $Re_V < 2100$ K_1	K_2	ψ_{Fl} (b) $Re_V \geq 2100$ K_3	K_4	$\psi_{Fl,m}$ (c)	$C_{Fl,0}$ [–]	μ-factor [–]	φ_P-factor [–]
SR-Pac by ENVIMAC	ceramic (size 2)	65	50	5.0	4.2	105.7	0.802	437	6.6468	−0.205	1.6302	−0.022	1.17	0.566	0.370	0.600
Tellerette	plastic (size 1)	47	18	1.5	–	190	0.930	63	4.86	−0.159	2.142	−0.052	1.38	0.566	0.300	0.64
by Ceilcote	(PP) (size 2)	70	25	3.0	12.0	110	0.930	63	1.65	−0.041	1.650	−0.0418	1.15	0.566	0.270	0.661
Nor-Pac rings	plastic	17	15	1.0	200.0	300.0	0.915	76.5	7.63	−0.256	1.825	−0.069	1.025	0.566	0.300	0.694
by NSW	(PP, PVDF)	28	28	2.0	44.5	180.0	0.927	82	7.63	−0.256	1.825	−0.069	1.025	0.566	0.200	0.694
		38	38	2.5	18.0	121.6	0.940	54	7.63	−0.256	1.825	−0.069	1.025	0.566	0.200	0.694
		50	50	3.0	7.3	90.0	0.952	43.2	7.63	−0.256	1.825	−0.069	1.025	0.566	0.200	0.694
		22	27	2.0	64.0	230.0	0.920	72	7.63	−0.256	1.825	−0.069	1.025	0.566	0.260	0.694

(a) $\psi_{Fl} = K_1 \cdot Re_V^{K_2}$ *for* $Re_V < 2100$; (b) $\psi_{Fl} = K_3 \cdot Re_V^{K_4}$ *for* $Re_V \geq 2100$; (c) $Re_V \in \{2100 - 10000\}$

prod. = production year

V4A ≡ stainless steel 1.4571 (SS 316Ti)
V2A ≡ stainless steel 1.4301 (SS 304)

Ceilcote Luftreinhaltung Air-Cure GmbH – Pfungstadt (Germany)
ENVIMAC Engineering GmbH – Oberhausen (Germany)
NSW ≡ Norddeutsche Seekabelwerke GmbH – Nordenham (Germany)
Raschig GmbH – Ludwigshafen (Germany)
rvt = RVT Process Equipment GmbH – Steinwiesen (Germany)
VFF ≡ Vereinigte Füllkörper-Fabriken GmbH & Co. KG – Ransbach-Baumbach (Germany)

Table 6-1c. List of technical data of structured, stacked packings and tube columns (TC) made of metal sheet, plastic and ceramic. Constants Ki for determination of resistance coefficient ψ_{Fl} acc. to Eq. (3-14) or form factors φ_P for determination of resistance coefficient of dry packing ψ acc. to Eq. (3-26), gas velocity at flooding point $u_{V,Fl}$ acc. to Eq. (2-69) and pressure drop $\Delta p/H$ acc. to Eqs. (4-48) and (4-54). List of form factors μ for determination of pressure drop $\Delta p/H$ acc. to ψ_{VL} – model; Eq. (5-9, 5-10)

Structured packing	Type	d_S [m]	a [$m^2 m^{-3}$]	ε [$m^3 m^{-3}$]	G [kg/m^3]	ψ_{Fl} (a) $Re_V < 2100$		ψ_{Fl} (b) $Re_V \geq 2100$		$\psi_{Fl,m}$ (c)	$C_{Fl,0}$ [–]	μ-factor [–]	φ_P-factor [–]
						K_1	K_2	K_3	K_4				
Montz-Pak metal sheet	**B1-200 Y**	0.22	200	0.978	176	4.70	−0.182	3.130	−0.133	1.027	0.566	0.120	0.691
	B1-300 Y	0.22	300	0.972	224	10.50	−0.321	2.490	−0.133	0.816	0.566	0.120	0.760
	B2-500 X	0.22	500	0.960	320	3.92	−0.321	0.929	−0.133	0.304	0.693	–	0.910
	B2-500 Y	0.22	500	0.950	400	8.19	−0.321	1.936	−0.133	0.630	0.566	0.152	0.815
	S1-180 X	0.22	180	0.970	240	9.81	−0.250	2.574	−0.075	1.370	0.630	–	0.598
	B1-100 Y	0.30	100	0.987	104	5.84	−0.163	5.835	−0.162	1.500	0.566	0.120	0.613
	B1-200 Y	0.30	200	0.978	176	4.36	−0.182	3.000	−0.133	0.983	0.566	0.120	0.708
	B1-300 Y	0.30	300	0.972	224	9.65	−0.321	2.290	−0.133	0.750	0.566	0.120	0.768
	B1-300 Y	0.45	300	0.972	224	8.97	−0.312	2.130	−0.133	0.698	0.566	0.120	0.795
	B2-500 X	1.00	500	0.955	360	2.60	−0.321	0.618	−0.133	0.202	0.693	–	0.940
Montz-Pak expanded metal sheet	**BS-460**	0.30	420	0.983	136	1.90	−0.1356	0.978	−0.050	0.624	0.566	–	0.812
Montz-Pak plastic	**C2-200 (PP)**	0.30	200	0.955	40.5	7.00	−0.215	3.728	−0.133	1.220	0.566	0.230	0.642
	C1-300 (Teflon)	0.30	300	0.900	90	4.76	−0.238	1.642	−0.099	0.710	0.566	0.230	0.790
Montz-Pak metal gauze	**A-2**	0.22	400	0.963	296	1.25	−0.14	1.250	−0.140	0.386	0.693	–	0.887
	A-3	0.22	500	0.950	400	1.45	−0.14	1.450	−0.140	0.437	0.693	–	0.870
	A-3	0.45	500	0.950	400	1.21	−0.14	1.210	−0.140	0.374	0.693	0.077	0.890

Table 6-1c. (continued)

Structured packing	Type	d_S [m]	a [m^2m^{-3}]	ε [m^3m^{-3}]	G [kg/m^3]	ψ_{Fl} (a) $Re_V < 2100$ K_1	K_2	ψ_{Fl} (b) $Re_V \geq 2100$ K_3	K_4	$\psi_{Fl,m}$ (c)	$C_{Fl,0}$ [−]	μ-factor [−]	φ_P-factor [−]
Mellapak	**250 Y**	0.15	250	0.960	320	11.42	−0321	2.710	−0.133	0.888	0.565	0.152	0.716
metal sheet	**250 Y**	0.22	250	0.960	320	8.19	−0.321	1.936	−0.133	0.630	0.566	0.091	0.816
by Sulzer													
	125 Y	1.00	125	0.985	120	5.76	−0.321	1.366	−0.133	0.448	0.566	–	0.865
	125 X	1.00	125	0.985	120	4.40	−0.321	1.044	−0.133	0.342	0.693	–	0.899
	250 Y	1.00	250	0.975	200	6.50	−0.321	1.537	−0.133	0.504	0.566	–	0.852
	250 X	1.00	250	0.980	160	2.60	−0.321	0.618	−0.133	0.202	0.693	–	0.940
	350 Y	1.00	350	0.965	280	5.76	−0.321	1.365	−0.133	0.448	0.566	–	0.868
	350 X	1.00	350	0.965	280	2.60	−0.321	0.618	−0.133	0.202	0.693	–	0.940
	500 Y	1.00	500	0.955	360	5.76	−0.321	1.365	−0.133	0.448	0.565	–	0.868
	500 X	1.00	500	0.955	360	2.60	−0.321	0.618	−0.133	0.202	0.693	–	0.940
Mellapak plastic	**250 Y (PP)**	0.30	250	0.955	40.5	1.62	−0.13	1.000	−0.069	0.564	0.566	–	0.834
Sulzer packing	**BX**	0.22	500	0.95	45	1.25	−0.14	0.7043	−0.065	0.408	0.693	–	0.880
metal gauze		0.50	500	0.95	45	1.21	−0.14	1.210	−0.14	0.374	0.693	–	0.890
Ralu Pak	**250YC**	0.22	250	0.963	296	11.06	−0.383	4.000	−0.250	0.495	0.566	0.080	0.855
metal sheet		0.30	250	0.963	296	10.09	−0.383	3.656	−0.250	0.452	0.566	0.080	0.856
by Raschig		0.45	250	0.963	296	9.74	−0.383	3.520	−0.250	0.436	0.566	0.080	0.872
Rombopak B		0.40	230	0.97	240	19.14	−0.42	19.14	−0.420	0.500	0.693	0.098	0.850
metal sheet		0.25	230	0.97	240	3.22	−0.206	1.023	−0.051	0.670	0.693	0.098	0.800
by Kühni													
Gempak	**A2-T-304**	0.22	202	0.97	240	12.45	−0.321	4.500	−0.188	0.932	0.566	0.120	0.725
metal sheet		0.30	202	0.97	240	10.51	−0.321	3.800	−0.188	0.787	0.566	0.120	0.754
by Koch-Glitsch													

Table 6-1c. (continued)

Structured packing	Type	d_S [m]	a [m^2m^{-3}]	ε [m^3m^{-3}]	G [kg/m^3]	ψ_{Fl} (a) $Re_V < 2100$		ψ_{Fl} (b) $Re_V \geq 2100$		$\psi_{Fl,m}$ (c)	$C_{Fl,0}$ [–]	μ-factor [–]	φ_P-factor [–]
						K_1	K_2	K_3	K_4				
Honey comb packing **plastic**		0.25	209.4	0.817	170	2.55	−0.35	0.551	−0.150	0.157	0.780	–	0.955
Fi-Pak **metal sheet** by ENVIMAC	**Fi-13F**	0.22	198	0.948	416	1.38	−0.029	1.375	−0.029	1.077	0.739	–	0.684
Nor-Pac Kompakt **plastic (PP)** by NSW	**size 1** **size 2**	0.30 0.30	240.0 117.3	0.904 0.934	768 3680	3.43 3.9	−0.084 −0.084	2.677 3.056	−0.052 −0.052	1.725 1.970	0.566 0.566	0.450 0.450	0.475 0.44
Nor-Pac swirl packing **plastic (PP)** by NSW	**size 3**	0.30	130.0	0.845	139.5	6.1	−0.458	1.027	−0.224	0.158	0.693	–	0.953
FX packing **plastic (PP)** by VFF	**FX-260-P**	0.45	250	0.962	34.2	0.76	−0.069	0.760	−0.069	0.425	0.63	0.062	0.875
Impuls packing **ceramic** by rvt	**50-IP-C**	0.50	102	0.83	374	2.61	−0.115	2.610	−0.115	1.000	0.634	0.400	0.706
Euroform packing **plastic** by Munters	**50-EP-P**	0.30	110	0.936	57.6	1.75	−0.104	1.750	−0.104	0.730	0.566	0.145	0.802
Durapak **glass** by Schott	**DUPA 280** **DUPA 400**	– –	280 400	0.82 0.72	450 700	8.19 8.19	−0.321 −0.321	1.936 1.936	−0.133 −0.133	0.630 0.630	0.566 0.566	– –	0.815 0.815
NET packing **plastic (PP)** by GEA 2H	**NET** **38-150**	0.60	80	0.959	37.7	3.0656	−0.2055	0.9424	−0.0513	0.616	0.628	–	0.820
FKP packing **plastic (PP)** by GEA 2H	**FKP 319** **FKP 327**	0.60 0.60	150 125	0.945 0.945	50.0 50.0	1.2930 2.4527	−0.2055 −0.2055	0.3975 0.754	−0.0513 −0.0513	0.259 0.492	0.693 0.693	– –	0.924 0.856

Table 6-1c. (continued)

Stacked packing	Type	N $[1m^{-3}]$	a $[m^2m^{-3}]$	ε $[m^3m^{-3}]$	G $[kg/m^3]$	ψ_{Fl} (a) $Re_V < 2100$		ψ_{Fl} (b) $Re_V \geq 2100$		$\psi_{Fl,m}$ (c)	$C_{Fl,0}$ [–]	μ-factor [–]	φ_P-factor [–]
						K_1	K_2	K_3	K_4				
stacked Białecki rings plastic	**50-BR-P-S**	8500	147	0.910	81	4.908	−0.14	3.50	−0.0958	1.525	0.693	0.400	0.541
stacked Białecki rings metal	**35-BR-M-S**	20000	183.5	0.941	53.1	12.50	−0.291	2.663	−0.0880	1.271	0.693	–	0.620
	50-BR-M-S (1974)	6174	113.4	0.969	248	6.280	−0.18	3.30	−0.0958	1.478	0.693	0.206	0.560
	25-BR-M-S (1977)	63500	275	0.933	536	4.370	−1.8	2.31	−0.0966	1.030	0.693	0.185	0.712
	50-BR-M-S (1991)	7650	128	0.965	280	4.675	−0.180	2.475	−0.0966	1.100	0.566	0.185	0.676
stacked PSL rings metal	**25-PSL-M-S**	60000	292	0.963	296	4.719	−0.134	3.544	−0.0966	1.574	0.566	–	0.530
	50-PSL-M-S	7650	142	0.968	256	4.030	−0.153	2.340	−0.082	1.174	0.566	0.200	0.650
stacked 50 mm Hiflow rings ceramic	**50 Hiflow-C-S**	5683	102.36	0.7582	532	0.541	−0.075	0.541	−0.075	0.288	0.693	0.089	0.906
stacked 50 mm Hiflow rings plastic (hydrophilized and non hydrophilized)	**50 Hiflow-PP-S**	8150	140.1	0.91	83.2	2.493	−0.2055	0.6113	−0.022	0.510	0.74	0.124	0.847
stacked 50 mm Raschig rings ceramic	**50-RR-C-S**	8768	137	0.697	666.6	5.530	−0.307	1.1	−0.0958	0.492	0.692	0.140	0.852
stacked 50 mm Pall rings ceramic	**50-PR-C-S**	7500	140	0.784	475.2	2.0932	−0.18	1.639	−0.089	0.754	0.740	0.275	0.773
stacked 50 mm SR-Pac ceramic	**50 SR-Pac-C-S**	–	115.8	0.783	477.4	2.779	−0.142	2.779	−0.142	0.845	0.693	–	0.750

Table 6-1c. (continued)

Tube column	Type		a $[m^2 m^{-3}]$	ε $[m^3 m^{-3}]$	G $[kg/m^3]$	ψ_{Fl} (a) $Re_V < 2100$ K₁	ψ_{Fl} (a) $Re_V < 2100$ K₂	ψ_{Fl} (b) $Re_V \geq 2100$ K₃	ψ_{Fl} (b) $Re_V \geq 2100$ K₄	$\psi_{Fl,m}$ (c)	$C_{Fl,0}$ [–]	μ-factor [–]	φ_P-factor [–]
Białecki rings tube column metal	**25 BR-TC-M**	75377	339.1	0.924	608	1.550	0	1.550	0	1.550	0.693	–	0.540
	50 BR-TC-M 1 tube bundle	10193	171.0	0.950	400	11.52	−0.205	4.453	−0.081	2.260	0.600	0.450	0.352
	52 BR-TC-M 7 tube bundles	9422	154.6	0.96	320	6.420	−0.14	3.2784	−0.052	2.060	0.625	0.320	0.380
	53 BR-TC-M 27 tube bundles	8740	156.4	0.950	400	4.164	−0.180	2.200	−0.097	0.978	0.780	–	0.705
25 mm Pall rings tube column metal	**25 PR-TC-M**	81528	343.7	0.929	568	0.409	−0.133	0.409	−0.133	0.168	0.780	0.025	0.950

(a) $\psi_{Fl} = K_1 \cdot Re_V^{K_2}$ *for* $Re_V < 2100$; (b) $\psi_{Fl} = K_3 \cdot Re_V^{K_4}$ *for* $Re_V \geq 2100$; (c) $Re_V \in \{2100 - 10000\}$
By determination of pressure drop:
- K=1 for structured packings and tube columns
- K≠1 for stacked packing bed

prod. = production year

V4A ≡ stainless steel 1.4571 (SS 316Ti)
V2A ≡ stainless steel 1.4301 (SS 304)

ENVIMAC Engineering GmbH – Oberhausen (Germany)
GEA 2H Water Technologies GmbH – Wettringen (Germany)
Kühni AG – Allschwil (Switzerland)
Montz GmbH – Hilden (Germany)
Munters Euroform GmbH – Aachen (Germany)
NSW ≡ Norddeutsche Seekabelwerke GmbH – Nordenham (Germany)
Raschig GmbH – Ludwigshafen (Germany)
Schott AG – Mainz (Germany)
Sulzer AG – Winterthur (Switzerland)
VFF ≡ Vereinigte Füllkörper-Fabriken GmbH & Co. KG – Ransbach-Baumbach (Germany)

Part 2

Principles of the Fluid Dynamic Design of Packed Columns for Liquid/Liquid Systems

List of Symbols for Part 2

Formula Variables, Latin letters

a	$m^2\ m^{-3}$	geometric packing surface area per unit volume
a_e	$m^2\ m^{-3}$	effective mass transfer area per unit volume
C_1	sm^{-1}	constant for a given system
C, C_0, C_T	–	constants
c	kmol m^{-3}	concentration of solute
D	$m^2\ s^{-1}$	diffusion coefficient of dissolved substance
D_{ax}	$m^2\ s^{-1}$	dispersion coefficient
d_E	m	diameter of the largest stable droplet
d_F	m	packing diameter
d_h	m	hydraulic diameter of packed bed
d_S	m	column diameter
d_T	m	diameter of droplet according to Sauter, $d_T = \sum n_i \cdot d_i^3 / \sum n_i \cdot d_i^2$
E	–	extraction factor, $E = m \cdot \lambda \cdot (\rho_D / \rho_C)$
g	ms^{-2}	local acceleration due to gravity
h	m	height of individual packing element
H	m	height of packed bed
m	–	model parameter in Eq. (7-21)
m_{CC}	kg/kg	slope of equilibrium line
N	m^{-3}	packing density of packed bed
u_C, u_D	$m^3\ m^{-2}\ s^{-1}$	specific flow rate of the continuous phase, specific flow rate of the dispersed phase
u_K	ms^{-1}	characteristic droplet velocity
u_R	$m^3\ m^{-2}\ s^{-1}$	slip velocity
$\dot{V}$	$m^3\ h^{-1}$	volumetric flow rate

w_E	ms^{-1}	end rising or falling velocity of droplets in empty column according to Mersmann in Eq. (7-16) or Levich in Eq. (7-15)
$\bar{w}_S$	ms^{-1}	middle rising or falling velocity of single droplet in packing bed
x, x^0	$m^3\ m^{-3}$	dispersed phase ratio or liquid hold-up of dispersed phase to empty column volume V_S or to effective column volume $x^0 = x/\varepsilon$

Formula Variables, Greek Letters

α	deg	flow channel angle in packed bed, see Fig. 1-2b
δ	%	relative error, based on experimental value
ε	$m^3\ m^{-3}$	void fraction of packed bed
λ	$\lambda = u_D/u_C$	phase flow ratio
η	$kgm^{-1}\ s^{-1}$	viscosity
ρ	kgm^{-3}	density
$\Delta\rho$	kgm^{-3}	difference in densities of immiscible liquid phases
σ	Nm^{-1}	interfacial tension
ψ	–	resistance coefficient for single-phase flow

Dimensionless Numbers

$Bo = \frac{u_C \cdot H}{D_{ax}}$	Bodenstein number
$Re_T = \frac{w_E \cdot d_T \cdot \rho_C}{\eta_C}$	Reynolds number of individual droplets
$Re = \frac{u \cdot d_T \cdot \rho_C}{\eta_C}$	Reynolds number of individual droplets

Indices

cal	calculated
C	continuous phase
D	dispersed phase
e	effective
exp	experimental
In/Out	Inlet/Outlet
Fl	flooding point
m	middle value based on Table 6-1a,b,c

Abbreviations

TC	tube column
BR	Białecki rings
M	metal
MIBK	methyl isobutyl ketone
PP/PVDF	polypropylene / polyvinylidene fluoride - plastics
Sy	used symbol
$D \rightarrow C$	mass transfer direction from dispersed phase D to continuous phase C
$C \rightarrow D$	mass transfer direction from continuous phase to C dispersed phase D
Chap.	chapter
Eq.	equation
s.	see

Basic Principles of Packed Column Design for Liquid/Liquid Systems

7.1 Introduction

Extractors used in industry can be operated with or without mechanical energy agitation, by pulsation or rotation [5, 10, 16, 17, 21].

Some of the simplest designs are spray columns without internals and packed columns containing randomly filled and stacked packing elements, tube columns or columns with structured packing, see Fig. 1-2a.

The basic design of liquid-liquid-extraction test plant with packed columns is shown in Fig. 7-1. It largely corresponds to the structure commonly used for gas/liquid systems. A distributor with bore holes is used for the pre-distribution of the dispersed phase, with the hole diameter roughly matching the expected droplet size. A distributor is also used for the pre-distribution of the continuous phase, in order to achieve an even flow of both phases in the packed extractor. Practical experience has shown that a poor pre-distribution of the continuous and dispersed phases results in strong back-mixing, which, in turn, leads to a significant reduction of the extractor's separation efficiency [3, 4, 10, 21].

If the dispersed phase, moving upwards in the extractor, is lighter than the continuous phase (i.e. $\rho_D < \rho_C$), it rises from the bottom to the top of the packing in counter-current to the heavier continuous phase. Phase separation then takes place at the top of the column, with the lighter phase being extracted from the column in the form of the continuous phase, see Fig. 7-1.

However, if the heavy phase is dispersed in the light phase at the top of the column (i.e. $\rho_D > \rho_C$), phase separation takes place in the lower column section, followed by extraction from the column.

The internals used in extraction columns consist of randomly filled or stacked packing elements or structured packings, mostly made of ceramic, metal or plastic, as shown in Fig. 1-2a or 1-2b.

To intensify mass transfer, pulsed packed columns are used, especially in cases where extremely high product purities must be achieved and energy-efficient pulsators can be used [7, 18, 21].

There are many areas of application for liquid/liquid extractors, where non-pulsed packed columns are preferred, e.g. for the separation of corrosive products, which must

J. Maćkowiak, *Fluid Dynamics of Packed Columns*,
Chemische Technik/Verfahrenstechnik, DOI 10.1007/b98397_7,

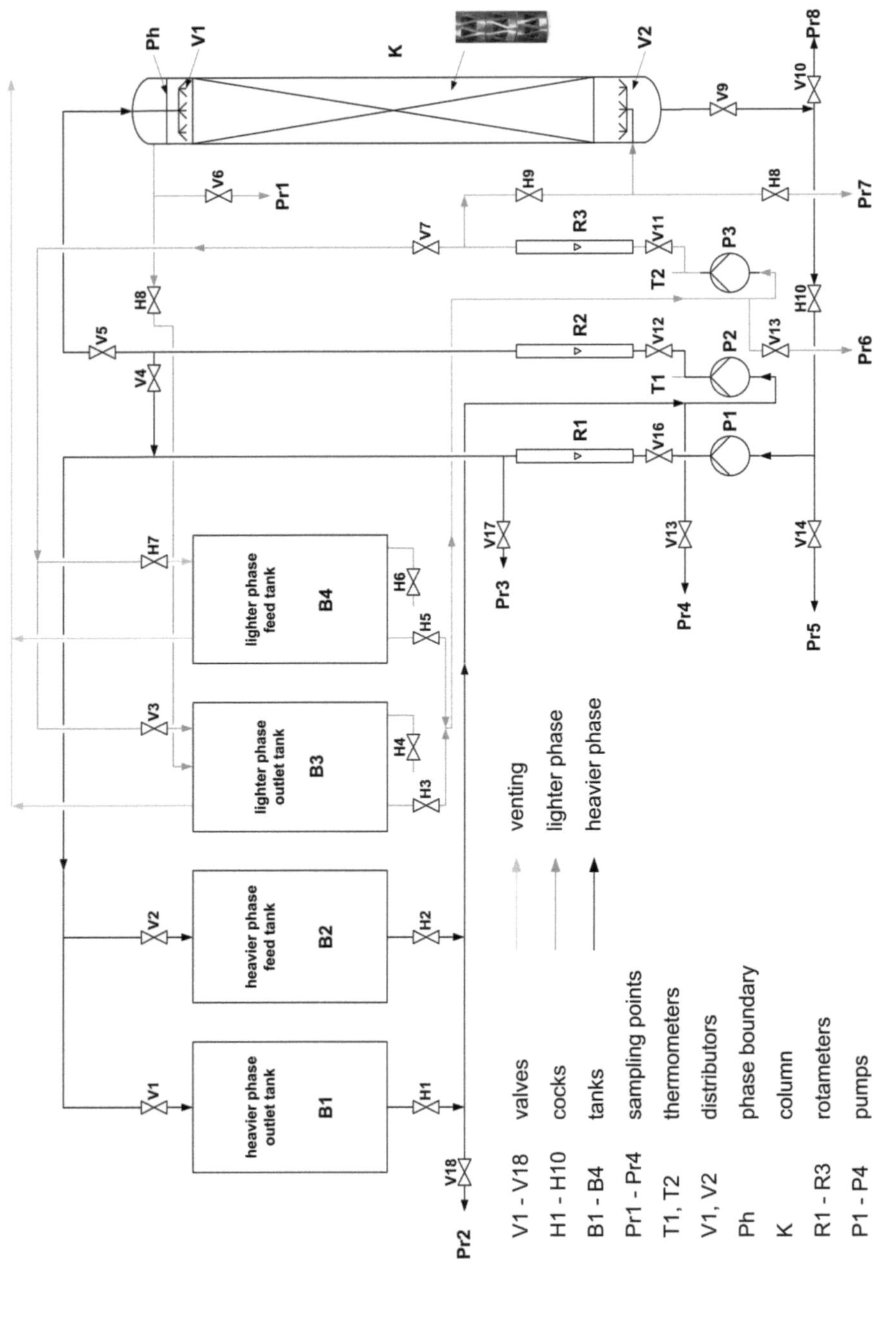

Figure 7-1. Schema of liquid-liquid-extraction test plant for the investigation of hydraulic and mass transfer behaviour of different packings

be extracted under higher temperatures and pressures, and for larger column diameters, where no additional energy input is possible due to technical reasons.

Low back-mixing is a particular advantage of tube columns or columns containing stacked Białecki rings, metal Fi-Pac and structured packings. Billet and Maćkowiak [3, 4] have carried out experiments to confirm that the mixing of the continuous phase in tube columns is low and that droplet flow can be assumed when determining the separation efficiency. The Bodenstein numbers Bo_C of various systems were found to be high, in the range of 30–40 [3, 4].

The mixing effects are not discussed further in this work. More information on this topic can be found in the relevant literature [3, 4, 10, 16, 21].

There are a number of studies and monographs available on the subject of designing packed columns. However, these are restricted to the empirical presentation of design documents, mostly for small ceramic Raschig rings, spheres or Berl saddles and occasionally for Pall rings or Białecki rings made of metal [10, 16, 21].

It was the objective of this book to present a uniform calculation model, based on previous publications [1, 2, 3, 6, 9, 11, 12, 13, 14, 15], which is applicable to any type of column internal and can be used to determine the main dimensions of extraction columns equipped with modern random or structured packings. They apply to randomly filled packing elements (Fig. 1-2a), stacked and structured packings or tube columns (Fig. 1-2b), which were tested using various systems, see Table 7-1.

7.2 Two-Phase Flow and Operating Ranges

For the design of non-pulsed extraction columns, it is necessary to ascertain the following parameters for a given flow rate of the raffinate phase: the flow rate of the extract phase, the concentration range of the transferring component, the specific flow rates u_D and/or u_C in the operating range of the extractor below the flooding point and the loading line.

It is then possible to determine the column diameter d_S, based on the volumetric flow rate of the continuous phase $\dot{V}_C$ and the specific flow rate u_C.

The following parameters are required for describing the operating range:

(a) dispersed phase hold-up

$$x = V_D/V_S \tag{7-1}$$

(b) Sauter droplet diameter d_T
(c) effective mass transfer area a_e
(d) mass transfer coefficients for mass transfer direction $D \rightarrow C$ or $C \rightarrow D$
(e) mixing effects

7.2.1 Dispersed Phase Hold-Up in Packed Columns Containing Random and Structured Packings

Figures 7-2, 7-3, 7-4, 7-5 and 7-6 illustrate the behaviour of the dispersed phase throughout the entire operating range of packed columns up to the flooding point.

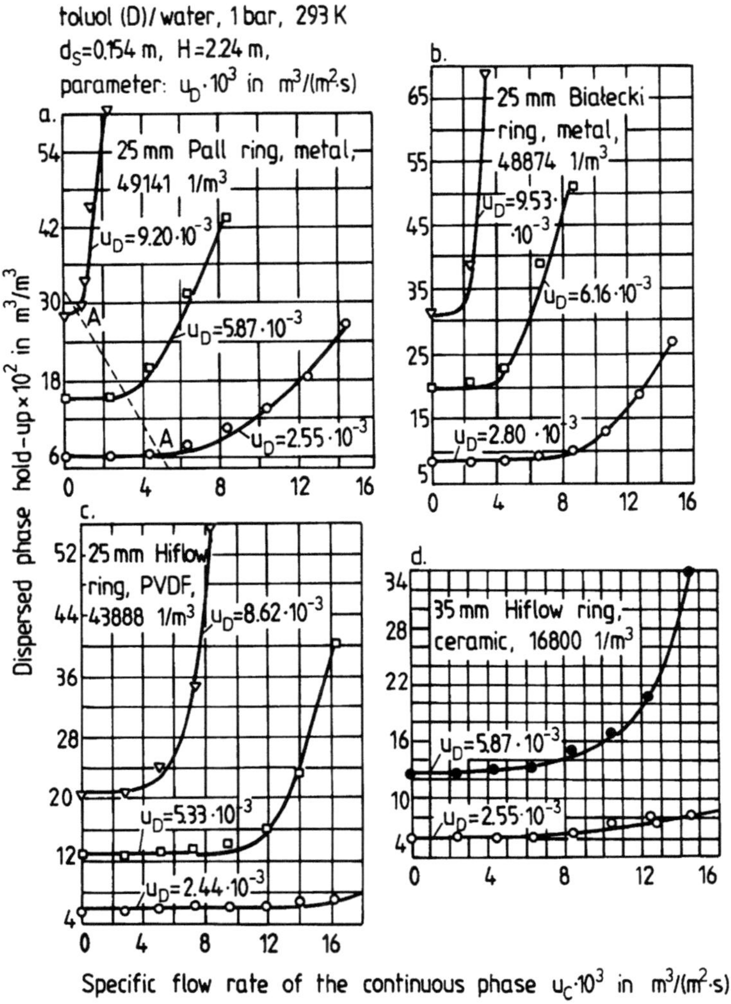

Figure 7-2. Dispersed phase hold-up x as function of specific flow rate of the continuous phase u_C, valid for various packings made of metal, plastic and ceramic [11–14]. Parameter: specific flow rate of dispersed phase u_D

Figures 7-2 and 7-3 show the dispersed phase hold-up x as a function of the specific flow rate u_C of the continuous phase, using various specific flow rates u_D of the dispersed phase as a parameter. The experimental data shown in Fig. 7-2 is applicable to different random packing elements, such as metal Pall rings, Białecki rings, Hiflow rings with a dimension of 25–38 mm, whereas the data shown in Fig. 7-3 is valid for 50 mm tube columns and other structured packings. The test system used for the experiments under normal conditions was toluol (D)/water, which has a high interfacial tension and is

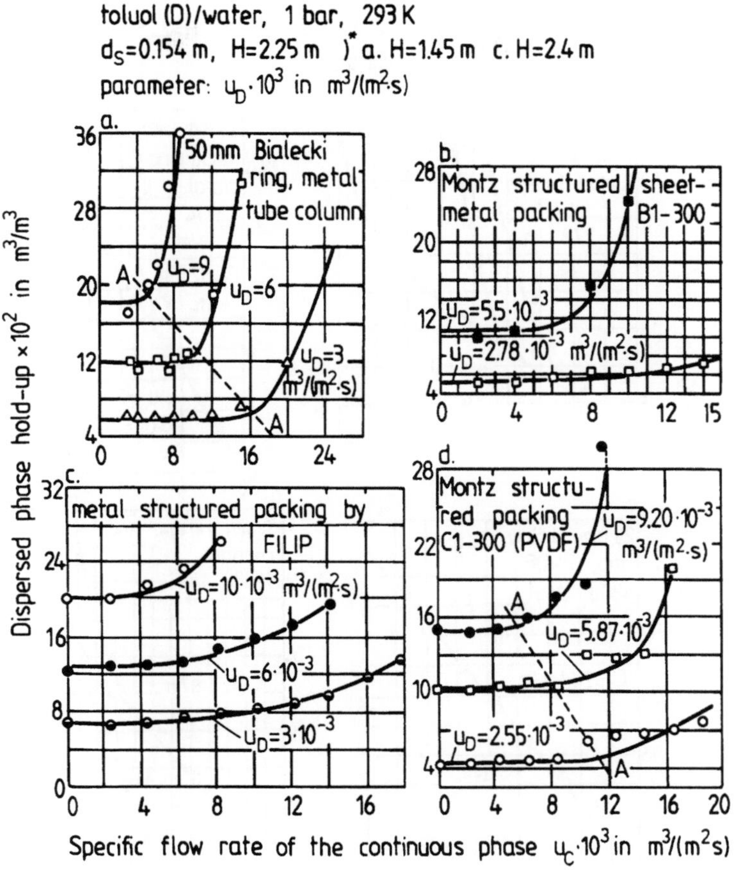

Figure 7-3. Dispersed phase hold-up x as function of specific flow rate of the continuous phase u_C, valid for various packings made of metal, plastic and ceramic [11–14]. Parameter: specific flow rate of dispersed phase u_D

recommended by the EFCE (European Federation of Chemical Engineering). The experiments were carried out using columns with a diameter of $d_S = 0.156$ m and/or 0.0532 m and a packing height of H = 2.4 m and/or 1.45 m [4, 11, 12]. The properties of the tested systems are listed in Table 7-1.

The figures show that all packings are characterised by a qualitatively equal course of the function of the dispersed phase hold-up x. Up to a certain distance from the flooding point $u_C < u_{C,Fl}$, Fig. 7-2, line A-A, the dispersed phase hold-up x is not dependent on the specific flow rate of the continuous phase u_C.

The dispersed phase hold-up x, as a function of the specific flow rate of the dispersed phase u_D, is now plotted in a graph, using the flow rate u_C of the continuous phase as a parameter, Figs. 7-4 and 7-5. The result is a linear relationship.

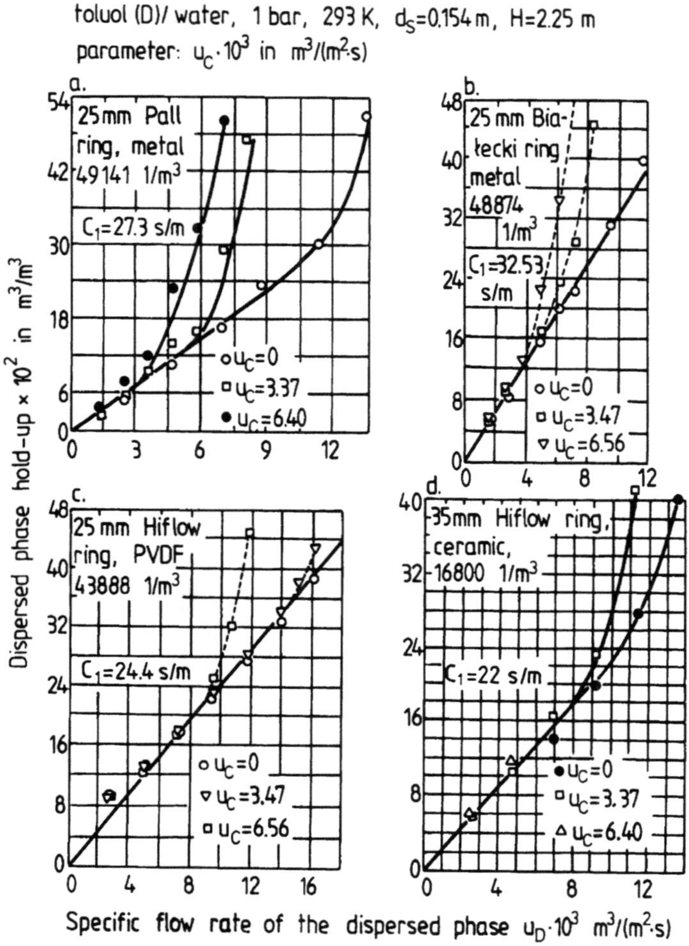

Figure 7-4. Dispersed phase hold-up x as function of specific flow rate of the dispersed phase u_D, valid for various random packings made of metal, plastic and ceramic [11–14]. Parameter: specific flow rate of continuous phase u_C

$$x = C_1 \cdot u_D \quad \text{and } x \neq f(u_C) \qquad [\mathrm{m^3 m^{-3}}] \tag{7-2}$$

between the dispersed phase hold-up x and the specific flow rate of the dispersed phase u_D, where C_1 is a system-specific packing constant.

Figures 7-4 and 7-6 illustrate the different behaviour of the dispersed phase in relation to different types of packings, e.g. the system constant C_1 in Eq. (7-2) is higher for random 25 mm metal Pall rings than it is for the 50 mm Białecki ring tube column. It can therefore be concluded that flooding occurs earlier in the random packing than it does in the tube column. An operating range, which is characterised by a linear relationship

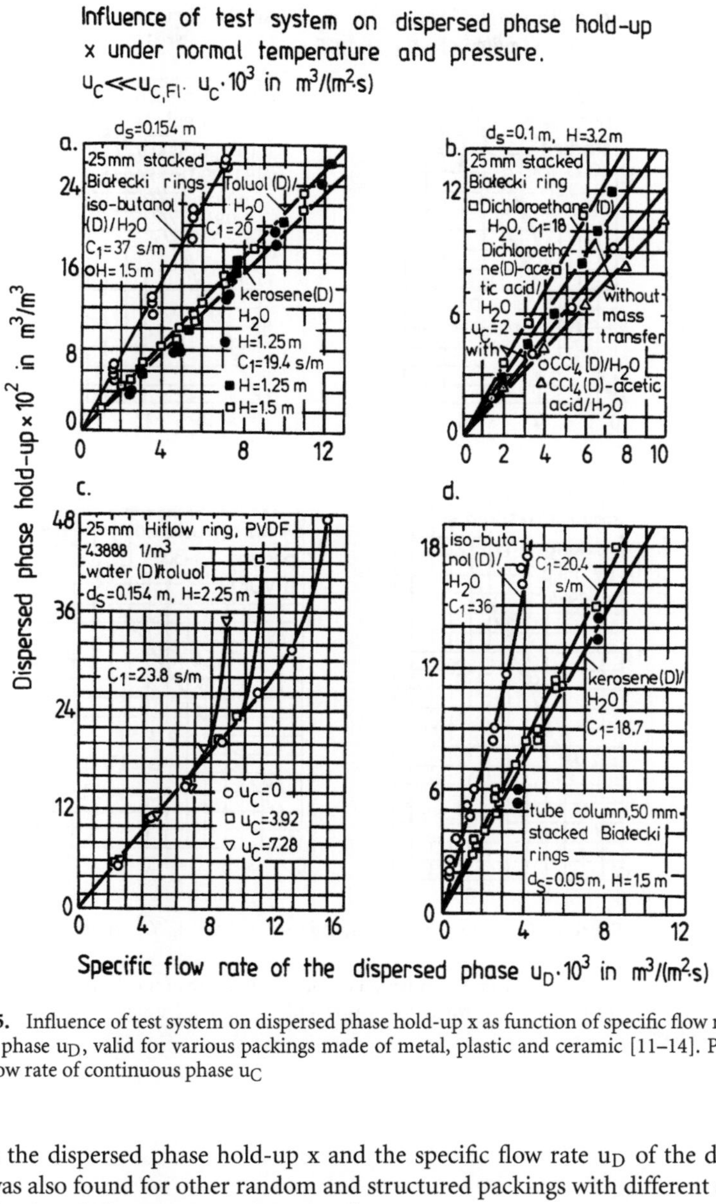

Figure 7-5. Influence of test system on dispersed phase hold-up x as function of specific flow rate of the dispersed phase u_D, valid for various packings made of metal, plastic and ceramic [11–14]. Parameter: specific flow rate of continuous phase u_C

between the dispersed phase hold-up x and the specific flow rate u_D of the dispersed phase, was also found for other random and structured packings with different systems, see Eq. (7-2) and Figs. 7-2, 7-3, 7-4, 7-5 and 7-6.

The linear relationship between the dispersed phase hold-up x, acc. to Eq. (7-2), can be seen in Figs. 7-2, 7-3, 7-4, 7-5 and 7-6. It therefore applies to liquid/liquid systems in the range below the loading line, i.e. for:

$$u_D \leq 0.65 \cdot u_{D,Fl} \tag{7-3a}$$

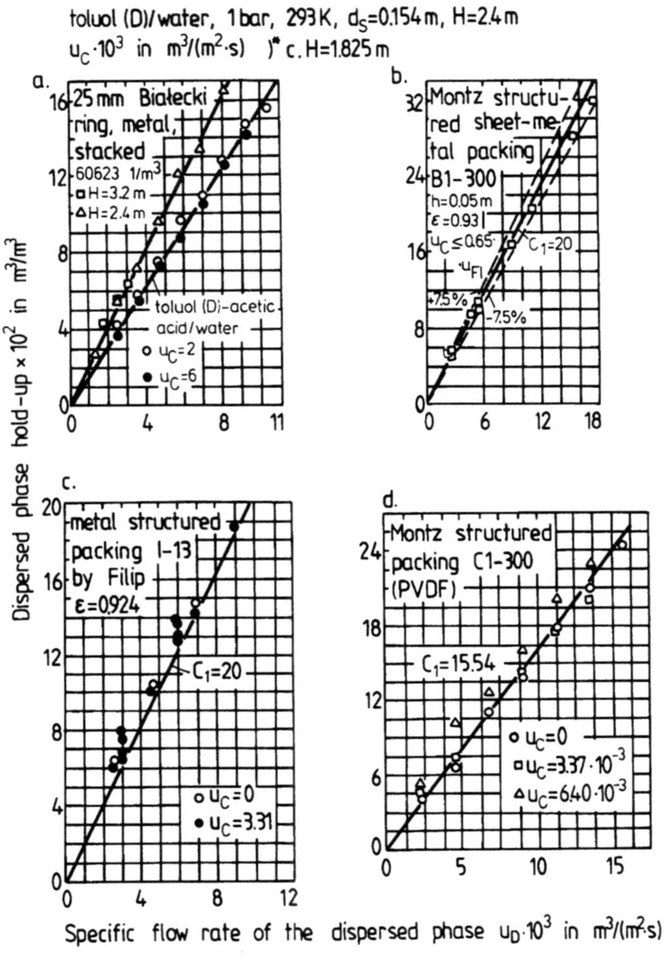

Figure 7-6. Dispersed phase hold-up x as function of specific flow rate of the dispersed phase u_D, valid for various test systems in the range, where $x = f(u_D)$ and $x \neq f(u_C)$. Parameter: specific flow rate of continuous phase u_C

and/or for

$$u_C \le 0.65 \cdot u_{C,Fl}. \tag{7-3b}$$

This behaviour of packed columns was already discussed in previous chapters in relation to gas/liquid systems.

Acc. to Fig. 7-5b and 7-6a, the dispersed phase hold-up x decreases for the mass transfer direction D → C and increases for C → D [11–13].

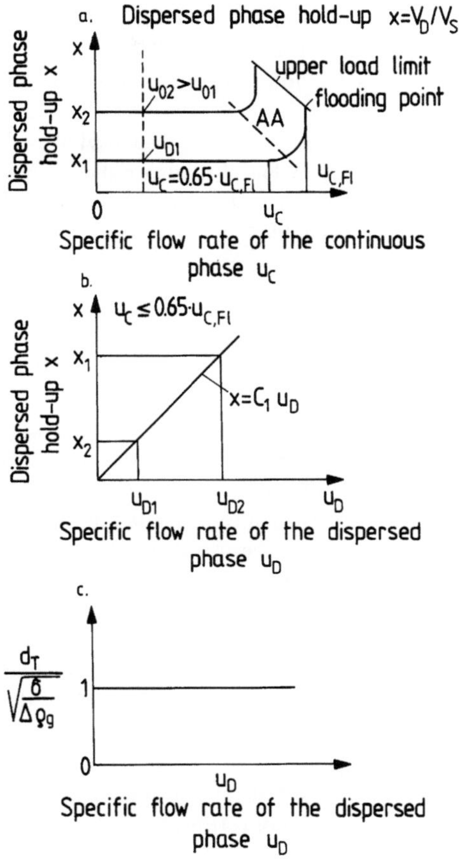

Figure 7-7. Schematic diagram of operating ranges in packed columns for liquid/liquid-extraction

The numerical values of the system constants C_1 for Eq. (7-2), which were experimentally determined for the packing elements tested, are summarised in Tables 7-2 and 7-3.

Based on the correlations $x = f(u_D, u_C)$, shown in Figs. 7-2, 7-3, 7-4, 7-5 and, 7-6 it can be concluded that there is a linear relationship between the dispersed phase hold-up x and the specific flow rate of the dispersed phase u_D in the operating range below the loading line, which for packed columns lies at approx. 65 % of the flooding point. This is shown by the schematic representation in Fig. 7-7.

The following approximate equation (7-4), derived in previous studies [11, 12], can be used to determine the dispersed phase hold-up x for any type of system:

$$x \cong C_1 \cdot u_D = \frac{1}{C_0 \cdot \varepsilon} \cdot \left[\frac{\rho_C^2}{4g \cdot \Delta\rho \cdot \sigma} \right]^{1/4} \cdot u_D \qquad \left[\mathrm{m^3 m^{-3}}\right], \qquad (7\text{-}4)$$

This equation can be used to verify the experimental data by calculation with an accuracy of approx. ±10 %.

Tables 7-2 and 7-3 contain a list of constants C_0 for a number of tested packings [11–13].

The experiments on the dispersed phase hold-up x, using systems with various properties shown in Fig. 7-6, also indicate that the variable C_0 in Eq. (7-4) is a system-independent constant, which is not dependent on the constructive design and size of the plant, see Table 7-2.

7.2.2 Droplet Diameter

Based on comprehensive experiments, using a tube column stacked with metal Białecki rings with a dimension of 25 and 50 mm as well as a structured packing of metal Białecki rings with a dimension of 25 mm, it was possible to determine the Sauter diameter d_T of the droplets by means of the photographic method [11–13]. The results are shown in Fig. 7-8. The experiments were carried out with the density difference of the liquids varying between 131.5 and 595.8 kgm^{-3} and an interfacial tension between 2.8 and 44.5 mNm^{-1}.

Using the photo-electric method developed by Pilhofer, it was possible to obtain comprehensive data material on the droplet diameter d_T in the 25 mm metal Białecki ring packing in a column with a diameter of $d_S = 0.10$ m and a packing height of 2.95 m [6, 11, 12, 14]. There were two test points in the packing, located at a packing height of $H = 0.75$ m and $H = 2$ m. It was at these points that the droplets were removed and the droplet size distribution was measured. The test systems, used under normal conditions, were toluol (D)/water [6, 11] as well as toluol (D)-acetic acid/water and toluol

TP	Column type	system at 1 bar, 293 K	$u_C \cdot 10^3$ $m^3/(m^2 \cdot s)$	$d_S \cdot 10^3$ [m]	H [m]
o	25 mm tube column	iso-butanol (D) / H_2O	1.5	25.5	1.25
ф	50 mm tube column		1.3	53.2	1.5
□		kerosene (D)/ H_2O	2.8	53.2	1.25
■		toluol (D)/H_2O	3	53.2	1.5

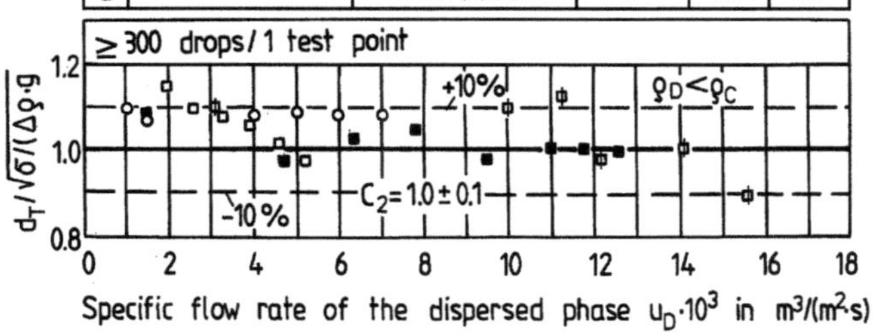

Figure 7-8. Quotient $d_T/\sqrt{\sigma/(\Delta\rho \cdot g)}$ as function of specific flow rate of the dispersed phase u_D, valid for 25 and 50 mm tube columns stacked aligned with metal Białecki rings made [11, 12]

(D)-acetone/water (C). During each test, the Sauter diameter d_T as well as the dispersed phase hold-up x were determined. The results are shown in Fig. 7-9a–c.

Using the test system toluol (D)/water, the droplets in the metal Białecki ring packing were found to have a diameter of $d_T = 5 \cdot 10^{-3} \pm 0.005$ m, irrespective of the location of the test point [6, 12].

In the range above the loading line, $u_C \leq 0.65 \cdot u_{C,Fl}$, in which the dispersed phase hold-up x, at a constant flow rate of the dispersed phase u_D, increases with the specific flow rate u_C of the continuous phase see Fig. 7-9b, the droplets were found to be larger. At the flooding point, the droplets were around 20 % larger, compared to the range below the loading line for $u_C \leq 0.65 \pm u_{C,Fl}$.

$$d_{T,Fl} \approx 1.2 \cdot d_T \qquad (7\text{-}5)$$

The Sauter droplet diameter d_T is therefore dependent on the physical properties of the relevant mixture:

$$d_T = f\left(\rho_D, \rho_C, \sigma, g\right) \qquad (7\text{-}6)$$

and is calculated using Eq. (7-7).

$$d_T = C_T \sqrt{\sigma / \Delta\rho \cdot g} \qquad [\text{m}] \qquad (7\text{-}7)$$

Based on Figs. 7-8 and 7-9, this equation is valid for pure binary mixtures and the mass transfer direction C → D for systems with low and high interfacial tension, acc. to Table 7-1. The following applies, based on Fig. 7-9a–c:

$$C_T = 1.00. \qquad (7\text{-}8a)$$

The mass transfer direction D → C has a significant influence on the system limit and leads to an increase in the droplet diameter d_T, compared to the pure test system toluol (D)/water under the same operating conditions, see Fig. 7-9a. The following applies:

$$C_T = 1.25 \qquad (7\text{-}8b)$$

for systems with a high interfacial tension for the mass transfer direction D → C and

$$C_T = 1.55 \qquad (7\text{-}8c)$$

for systems with low and moderate interfacial tension for the mass transfer direction D → C [4, 12, 19].

In the operating range, which is characterised by a linear relationship between the dispersed phase hold-up x and the specific flow rate u_D of the dispersed phase, the droplet diameter d_T is practically load-independent, see Eq. (7-2).

Figure 7-9. Sauter droplet diameter d_T as function of: (**a**) specific flow rate of the dispersed phase u_D; (**b**) relative column load $u_C/u_{C,Fl}$; (**c**) mean acetone concentration in water for mass transfer direction C → D [11–14]

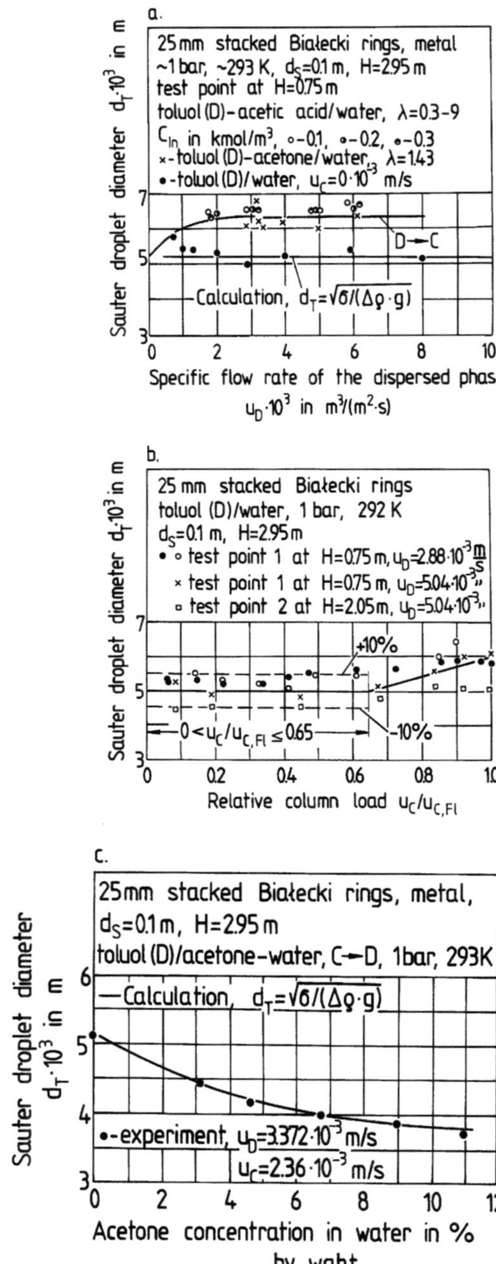

7.3 Determining the Flooding Point

7.3.1 Introduction

In order to determine the column diameter d_S of a packed column at given volumetric flow rates of the raffinate and extract phases, it is necessary to ascertain the flooding point. In extraction columns, flooding occurs if:

(a) the specific flow rate of the continuous phase u_C is too high or
(b) the specific flow rate of the dispersed phase u_D is too high

In the first case (a), droplet entrainment from the extraction column occurs at the flooding point. In the second case, the dispersed phase gathers at one end of the column and fills the entire column volume with droplets, which strongly obstructs the flow of the continuous phase, leading to its build-up.

The typical hydraulic characteristics of a packed column are shown in Fig. 7-7, plotting the correlation between the dispersed phase hold-up h_D and the specific flow rate of the dispersed phase u_D. The parameter here is the specific flow rate of the continuous phase u_C. This figure shows the course of the most important parameters of a packed column in a single diagram. Up to 65 % of the flooding capacity $u_{D,Fl}$, the flow rate of the continuous phase u_C has practically no influence on the dispersed phase hold-up.

The following Figs. 7-10, 7-11, 7-12 and 7-13 show the capacity diagrams for a number of different packed columns, using various systems, acc. to Table 7-1 [11, 12, 13, 14]. They form the basis for the development of the author's own model for determining the flooding capacities in non-pulsed extraction columns [15].

The models used in practice to determine the specific flow rate of the disperse $u_{D,Fl}$ and/or the continuous phases $u_{C,Fl}$ at the flooding point are usually based on the two-layer model, developed by Gayler, Roberts and Pratt [8].

According to this model, the relative velocity of the droplet flow of both phases u_R is linked to the dispersed phase hold-up x^0 and the falling or rising velocity of the individual droplet $\overline{w}_S$. The best known models for determining the relative velocity u_R for liquid/liquid systems are as follows:

$$u_R = \frac{u_D}{\varepsilon \cdot x^0} + \frac{u_C}{\varepsilon \cdot (1 - x^0)}. \tag{7-9}$$

and

$$u_R = \overline{w}_S \cdot \left(1 - x^0\right), \tag{7-10}$$

acc. to Thornton [20] or

$$u_R = \overline{w}_S \cdot e^{-b \cdot x}, \tag{7-11}$$

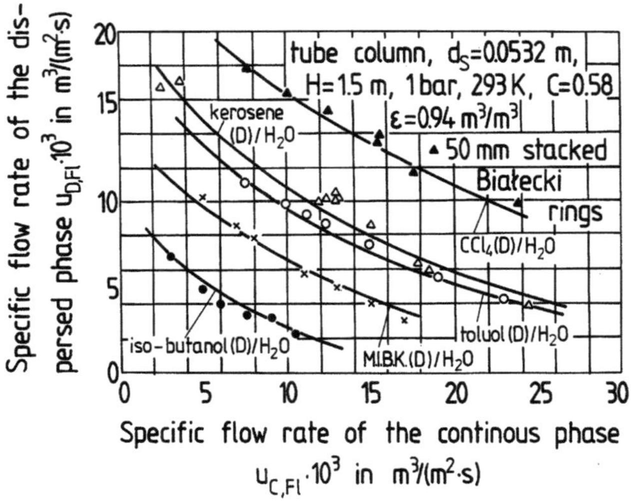

Figure 7-10. Capacity diagram for 50 mm tube column stacked aligned with 50 mm metal Białecki rings, valid for various test systems [2, 12]

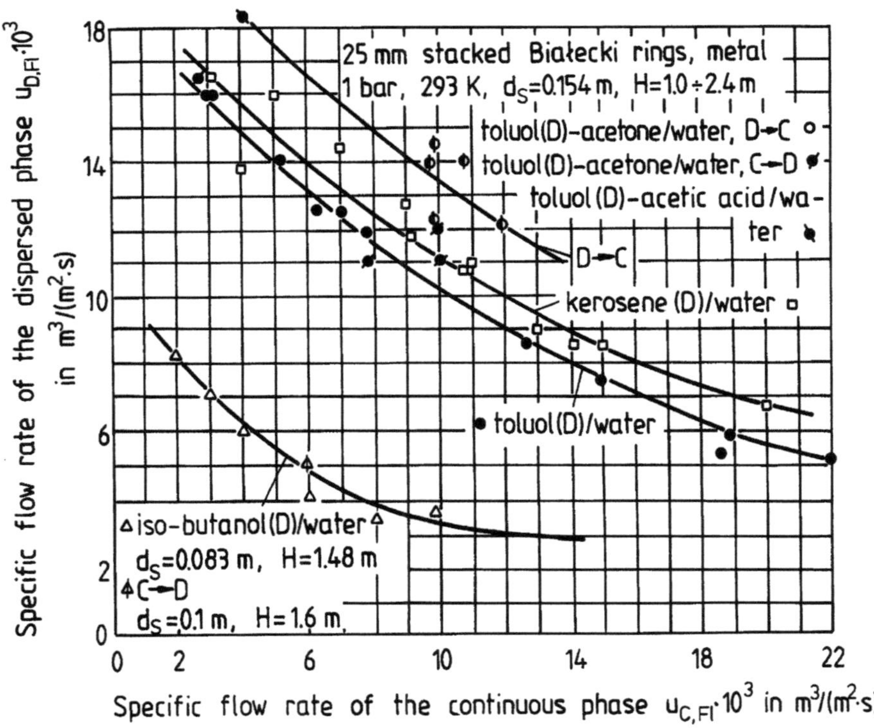

Figure 7-11. Capacity diagram for stacked 25 mm Białecki rings – valid for various test systems and for mass transfer directions C → D and D → C [11–14]

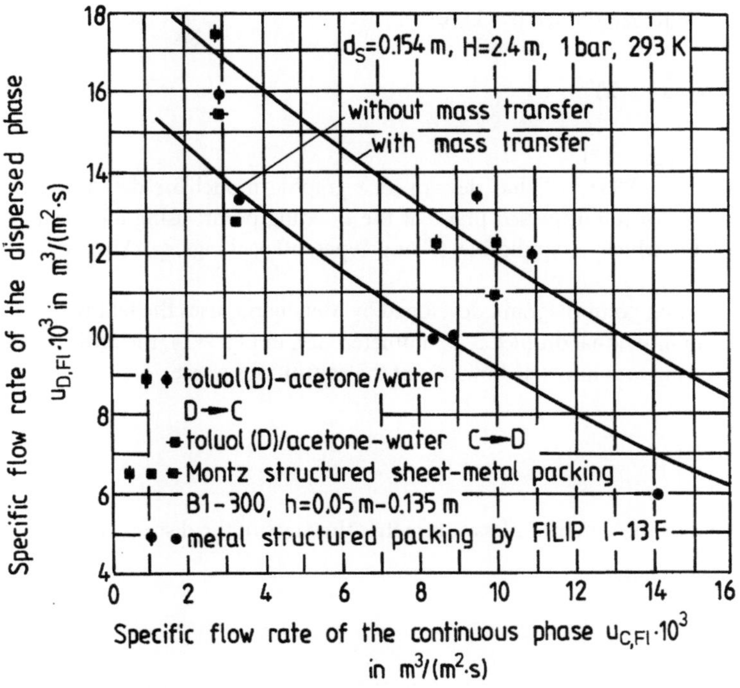

Figure 7-12. Capacity diagram for structured sheet-metal packings type Montz-B1-300Y and Fi-Pac-200, valid for system toluol (D)/water under normal conditions with and without mass transfer C → D and D → C [11, 12]

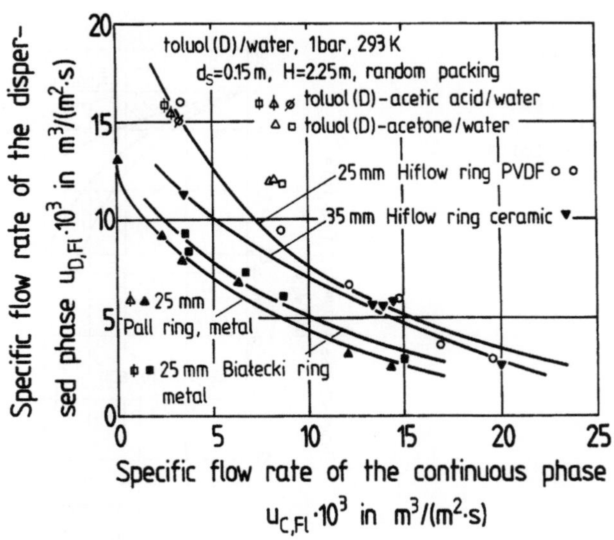

Figure 7-13. Capacity diagram for random packings, valid for system toluol (D)/water under normal conditions with and without mass transfer C → D and D → C [12–14]

acc. to Latan, Kehat, quoted by [2] or

$$u_R = \overline{w}_S \cdot (1 - x)^{m-1}\,, \tag{7-12}$$

acc. to Mersmann [17].

Mersmann (1980) [17] has developed a graphic model for determining the specific flow rate of the dispersed phase at the flooding point, using a capacity diagram, see Fig. 7-14, which is applicable to any types of packings, packing structures and materials.

The flooding point diagram, developed by Mersmann, uses the falling or rising velocity $\overline{w}_S$ of an individual droplet as a parameter. Acc. to Fig. 7-14, the ordinate axis shows the dimensionless volumetric flow rate of the continuous phase

$$\frac{u_{C,Fl}}{\overline{w}_S \cdot \varepsilon} \tag{7-13a}$$

plotted against the dimensionless volumetric flow rate of the dispersed phase

$$\frac{u_{D,Fl}}{\overline{w}_S \cdot \varepsilon} \tag{7-13b}$$

The experimental data range is based on experimental values mainly for small, ceramic packing elements with d ≤ 0.035 mm [1, 8, 10, 17, 20]. It is marked by the hatched area in Fig. 7-14.

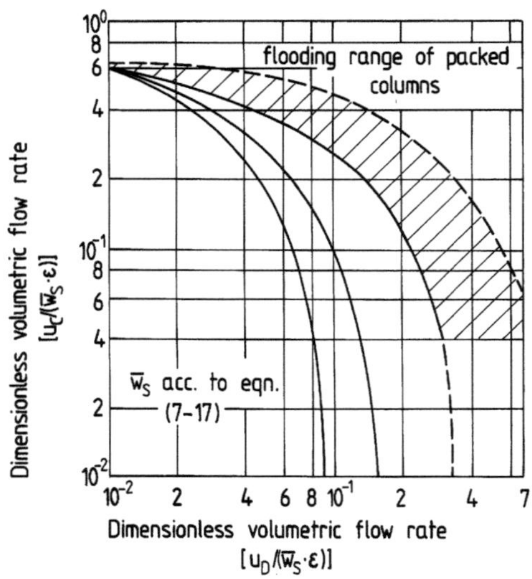

Figure 7-14. Flood load diagram according to Mersmann [17] for non-pulsed packed column. The experimental data is marked by the hatched area

7.3.2
Rising and Falling Velocity of Droplets in Packings — New Model

While the droplet is falling or rising through the packed bed, it collides with the individual packing elements, which means that the falling or rising velocity of the droplet $\overline{w}_S$ is lower than that of the empty column w_E. The studies of Maćkowiak and Billet [11–13] and of Billet, Maćkowiak and Pajak [2] have presented the following correlation for the individual droplet velocity $\overline{w}_S$:

$$\overline{w}_S = C \cdot w_E \quad \text{where } C < 1, \tag{7-14}$$

is different for each packing element.

The droplet velocity of an individual droplet in an empty column, acc. to Levich and quoted in [2], is given as:

$$w_E = 1.414 \cdot \sqrt[4]{\frac{\sigma \cdot \Delta\rho \cdot g}{\rho_C^2}} \qquad \left[\text{ms}^{-1}\right] \tag{7-15}$$

The only difference between the above and the correlation (7-16), developed by Mersmann [17], consists in the constant numerical value:

$$w_E = 1.55 \cdot \sqrt[4]{\frac{\sigma \cdot \Delta\rho \cdot g}{\rho_C^2}} \qquad \left[\text{ms}^{-1}\right] \tag{7-16}$$

Based on the experiments carried out by other authors, Mersmann [17] was able to derive the following empirical equation for determining the rising and falling velocity $\overline{w}_S$ of individual droplets in random packings, containing ceramic packing elements:

$$\overline{w}_S \cong \frac{w_E}{1 + 1.4 \cdot \left[\dfrac{w_E^2 \cdot \rho_D}{(d_F - d_E) \cdot \Delta\rho \cdot g}\right]} \qquad \left[\text{ms}^{-1}\right] \tag{7-17}$$

with a droplet diameter d_E acc. to Eq. (7-18):

$$d_E = 2.44 \cdot \sqrt{\frac{\sigma}{\Delta\rho \cdot g}} \qquad [\text{m}] \tag{7-18}$$

and d_F as the nominal packing diameter. Equations (7-16) and (7-17) are applicable to large, deformed droplets in excess of 10^{-3} m [17] in the range, in which the rising or falling velocity of the individual droplet is not dependent on the droplet size.

In the case of structured or random packings with the same diameter d_F but different geometric data a and ε, the droplet velocities $\overline{w}_S$ are likely to differ. However, this is not taken into account by Eq. (7-17).

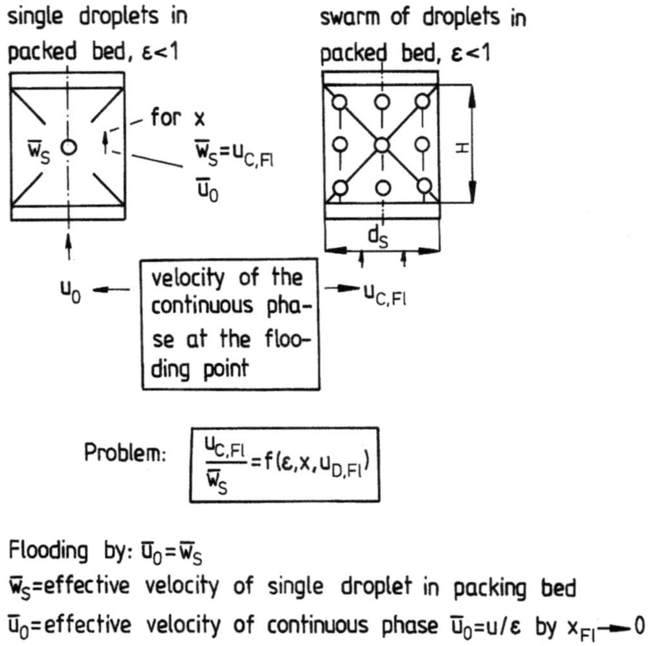

Figure 7-15. Model of suspended bed of droplets (SBD) for description of flooding point in liquid/liquid extraction column with packed bed

Based on the suspended bed of droplets (SBD) model, presented in Chap. 2, and the assumptions shown in Fig. 7-15, it possible to apply Eq. (2-57), derived in Chap. 2, to liquid/liquid systems. Assuming that u_T equals $\overline{w}_S$ ($u_T = \overline{w}_S$), the following correlation can be derived for determining the mean rising or falling velocity $\overline{w}_S$ of large, deformed droplets in liquids:

$$\overline{w}_S = u_T = 0.8 \cdot \cos\alpha \cdot \psi_{m,\,Fl}^{-1/6} \cdot \left(\frac{d_h}{d_T}\right)^{1/4} \cdot \left(\frac{d_T \cdot \Delta\rho \cdot g}{\rho_C}\right)^{1/2} \quad \left[\mathrm{ms}^{-1}\right] \qquad (7\text{-}19)$$

The following applies for $\alpha = 45°$:

$$\overline{w}_S = 0.566 \cdot \psi_m^{-1/6} \cdot \left(\frac{d_h}{d_T}\right)^{1/4} \cdot \left(\frac{d_T \cdot \Delta\rho \cdot g}{\rho_C}\right)^{1/2} \quad \left[\mathrm{ms}^{-1}\right]. \qquad (7\text{-}19a)$$

$\psi_{m,Fl}$ in Eq. (7-19) denotes the resistance coefficient of the continuous phase, which practically becomes constant in turbulent single-phase flow, $Re_C \geq 2100$, for a packing element or a family of packing elements, see Tables 6-1a–c. In the case of liquid/liquid systems, the resistance coefficient $\psi_{m,Fl}$ can be regarded as a packing-specific parameter, as packed columns used for extraction are operated in the range of low Reynolds numbers of the continuous phase, $Re_C < 100$.

d_h represents the hydraulic diameter of the packed bed, acc. to Eq. (7-20):

$$d_h = 4 \cdot \frac{\varepsilon}{a} \qquad [\mathrm{m}]\,. \tag{7-20}$$

Based on Eqs. (7-19), (7-9b) and (7-9c), the droplet velocities for the mass transfer direction D → C are predicted to be higher than for pure binary mixtures and systems with the mass transfer direction C → D.

Equation (7-19), which was derived for gas/liquid systems in view of determining the mean rising or falling velocity of droplets in random and structured packings for liquid/liquid systems, has been verified using a number of modern and classic, randomly filled and stacked packing elements and structured packings made of plastic, ceramic and metal, based on the data found in publications [2, 11–14]. The evaluation was based on experimental data relating to the following systems, acc. to Table 7-1:

- toluol (D)/water,
- water (D)/toluol,
- kerosene (D)/water,
- isobutyl alcohol (D)/water,
- MIBK (D)/water,
- CCl_4 (D)/water

The packings and systems used for the experiments as well as the mean $\overline{w}_S$ values are listed in Table 7-4. The table contains the $\overline{w}_S$ values calculated by Mersmann, acc. to Eq. (7-17), and presented by Maćkowiak [15], based on Eq. (7-19).

A comparison between the calculated and experimental values, see Table 7-4, shows good concurrence between both formulas and the experiment. An additional advantage of Eq. (7-19) lies in the fact that it allows a prediction of the individual droplet velocity in structured packings, stacked packing elements and small, randomly filled ceramic packing elements, which is sufficiently accurate for practical applications.

7.3.3
Modified Flooding Point Diagram [15]

The flooding point diagram, developed by Mersmann [17], is shown in Fig. 7-14. The hatched area covers the experimental results found in literature and evaluated by Mersmann [17]. A more accurate prediction of the loading line, however, requires a more accurate plotting of the loading curves, as the loading capacity of an extractor is highly dependent on the mass transfer direction [9–11]. For this purpose, the experimental data [2, 11, 14], together with new data [6, 9] is plotted in Fig. 7-16.

Equation (7-19) [15] was used for determining the mean droplet velocity $\overline{w}_S$, which is required as a variable. The properties of the system were varied, as shown in Table 7-7.

The experimental data plotted on the diagram in Fig. 7-16 are grouped around 3 curves [15].

Figure 7-16. Flood load diagram according to Mersmann [15], modified by Maćkowiak with own test points and [6, 9, 11–14]. Symbols see Table 7-4

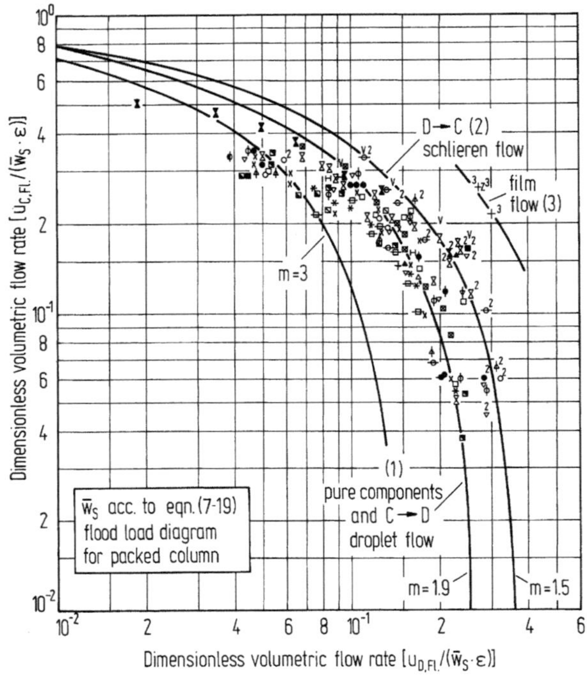

Case1: Curve 1 in Figure 7-16 – Pure Components and Ternary Systems C → D

The plotted test points for pure binary mixtures or ternary mixtures with the mass transfer direction C → D are grouped around curve 1, see Fig. 7-16. This curve is therefore valid for any types of column internals, Table 7-4, made of metal (with bare surface), ceramic and plastic (PP), if there is a droplet flow prevalent in the column.

Case2: Curve 2 in Figure 7-16 – Ternary Systems D → C

If, however, the organic phase is used as the dispersed phase, droplet flow in packing elements with an easily wettable surface can only be achieved for ternary mixtures for the mass transfer direction C → D. For the mass transfer direction D → C, the fast coalescence of droplets leads to schlieren formation [2, 6, 11–13]. In schlieren flow, the droplets are significantly larger than in droplet flow, which occurs in the mass transfer direction C → D. The loading capacity of the extractor is considerably higher, as $\overline{w}_S \approx d_T^{1/4}$, acc. to Eq. (7-19), and the falling velocity of the individual droplet is increased by the factor 1.1 for the mass transfer direction D → C, compared to the mass transfer direction C → D. This results in the second limit line in Fig. 7-16.

Limit line 2 was created by the plotting of experimental data, which was based on the mass transfer conditions for the mass transfer direction D → C, using the following ternary mixtures:

- isobutyl alcohol (D)-acetic acid-water
- toluol (D)-acetic acid-water
- toluol (D)-acetone-water
- water (D)-acetone-toluol

and various internals:

- 15–25 mm metal Pall rings
- 20–38 mm Hiflow rings made of plastic and ceramic
- 25 mm metal Białecki rings
- metal VSP rings, size 1
- stacked metal Białecki rings, 25 mm
- tube column with stacked Białecki rings, 25 and 50 mm
- sheet-metal packings produced by Montz B1-300, C1-300 and metal packing Fi-Pac

The experimental data points, which are valid for the binary system toluol (D)/water and for the plastic Montz packing C1-300, are also grouped around the second curve.

Case3: Curve 3 in Figure 7-16

If the packing surface is easily wet by the dispersed phase, as is the case with the random 25 mm Pall rings made of PVDF and the metal 1" VSP ring with a matt V2A surface, using the organic dispersed phase, the packing elements are covered with a film. In the case of film flow, the maximum flow rates in the extractor are found to be even higher. The few experimental values available are on the border of the hatched area in Mersmann's diagram, Fig. 7-14. In Fig. 7-16, they are grouped around curve 3.

7.3.4 Model for Determining the Specific Flow Rate of the Dispersed Phase at Flooding Point for Liquid/Liquid Systems

Mersmann [17] has developed the following approximate equation for calculating the specific flow rate of the dispersed phase at the flooding point:

$$\frac{u_{D,Fl}}{\overline{w}_S} = \frac{\varepsilon}{m} \cdot \left[1 - \left(\frac{u_C}{\overline{w}_S}\right)^{0.6}\right] \cdot \left[1 - \frac{\left[1 - \left(\frac{u_C}{\overline{w}_S}\right)^{0.6}\right]}{m}\right]^{m-1} - \frac{u_C}{\overline{w}_S} \cdot \frac{\frac{1}{m} \cdot \left[1 - \left(\frac{u_C}{\overline{w}_S}\right)^{0.6}\right]}{1 - \frac{1}{m} \cdot \left[1 - \left(\frac{u_C}{\overline{w}_S}\right)^{0.6}\right]} \tag{7-21}$$

In order to apply this equation, it is necessary to ascertain the droplet velocity $\overline{w}_S$ as well as the specific flow rate u_C of the continuous phase, the void fraction of the packing ϵ and the parameter m. The numerical values of this parameter were determined in [15]. However, the model parameters *m* are limited to cases *1* and *2*, described in Sect. 7.3.3, as they are particularly important for practical applications in liquid/liquid extractions.

Based on the experimental values shown in Figs. 7-11, 7-12 and 7-13, the parameter *m*, required for solving equation (7-21) for the mass transfer direction C → D and for pure binary mixtures, was found to be:

$$m = 1.9 \tag{7-22}$$

see solid line *1*. The experimental values for the mass transfer direction D → C, which are grouped around curve *2* in Fig. 7-16, can be described by Eq. (7-21), using the parameter:

$$m = 1.5 \tag{7-23}$$

Figure 7-16 shows an additional flooding point curve for the parameter:

$$m = 3.0 \tag{7-24}$$

For the higher values of the dimensionless flow rate of the continuous phase $u_C/(w_S \pm \varepsilon) > 0.25$, Fig. (7-16) shows considerably lower values of the dimensionless flow rate of the dispersed phase $u_D/(w_S \pm \varepsilon)$, compared to the values calculated using Eq. (7-21) with the parameter $m = 1.9$. The parameter ranges between $m = 1.9$ and *3*.

Here, it is recommended to perform the graphic estimation of the specific liquid flow rate of the dispersed phase at the flooding point $u_{D,Fl}$ using Fig. 7-16.

Figure 7-17 [15] shows a comparison between the experiment and the calculated flooding point data, Eq. (7-21), with the parameter $m = 1.9$ for pure binary mixtures and ternary mixtures C → D and for the mass transfer direction D → C with the parameter $m = 1.5$. As can be seen from the comparison, the experimental data has been verified by calculation with an accuracy of less than ±20 %. It was therefore possible to significantly consolidate and generalise the information available on the loading capacity of non-pulsed extractors, compared to Mersmann's flooding point diagram shown in Fig. 7-14.

Comparison of the Model Based on Eqs. (7-21) and (7-19) with Literature Data

Comprehensive studies on the operating characteristics of pulsed and non-pulsed packed columns have been carried out by Author [A, 2–4, 11–15], Bender, Berger, Leuckel and Wolf [1], Brandt, Reissinger, Schäfer [5] as well as Pilhofer [18].

Their experiments are particularly significant from a practical point of view, as they cover a wide range of constructive, operational and material parameters, with the packing height ranging from 3 m [5, 18] to 5 m [1], to name one example. It is also

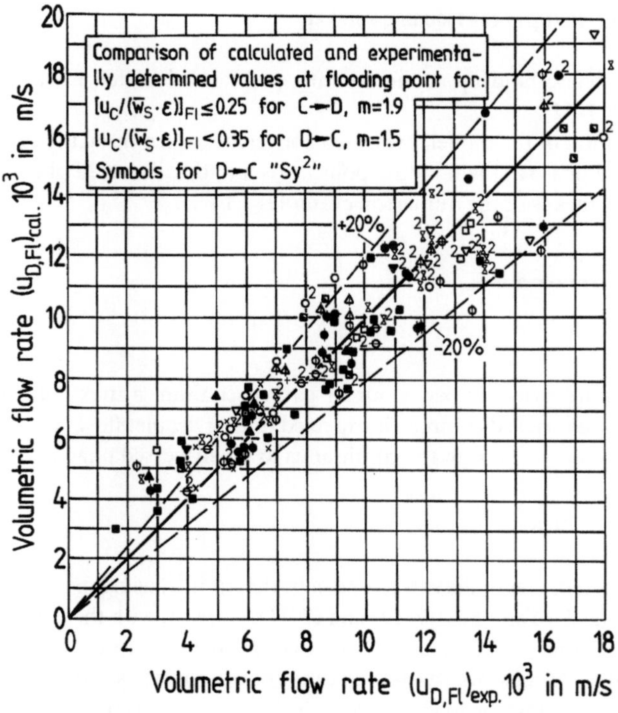

Figure 7.17. Comparison of experimentally determined specific flow rates at flooding point $u_{D,Fl}$ and values calculated according to Eq. (7-21) using Eq. (7-19) and model parameter based on [15]: $m = 1.9$ for mass transfer direction C → D; $m = 1.5$ for mass transfer direction D → C

the application of modern experimental methods for determining the flooding point by means of differential pressure techniques [18] as well as the long-term experience of Berger et al. [1], which make the comparison with the model based on Eqs. (7-12) and (7-21) particularly interesting.

Table 7-5 shows a comparison between the values calculated throughout this work and the literature data for 10–25 mm Pall rings and 15–25 mm Raschig rings made of metal and ceramic. The accordance between the experimental data and the calculated values is quite satisfactory.

The same goes for the comparison between the calculated values and the new experimental data presented by Pilhofer [18] for 15 mm ceramic Berl saddles, see Table 7-6. In the case of systems with the ratio $d_h/d_T > 1$, the prediction of the maximum capacity of the packing by means of Eq. (7-21) is very accurate. Table 7-6, point 4, shows that the droplet size occurring in the system toluol (D)/water corresponds to the size of the free channels d_h. Hence, the droplet fall in the packing is marked by a strong decelerating effect, which means that the experimental values $(u_D + u_C)_{max.}$ clearly fall below the values $(u_D{+}u_C)_{cal}$ calculated acc. to Eq. (7-21).

This can be explained by Eq. (7-19). The separation of systems with high interfacial tension requires the use of larger packing elements. The presented new model applicable is only if $d_h > d_T$.

Dispersed Phase Hold-Up at the Flooding Point x_{Fl}

The dispersed phase hold-up at the flooding point is calculated iteratively by solving Eqs. (7-9) and (7-12).

The application of the model, presented above, for determining the specific flow rate of the dispersed phase at the flooding point $u_{D,Fl}$ and the dispersed phase hold-up $x^0{}_{Fl}$ is illustrated by means of the numerical examples 7.1–7.3.

7.4 Conclusions

According to fluid dynamic behaviour of packed column, a maximum mass transfer area in non-pulsed packed columns is expected during droplet flow, mostly for the mass transfer direction C → D from the continuous to the dispersed phase.

$$a_e = 6 \cdot \frac{x}{d_T} \left[m^2/m^3\right] \qquad (7\text{-}25)$$

The column diameter of the extractor can be determined using the flooding point diagram, acc. to Fig. 7-16 and Eqs. (7-19) and (7-21). Curve 1, with the parameter $m = 1.9$, applies to the transfer direction C → D and to pure binary mixtures as well as to random and structured packings as well as tube columns of various types and sizes. The operation of packed column below the loading line $u_C \leq 0.65 \cdot u_{C,Fl}$ is recommended.

The flooding point diagram, Fig. 7-16, contains the falling or rising velocity of the individual droplet w_S as a variable, which is calculated using Eq. (7-19) and based on the SBD model, Chap. 2. This equation produces numerical values, which match the results of Mersmann's Eq. (7-17), see Table 7-4. Equation (7-19) covers a wider range of random and structured packings, compared to Eq. (7-17). The numerical values of the mean resistance coefficients $\psi_{m,Fl}$ for the packings not mentioned here can be found in Table 6-1a, b and c.

The current work shows that Eq. (7-19), which was derived for gas/liquid systems in Chap. 2, is also applicable to liquid/liquid systems. The range of constructive, operational and material parameters can be found in Table 7-7.

In the case of mass transfer of droplets into the continuous phase D → C, the loading lines are considerably higher, curve 2 Fig. 7-16, compared to the mass transfer direction C → D, curve 1, which is due to schlieren formation in the packing. The parameter m for Eq. (7-21) was found to be $m = 1.5$.

The loading lines are even higher in the case of packed columns, in which no droplet or schlieren flow occurs, curve, in Fig. 7-16. Packing elements made with easily wettable materials, such as PVDF, treated PP and matt, metal surfaces, are characterised by film flow. This was the result of studies with the systems toluol (D)/water, using 25 mm Pall rings made of PVDF.

Figure 7-1 shows a diagram of a test plant used for determining the hydraulic and mass transfer parameters of packed columns for liquid/liquid extraction. A photo of this plant can be seen in Fig. 7-18.

Figure 7-18. View of test plant $d_S = 0.15$ m for investigating fluid dynamics and mass transfer of packings for liquid/liquid extraction, filled with stacked 25 mm metal Białecki rings

The following numerical examples illustrate the application of the correlations, presented above, for the design of packed columns for liquid/liquid systems.

Numerical Examples Chapter 7

Numerical Example 7.1

The aim is to calculate the specific flow rate of the dispersed phase at the flooding point $u_{D,Fl}$, using Eq. (7-21), and to determine the dispersed phase hold-up at the flooding point x_{Fl} for 38 mm ceramic Hiflow rings in an extraction column with a diameter of 0.156 m for the toluol (D)/water system.

The column is operated at a specific flow rate of the continuous phase of $u_C = 3.37 \pm 10^{-3}\ m^3 m^{-2} s^{-1}$.

The experimentally determined value is given as $(u_{D,Fl})_{exp} = 11.42 \cdot 10^{-3}\ m^3 m^{-2} s^{-1}$.

Solution

Physical properties:

$$\rho_D = 998.2\,\text{kgm}^{-3};\quad \rho_D = 866.7\,\text{kgm}^{-3};\quad \sigma = 0.0351\,\text{Nm}^{-1}$$

Packing data acc. to Table 6-1b (38 mm Hiflow ring, ceramic):

$$a = 110\,\text{m}^2\,\text{m}^{-3};\quad \varepsilon = 0.831\,\text{m}^3\,\text{m}^{-3}\quad \psi_m \cong 1.725$$

Based on Eq. (7-20), the hydraulic diameter is given as:

$$d_h = 4 \cdot \frac{0.831}{110} = 0.0302\ \text{m}$$

and, using Eq. (7-8a), the mean droplet diameter d_T, based on Eq. (7-7), is given as:

$$d_T = 1.00 \cdot \sqrt{\frac{0.0351}{(998.2 - 866.7) \cdot 9.81}} = 0.00522\ \text{m}$$

Hence, the falling velocity of the individual droplets for the contact angle of the packed bed $\alpha = 45°$ and $0.8 \cdot \cos\alpha = 0.566$, based on Eq. (7-19), is as follows:

$$\overline{w}_S = 0.566 \cdot 1.725^{-1/6} \cdot \left(\frac{30.2}{5.22}\right)^{1/4} \cdot \sqrt{\frac{0.00522 \cdot 131.5 \cdot 9.81}{988.2}} = 0.0657\ \text{ms}^{-1}$$

The specific flow rate of the dispersed phase at the flooding point, acc. to Eq. (7-21), with $m = 1.9$, acc. to Eq. (7-22), has the numerical value of $u_{D,Fl} = 11.6 \cdot 10^{-3}\ \text{ms}^{-1}$ for $u_C = 3.37 \pm 10^{-3}\ \text{ms}^{-1}$.

Hence, the deviation from the experimentally derived value is:

$$\delta\left(u_{D,Fl}\right) = \left(\frac{11.42 - 11.6}{11.42}\right) \cdot 100\,\% = -1.6\,\%$$

The dispersed phase hold-up x_{Fl} at the flooding point is determined iteratively, using Eqs. (7-10) and (7-12) for $m = 1.9$. Trial and error led to the value $x = 0.40\ m^3m^{-3}$. The following applies, based on Eq. (7-9):

$$u_{R,Fl} = \frac{u_{D,Fl}}{\varepsilon \cdot x^0} + \frac{u_{C,Fl}}{\varepsilon \cdot (1 - x^0)} = \frac{u_{D,Fl}}{x} + \frac{u_{C,Fl}}{\varepsilon - x}$$

$$= \frac{11.6 \cdot 10^{-3}}{0.4} + \frac{3.37 \cdot 10^{-3}}{0.841 - 0.4}$$

$$= 3.66 \cdot 10^{-2}\ ms^{-1}$$

and, based on Eq. (7-12):

$$u_R = \overline{w}_S \cdot (1 - x)^{m-1}$$

$$= 0.0657 \cdot \left(1 - \frac{0.4}{0.537}\right)^{0.9}$$

$$= 3.64 \cdot 10^{-2}\ \mathrm{ms}^{-1}$$

Hence, the numerical value u_R, acc. to Eq. (7-9), matches the u_R value, acc. to Eq. (7-12). The experimental value x_{Fl}, based on Fig. 7-4d, is given as $x_{Fl,exp} = 0.41\ m^3m^{-3}$.

The relative deviation is now as follows:

$$\delta\left(x_{Fl}\right) = \left(\frac{0.41 - 0.40}{0.41}\right) \cdot 100\,\% = +2.43\,\%.$$

Numerical Example 7.2

Assuming a specific liquid flow rate of the dispersed phase of $u_D = 6 \times 10^{-3}\ m^3m^{-2}s^{-1}$, the aim is to determine the dispersed phase hold-up, based on the data presented in the numerical example 7.1.

Solution

The relative column load for the flow rate u_D is given as:

$$\frac{u_D}{u_{D,Fl}} = \frac{6 \cdot 10^{-3}}{11.6 \cdot 10^{-3}} = 51.7 \cdot 10^{-2} = 51.7\,\%$$

The extractor is therefore operated below the loading line. Here, correlation (7-4) can be applied, with $C_0 = 0.47$, acc. to Table 7-3:

$$x = \frac{1}{0.47 \cdot 0.831} \cdot \left[\frac{998.2^2}{4 \cdot 9.81 \cdot (998.2 - 866.7) \cdot 0.0351} \right]^{1/4} \cdot 0.006$$
$$= 13.2 \cdot 10^{-2}\ \mathrm{m^3 m^{-3}} \qquad \text{and } C_1 = 22.05\ \mathrm{sm^{-1}}$$

The experimentally derived value, based on Fig. 7-4d, is as follows:

$$x_{\mathrm{exp}} = 13.0 \cdot 10^{-2}\ \mathrm{m^3 m^{-3}}$$

The relative deviation is therefore:

$$\delta(x) = \left(\frac{13.0 - 13.2}{13.0} \right) \cdot 100\ \% = -1.54\ \%.$$

Numerical Example 7.3

The aim is to extract acetic acid from toluol by means of pure water in an extraction column with a diameter of 0.154 m.

The column containing Montz packing, type B1-300Y, is operated at a specific liquid flow rate of the continuous phase of $u_C = 3.18 \times 10^{-3}\ \mathrm{ms^{-1}}$ ($11.45\ \mathrm{m^3 m^{-2} h^{-1}}$). What is the predicted flooding capacity of the dispersed phase $u_{D,Fl}$?

Solution

Physical properties:

$$\rho_D = 998\,\mathrm{kgm^{-3}}; \quad \rho_D = 862\,\mathrm{kgm^{-3}}; \quad \sigma = 0.026\,\mathrm{Nm^{-1}}$$

Technical data of the Montz packing, acc. to Table 6-1c:

$$a = 300\,\mathrm{m^2\,m^{-3}}; \quad \varepsilon = 0.972\,\mathrm{m^3\,m^{-3}}; \quad \psi_{\mathrm{Fl,m}} \cong 0.888$$

1. Determining the falling velocity of the droplet $\overline{w}_S$ acc. to Eq. (7-19):

$$d_h = 4 \cdot \frac{\varepsilon}{a} = 4 \cdot \frac{0.972}{300} = 0.01296\ \mathrm{m}$$

and for D → C with Eq. (7-8b), Eq. (7-7) leads to a droplets diameter d_T of:

$$d_T = 1.25 \cdot \sqrt{\frac{\sigma}{\Delta \rho g}} = 1.25 \cdot \sqrt{\frac{0.026}{136.9 \cdot 9.81}} = 5.5 \cdot 10^{-3}\ \mathrm{m}$$

Based on Eq. (7-19), the falling velocity of the droplet, with $\alpha = 45°$, is given as:

$$\overline{w}_S = 0.566 \cdot 0.888^{-1/6} \cdot \left(\frac{12.96}{5.5}\right)^{1/4} \cdot \sqrt{\frac{5.5 \cdot 10^{-3} \cdot 136 \cdot 9.81}{988}}$$

$$= 6.16 \cdot 10^{-2}\ \mathrm{ms}^{-1}$$

Equation (7-21), for $u_C = 3.18 \times 10^{-3}\ \mathrm{ms}^{-1}$ and for $D \rightarrow C$ and $m = 1.5$, leads to:

$$u_{D,Fl} = 18.25 \cdot 10^{-3}\ \mathrm{ms}^{-1}$$

The experimentally derived value, acc. to [10, 11], is as follows (see Fig. 7-11):

$$u_{D,Fl} = 18.3 \cdot 10^{-3}\ \mathrm{ms}^{-1}$$

Hence, the relative deviation is:

$$\delta\left(u_{D,Fl}\right) = \left(\frac{18.3 - 18.25}{18.3}\right) \cdot 100\ \% = +0.27\ \%.$$

Annex Chapter 7

Tables Chapter 7

Table 7-1. Physical properties of the test systems investigated, valid for 293 K, 1 bar [11–14]

Test system	$\Delta\rho$ $[kgm^{-3}]$	ρ_C $[kgm^{-3}]$	$\sigma \cdot 10^3$ $[Nm^{-1}]$	Transferred component	$D_D \cdot 10^9$ $[m^2s^{-1}]$	$D_C \cdot 10^9$ $[m^2s^{-1}]$	$\eta_C \cdot 10^9$ $[kgm^{-1}s^{-1}]$	$\eta_D \cdot 10^9$ $[kgm^{-1}s^{-1}]$
toluol (D)/water	131.5	998.2	35.1	acetone	2.79	1.156	1.075	0.586
	131.5	998.2	35.1	acetic acid	2.52	0.88	1.075	0.586
iso-butanol (D)/water	148.5	985.6	2.8	acetic acid	0.926	0.353	1.46	4.703
paraffin (D)/water	198.2	998.2	32.0	–	–	–	–	–
CCl_4 (D)/water	595.8	998.2	44.5	acetic acid	1.472	–	1.0	0.969
MIBK (D)/water	190.8	995.6	10.0	–	–	–	1.07	0.582
p-xylene (D)/water	137.2	998.2	33.7	–	–	–	–	–
dichloroethane (D)/water	253.8	998.2	28.0	acetic acid	2.35	0.88	1.0	0.835
water (D)/toluol	131.5	866.7	35.1	–	–	–	0.586	1.0

Table 7-2. List of the experimentally determined constants C_1 for Eq. (7-2) and C_0 for Eq. (7-4), valid for 25 and 50 mm tube columns with stacked Białecki rings and for metal structured packing made of 25 mm stacked Białecki rings for various systems in range below the loading line $u_C \leq 0.65 \cdot u_{D,Fl}$ [2, 11–13]

Column type	System 1 bar, 293 K	$d_S \cdot 10^3$ [m]	H [m]	C_1 [sm^{-1}]	C_0	$Re = \frac{u \cdot d_T \cdot \rho_C}{\eta_C}$
50 mm Białecki rings tube column	iso-butanol (D)/water	53.2	1.5	36.0	0.463	62.8
25 mm Białecki rings metal, stacked		100	3.2	38.5	0.470	
		83	1.5	37.0	0.458	
25 mm Białecki rings tube column	p-xylene (D)/water	25.5	0.5	20.4	0.464	526.7
	MIBK (D)/water	25.5	0.625	24.0	0.465	230.9
50 mm Białecki rings tube column	paraffin (D)/water	53.2	1.5	18.7	0.4525	504
25 mm Białecki rings metal, stacked		154	1.25	19.4	0.444	
50 mm Białecki rings tube column	CCl_4 (D)/water	53.2	1.5	12.6	0.469	313.7
25 mm Białecki rings metal, stacked		220	2.4	13.8	0.4344	
		100	3.2	12.34	0.4858	
	dichloroethane (D)/water	100	3.2	16.8	0.493	34.76
		100	2.95	18.0	0.465	

Table 7-3. List of the experimentally determined constants C_1 and C_0 for Eqs. (7-2) and (7-4) for structured and random packings investigated, valid for the range the below loading line $u_C \leq 0.65 \cdot u_{D,Fl}$ [11–12]

Column type	ε [$m^3 m^{-3}$]	N [m^{-3}]	$d_S \cdot 10^3$ [m]	H [m]	$C_{1,exp.}$ [sm^{-1}]	$C_{0,exp.}$
(a) System: toluol (D)/water, 1 bar, 293 K						
50 mm tube column, metal	0.94	9000	53.2	1.45	19.4	0.4722
					20.4	0.4491
25 m Białecki rings metal, stacked	0.928	60623	154	2.4	19.9	0.465
			220	2.4	19.9	
Fi-Pac by Filip metal	0.94	–	154	2.4	22.0	0.470
sheet metal packing B1-300 by Montz	0.93	–	154	2.4	20.6	0.475
Montz packing PVDF, C1-300	0.897	–	154	2.4	15.54	0.618
25 mm Pall rings metal, random	0.94	49149	154	2.25	27.3	0.338
25 mm Białecki rings metal, random	0.94	48874	154	2.25	32.5	0.284
25 mm Hiflow rings PVDF, random	0.924	43888	154	2.25	24.4	0.385
35 mm Hiflow rings ceramic, random	0.83	16800	154	2.25	22.0	0.47
(b) System: water (D)/toluol, 1 bar, 293 K						
25 mm Hiflow rings PVDF, random	0.924	43888	154	2.25	23.8	0.365
Montz packing PVDF, C1-300	0.897	-	154	2.4	14.54	0.615

Table 7-4. List of mean velocities of single droplets in packed bed $\overline{w}_S$ for packings and systems investigated, symbols for Fig. 7-16

Sy	$d \cdot 10^3$ [m]	Packing or column type	a [m^2m^{-3}]	ε [m^3m^{-3}]	System no.	$\overline{w}_{S,exp.}$	$\overline{w}_{S,calc.}$ acc. to eqn. (7-19) [ms^{-1}]	$\overline{w}_{S,calc.}$ acc. to eqn. (7-17) [ms^{-1}]	$\Psi_{h,Fl}$ s. table 6-1	$C_{Fl,0}$
ɸ	28	Hiflow ring, PP, random	185.3	0.922	1	0.61	0.060	0.060	1,60	0.566
-●-					2	0.66	0.064	0.059		
●	38	Hiflow ring, ceramic, random	110	0.831	1	0.066	0.0654	0.071	1.725	0.566
○	25	Pall ring, metal, random	215	0.942	1	0.051	0.054	0.056	2.42	0.566
■	25	Intalox saddle, ceramic, random	207	0.69	1	0.041	0.0516	0.056	2.1	0.566
▲	20	Hiflow ring, ceramic, random	283	0.763	1	0.033	0.0510	0.045	1.60	0.566
H	32	VSP ring, metal (bright), random	200	0.972	1	0.062	0.060	0.065	2.2	0.566
					3	0.062	0.056	0.065		
⧗	25	Białecki rings, metal, stacked, N = 63,300 m^{-3}	275	0.933	1	0.077	0.071	–	1.030	0.693
⧗					4	0.047	0.050	–		
I					5	0.12	0.112	–		
+	25	Pall ring, PVDF, random	182	0.909	3	0.0546	0.0557	0.056	2.42	0.566
□	–	Fi-Pac	200	0.944	1	0.069	0.0825	–	1.077	0.739
x	25	tube column (7 tubes) with 25 mm metal Białecki rings	339.1	0.924	4	0.032	0.0412	–	1.55	0.693
*					5	0.078	0.0888	–		
Φ	–	Montz packing C1-300, PVDF	300	0.95	1	0.068	0.060	–	0.71	0.566
▽	–	Montz packing B1-300 h = 0.050 m	300	0.972	1	0.066	0.062	–	–	0.566
▼	–	Montz packing B1-300 h = 0.135 m	300	0.972	6	0.066	0.062	–	–	0.566
ɸ	–	Montz packing C1-300, PVDF	300	0.9	1	0.068	0.0654	–	0.71	0.566
⊠	53	tube column (1-7 tubes) with 53 mm metal Białecki rings	156.4	0.95	7	0.067	0.0635	–	1.55	0.693
◪					8	0.073	0.0735	–		
◩					9	0.055	0.064	–		
⊟					5	0.104	0.1081	–		
-⊖-	15	Pall ring, metal, random	354	0.95	10	–	0.036	0.0260	2.42	0.566
V	32	VSP ring, metal (matt), random	200	0.972	11	–	0.060	–	2.2	0.566
⍋	25	Białecki rings metal, random N = 52,000 m^{-3}	225	0.945	1	0.0533	0.0528	0.056	2.66	0.566

u_D, u_C = i

$\delta(i) = (i_{exp} - i_{cal})/i_{exp} \cdot 100$ %

System no. 1: toluol (D) / water

System no. 2: water (D) / toluol + acetic acid, C → D

System no. 3: water (D) / acetone – toluol, C → D

System no. 4: iso-butanol (D) / water

System no. 5: CCl_4 (D) / water

System no. 6: toluol (D) / acetone – water, C → D

System no. 7: water (D) / toluol

System no. 8: kerosene (D) / water

System no. 9: MIBK (D) / water

System no. 10: iso-butanol (D) – acetic acid / water, C → D

System no. 11: water (D) / acetone – toluol, D → C

Sy^2 = symbol with ^{2} valid for ternary mixtures and mass transfer direction D → C; transferred components: acetic acid, acetone and cinnamic acid. For calculation of $\overline{w}_S$ according to eqn. (7-19), experimental values were used for d_T.

Table 7-5. Comparison of calculated values acc. to Eqs. (7-21) and (7-19) of this work with experimentally determined values by Bender, Berger, Leuckel and Wolf [1] for maximum total load in non-pulsed packed columns

	Packing	d_S/H [m] [m]	a/ε $[m^2 m^{-3}]$ $[m^3 m^{-3}]$	$(u_D+u_C)_{Fl}$ $[m^3 m^{-2} h^{-1}]$		$\delta_i(u_D+u_C)_{Fl}$ [%]	System
				exp.	cal.		***M***
1	25 mm Pall rings metal $\lambda = 0.7$ $\psi = 2.42$	$\frac{0.15}{5}$	$\frac{215}{0.942}$	65	62	+4.6	No. 1, m = 1.9
2	25 mm Raschig rings ceramic $\lambda = 0.7$ $\psi = 3$	$\frac{0.15}{5}$	$\frac{177}{0.693}$	42.8 41.0	41.6 41.0	+2.8 0	No. 1, m = 1.9
3	15 mm Pall rings metal $\lambda = 0.7$ $\psi = 2.42$	$\frac{0.1}{5}$	$\frac{350}{0.93}$	39.0 49.0 34	45.4 56.7 38.5	−16.4 −15.7 −13.2	No. 1, m = 1.9, D → C No. 2, C → D No. 3, m = 1.9
4	$\lambda = 1.2$			72	60.9	+15.4	No. 4, m = 1.5, D → C
5	10 mm Pall rings metal $\lambda = 1.2$ $\psi = 2.42$	$\frac{0.1}{5}$	$\frac{515}{0.92}$	58	51.5	+11.5	No. 4, m = 1.5, D → C
6	15 mm Raschig rings metal $\lambda = 1.2$ $\psi = 8.5$	$\frac{0.1}{5}$	$\frac{350}{0.92}$	42.1	48.5	−15.2	No. 4, m = 1.5, D → C

$u_D, u_C = i$
$\delta(i) = (i_{exp} - i_{cal})/i_{exp} \pm 100\ \%$
System no. 1: toluol (D)/water
System no. 2: toluol (D)-acetone/water
System no. 3: toluol (D)/acetone/water C → D
System no. 4: ethyl hexanol (D)-acetic acid/water D → C

Table 7-6. Comparison of calculated values acc. to Eqs. (7-19) and (7-21) of this work with experimentally determined values by Pilhofer [18] for maximum total load in non-pulsed packed columns

	System	$(u_D+u_C)_{Fl}$ $[m^3m^{-2}h^{-1}]$		$\delta_i(u_D+u_C)_{Fl}$
		exp.	cal.	[%]
1	methylene chloride (D)/water	40.3	41.0	−1.74
2	MIBK (D)/water	31.7	30.6	+3.5
3	ethyl acetate (D)/water	24.4	22.4	+8.2
4	toluol (D)/water	20.2	28.4	−40.6
		24.0	28.9	−20.4

$u_D, u_C = i$
$\delta(i) = (i_{exp} - i_{cal})/i_{exp} \cdot 100\ \%$

Table 7-7. Range of various materials, designs and operating parameters, valid for the correlation for determining the specific flow rate of the dispersed phase at the flooding point, based on Eqs. (7-21) and (7-19) [15] and investigated packings

0.7	≤	ψ_{Fl}	≤	8.5 [−]
0.025	≤	d_S	≤	0.22 m
0.696	≤	ε	≤	0.972 m^3m^{-3}
110	≤	a	≤	515 m^2m^{-3}
1.25	≤	H	≤	5.0 m
866	≤	ρ_C	≤	1260 kgm^{-3}
800	≤	ρ_D	≤	1594 kgm^{-3}
99.5	≤	$\Delta\rho$	≤	596 kgm^{-3}
1	≤	σ	≤	44.5 mNm^{-1}
0.596	≤	η_C	≤	1.5 mPas
0.596	≤	η_D	≤	9.28 mPas
		d_h/d_T	>>	1

10–25 mm	Pall rings made of metal and plastic
12–25 mm	metal Białecki rings
20–38 mm	Hiflow rings made of ceramic and plastic
20–25 mm	Intalox saddles
15 mm	metal Raschig rings
15 mm	Berl saddles made of ceramic
15 mm	Raschig rings made of ceramic
25, 50 mm	tube column with Białecki rings
25 mm	packing with metal Białecki rings
Fi-Pac 200	sheet-metal packing
B1-300, C1-300	Montz packing
10–25 mm	Pall rings made of metal and plastic
12–25 mm	metal Białecki rings
20–38 mm	Hiflow rings made of ceramic and plastic
20–25 mm	Intalox saddles
15 mm	metal Raschig rings
15 mm	Berl saddles made of ceramic
15 mm	Raschig rings made of ceramic
25, 50 mm	tube column with Białecki rings
25 mm	packing with metal Białecki rings
Fi-Pac 200	sheet-metal packing
B1-300, C1-300	Montz packing

References Chapter 7

1. Bender E, Berger R, Leuckel W, Wolf D. Untersuchungen zur Betriebscharakteristik pulsierter Füllkörperkolonnen für die Flüssig/Flüssig-Extraktion. Chem.-Ing. Techn. 51(1979) No. 3, p. 192–199
2. Billet R, Maćkowiak J, Pajak M. Hydraulics and Mass Transfer in Filled Tube Columns. Chem.-Eng.-Process. 19 (1985) p. 39–47
3. Billet R, Landau V, Maćkowiak J. Rückvermischung der kontinuierlichen Phase in Rohrkolonnen. Chemie Technik 13 (1984), No. 10, p. 88–94
4. Billet R, Maćkowiak J. Relation Between Axial Mixing in the Continuous Phase and Hold-up in Tube Columns Filled with Białecki-Rings for Liquid/Liquid-Extraction. Chem. Biochem. Eng. 2(2) (1988), p. 91–97
5. Brandt H, Reissinger KH, Schröter J. Moderne Flüssig/Flüssig-Extraktoren - Übersicht und Auswahlkriterien. Chem. Ing. Techn. 50 (1978) No. 5, p. 345–354
6. Braun Chr. Untersuchungen zur Fluiddynamik begaster Gegenstromextraktion. Dissertation TU-Bochum (1991) VDI-Verlag No. 261/ Reihe 3
7. Pendulum Pulsator. Technical Information – ENVIMAC Engineering GmbH, 46149 Oberhausen
8. Gayler R, Roberts NW, Pratt HRC. Liquid-Liquid Extraction: Part IV. A Father Study of Hold-Up in Packed Columns. Trans. Inst. Chem. Engrs. 31(1953), p. 57–68
9. Gosch A,. Einfluss von Verunreinigungen auf die Wirksamkeit und die Belastbarkeit von statischen Gegenstromextraktoren. Dissertation- TU- Bochum (1988)
10. Laddha GS, Degaleesan TE. Transport phenomena in liquid extraction. Tata Mc Graw-Hill Publishing C. Ltd. New Dehli (1976)
11. Maćkowiak J. Grundlagen der Auslegung von Kolonnen mit regellosen Füllkörpern und Packungen. Presentation at GVC-VDI annual meeting in Munich, 19/21 September 1984
12. Maćkowiak J, Billet R. New Method of Packed Column Design for Liquid/Liquid Extraction Processes with Random and Stacked Packings. Ger. Chem. Eng. 1(1986) p. 48–64
13. Maćkowiak J, Billet R. Podstawy projektowania kolumn wypełnionych do procesów ekstrakcji ciecz-ciecz (orig. Polish). Inż. Chem. i Procesowa 3 (1986), p. 351–371
14. Maćkowiak J, Billet R. Beitrag zur Auslegung von Flüssig/ Flüssig- Extraktoren mit regellos geschütteten Füllkörpern für Systeme großer Grenzflächenspannung. Chem.Techn. 40 (1988) No. 8, p. 339–344
15. Maćkowiak J. Grenzbelastung von unpulsierten Füllkörperkolonnen bei der Flüssig/Flüssig- Extraktion. Chem.-Ing.-Tech. 65 (1993) No. 4, p. 423/429
16. Mersmann A. Thermische Verfahrenstechnik. Springer-Verlag (1980)
17. Mersmann A. Zum Flutpunkt in Flüssig/Flüssig-Gegenstromkolonnen. Chem.-Ing. Techn. 52 (1980) No. 12, p. 933–942
18. Pilhofer Th. Belastungsgrenzen von pulsierten Füllkörper-Extraktionskolonnen. Chem.-Ing. Techn. 62 (1990) No. 8, p. 661–663
19. Seibert AF, Fair JR. Hydrodynamics and Mass Transfer in Spray and Packed Columns. Ind.- Eng. Chem. Res. 27 (1988), p. 470/481
20. Thornton J D. Spray liquid-liquid extraction columns: Prediction of limiting hold-up and flooding rates. Chem. Eng. Sci. 5 (1956) p. 201/208
21. Ziołkowski Z. Ekstrakcja cieczy w przemyśle chemicznym (orig. Polish) WNT, Warszawa (1980)

Index

MIX
Papier aus verantwortungsvollen Quellen
Paper from responsible sources
FSC® C105338

If you have any concerns about our products,
you can contact us on
ProductSafety@springernature.com

In case Publisher is established outside the EU,
the EU authorized representative is:
Springer Nature Customer Service Center GmbH
Europaplatz 3, 69115 Heidelberg, Germany

Printed by Libri Plureos GmbH
in Hamburg, Germany